Liver Carcinogenesis

The Molecular Pathways

NATO ASI Series

Advanced Science Institutes Series

A series presenting the results of activities sponsored by the NATO Science Committee, which aims at the dissemination of advanced scientific and technological knowledge, with a view to strengthening links between scientific communities.

The Series is published by an international board of publishers in conjunction with the NATO Scientific Affairs Division

A Life Sciences B Physics	Plenum Publishing Corporation London and New York
C Mathematical and Physical Sciences D Behavioural and Social Sciences E Applied Sciences	Kluwer Academic Publishers Dordrecht, Boston and London
F Computer and Systems Sciences G Ecological Sciences H Cell Biology I Global Environmental Change	Springer-Verlag Berlin Heidelberg New York London Paris Tokyo Hong Kong Barcelona Budapest

NATO-PCO DATABASE

The electronic index to the NATO ASI Series provides full bibliographical references (with keywords and/or abstracts) to more than 30000 contributions from international scientists published in all sections of the NATO ASI Series. Access to the NATO-PCO DATABASE compiled by the NATO Publication Coordination Office is possible in two ways:

- via online FILE 128 (NATO-PCO DATABASE) hosted by ESRIN, Via Galileo Galilei, I-00044 Frascati, Italy.
- via CD-ROM "NATO Science & Technology Disk" with user-friendly retrieval software in English, French and German (© WTV GmbH and DATAWARE Technologies Inc. 1992).

The CD-ROM can be ordered through any member of the Board of Publishers or through NATO-PCO, Overijse, Belgium.

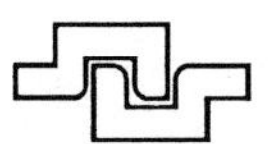

Series H: Cell Biology, Vol. 88

Liver Carcinogenesis

The Molecular Pathways

Edited by

George G. Skouteris

Deutsches Krebsforschungszentrum
Department of Applied Tumor Virology
Laboratory of Virus-Host Cell Interactions
Im Neuenheimer Feld 242
D-69120 Heidelberg, Germany

Springer-Verlag
Berlin Heidelberg New York London Paris Tokyo
Hong Kong Barcelona Budapest
Published in cooperation with NATO Scientific Affairs Division

Proceedings of the NATO Advanced Study Institute on molecular aspects of liver carcinogenesis, held at City of Delphi, Greece, January 8–18, 1994

ISBN 3-540-58371-8 Springer-Verlag Berlin Heidelberg New York
ISBN 0-387-58371-8 Springer-Verlag New York Berlin Heidelberg

CIP data applied for

Printed in Germany

Typesetting: Camera ready by authors
SPIN 10134843 31/3130 - 5 4 3 2 1 0 - Printed on acid-free paper

To Mary

FOREWORD

The investigation of the mechanisms controlling liver growth has led to the identification of many regulators such as the liver-acting growth modulators (TGF-α, HGF, TGF-β) and the activation of signal transduction pathways (inositol phosphates, tyrosine phosphorylation, PLC etc.).

Although the progress in understanding liver growth is extraordinary, very little is known about the regulatory steps governing the development of liver cancer. In this Advanced Study Institute held at the very stimulating environment of the European Cultural Center of Delphi, various aspects of liver carcinogenesis were discussed. In this meeting we addressed various aspects but not all the etiologies of the development of hepatocellular carcinoma. More emphasis was given to link the normal liver growth with the molecular phenomena leading to the development of a malignant phenotype.The functions of the newly discovered Hepatocyte Growth Factor (HGF) and its receptor (c-met) were discussed in detail. The evidence that HGF plays a major role in liver regeneration is now converging from data on animal and human material, although its role in liver neoplasia is not yet clear as pointed out by Michalopoulos (Pittsburgh, USA), Matsumoto (Osaka, Japan) and Comoglio (Torino, Italy). Schulte-Hermann (Vienna, Austria) pointed out the mechanisms of promotion and initiation in liver tumors, as well as the pathways of active cell death (apoptosis) in liver regeneration and neoplasia. Emphasis was also given on the action of non-genotoxic agents in super-promoting clonal growth of spontaneously initiated hepatocytes.

The activation of the hepatic stem cell compartment in the liver by HGF was presented by Alison (London, UK) and Thorgeirsson (Bethesda, USA). Shafritz (Ney York, USA) also discussed the activation of the hepatocyte progenitor cell in D-galactosamine-induced hepatic injury.
The role of the proto-oncogene *c-myc* was analyzed in liver growth biology by Skouteris (Heidelberg, Germany). Although this oncogene is over-expressed in most hepatocellular carcinomas, it is also expressed in normal hepatocyte cultures and it seems to be essential for initiation of hepatocyte DNA synthesis. In depth analysis of the action of norepinephrine, angiotensin and vasopressin on hepatocyte proliferation (Christoffersen, Oslo, Norway) revealed that these compounds mediate their effects not through PIP2 turnover but rather through other pathways such as that of arachidonic acid or phosphatidylcholine.

An interesting subtractive-hybridization procedure was described by Hakvoort (Amsterdam, Netherlands) to be used for the identification of new gene products in the liver following a regenerating stimulus. Siddiqui (Colorado, USA) and Andrisani (West Lafayette, USA) discussed the interactions of the hepatitis B transactivator protein (HBx) with cellular targets such as CREB and HBV I and II enhancers. Hepatocytes were presented as a model system for studying HBV infection (Guguen-Guillouzo, Rennes, France). A duck model of hepatitis infection in relation to the development of hepatocellular carcinoma was presented by Cova (Lyon, France). The various methodologies in preparing animal hepatocytes were analyzed in detail by Skett (Glasgow, UK) and their uses summarized by Rogiers (Brussels, Belgium). Similar details were discussed for human hepatocytes and other cells of the liver, focussing in particular on their *in vitro* growth requirements by Strain (Birmingham, UK).

The various technologies used to deliver foreign genes in hepatocytes were presented in the last part of the meeting. The use of asialoglycoprotein-gene complexes to target foreign genes in hepatocytes via AsG receptors, was analyzed by Wu (Connecticut, USA). This study also underlined the correlation between the disruption of the microtubules (using colchicine) and the persistence of the expression of the introduced gene. A new highly promising approach to deliver foreign genes using the receptor-mediated endocytosis route was analyzed by Zatloukal (Graz, Austria). The "transferrinfection" is used to transfer foreign genes into a variety of cells and in this technique the virus serves only as an endosome-lysis agent to ensure the release of the endocytosed DNA from the endosome, thus drastically augmenting the transfer efficiency.

I would like to thank the other two co-organizers, Prof.G.Michalopoulos and Dr.P.Skett for efficient support and would also like to thank the NATO Scientific Affairs Division for the kind and generous support of this project. I would also like to thank Flora Poloway for her fierce contribution in the organization of this ASI and for valuable editorial assistance.

G.G.Skouteris
Editor

Contents

Molecular pathways in hepatocyte proliferation 1
G.G.Skouteris

HGF and liver growth regulation 13
G.K.Michalopoulos

Structure, pleiotropic actions and organotrophic roles of hepatocyte growth factor 33
K.Matsumoto and T.Nakamura

Hepatocyte Growth Factor and its variant with a deletion of five amino-acids are distinguishable in biological, physicochemical and immunochemical properties 55
N.Shima and K.Higashio

Growth stimulation mediated by G protein-coupled receptors in hepatocytes: Synergism with epidermal growth factor and mechanisms of signal transduction 71
T.Christoffersen, G.H.Thoresen, O.F.Dajani, D.Sandnes and M.Refsnes

Protein tyrosine phosphatases in signal transduction 93
S-H.Shen and D.Banville

C-myc is essential for initiation of DNA synthesis in EGF-stimulated hepatocyte cultures 109
G.G.Skouteris

Cellular biology of the rat hepatic stem cell compartment 129
S.S.Thorgeirsson, R.P.Evarts, K.Fujio and Z.Hu

Changes in protein expression during oval cell proliferation in the liver 147
T.W.Jordan, I.E.Nickson and H.Feng

Stem cell activation in the acetylaminofluorene-treated regenerating rat liver: A bile ductular reaction? 163
T.V.Anilkumar, M.Golding, C.Sarraf, E-N.Lalani, R.Poulsom and M.Alison

Apoptosis and its role in hepatic carcinogenesis by non-genotoxic agents 181
R.Schulte-Hermann, W.Bursch, B.Grasl-Kraupp, W.Huber, B.Ruttkay-Nedecky and A.Wagner

Cell proliferation and cell death in the liver: Patterns, mechanisms and measurements 197
M.Alison, M.Golding and C.Sarraf

Preneoplastic changes during non-genotoxic hepatocarcinogenesis 215
C.J.Powell, B.Secretan and S.Cottrell

mRNA composition of rat liver tumors initiated by aromatic amines 231
A.Bitsch, M.Jost and H.Richter

Liver γ-Glutamyl transpeptidase activity after cyclosporine A and amlodipine treatment 249
J.G.Maj, J.J.Tomaszewski and A.E.Haratym

Enhanced expression of the 27KDa heat-shock protein during DENA-induced hepatocarcinogenesis in rat and in human neoplastic and non-neoplastic liver tissues 261
N.Mairesse, M.Delhaye, B.Gulbis and P.Galand

Hepatic recovery of dimethylnitrosamine-cirrhotic rats after injection of the liver growth factor 275
J.J.Diaz-Gill, C.Rua, C.Machin, M.Rosa Cereceda, M.Carmen Guijarro and P.Escartin

Cell contact-mediated regulation of hepatocyte differentiation/proliferation: Role of LRP(s) 287
A.Corlu, P.Loyer, S.Cariou, G.Ilyin, I.Lamy, M.Corral-Debrinski and C.Guguen-Guillouzo

Interactions of human hepatocytes with hepatitis B virus. 301
S.Rumin, P.Gripon, M.Corral-Debrinski, J.Gillard and C.Guguen-Guillouzo

Liver-specific aspects of hepatitis B virus gene expression 311
M.J.Kosovsky, H.F.Maguire, B.Huan and A.Siddiqui

Transcriptional involvement of the hepatitis B virus protein in cellular transduction systems. Protein-protein interactions with bZip transactivators 341
O.M.Andrisani

Role of duck hepatitis B virus infection, aflatoxin B_1, and p53 mutation in hepatocellular carcinomas of ducks 359
L.Cova

The preparation and culturing of rat hepatocytes 373
P.Skett

Isolation and growth of hepatocytes and biliary epithelial cells from normal and diseased human livers 389
A.J.Strain, L.Wallace, R.Joplin, J.Neuberger and D.Kelly

A novel strategy for isolating HEPR, a human small intestinal cytidine deaminase 411
C.Hadjiagapiou, F.Giannoni, T.Funahashi and N.Davidson

Regenerating liver: Isolation of up- and down-regulated gene products 421
T.B.M.Hakvoort and W.H.Lamers

Hepatotrophic and renotrophic activities of HGF *in vivo:* Possible application of HGF for hepatic and renal diseases 435
K.Matsumoto and T.Nakamura

Targeted gene delivery and expression 449
G.Y.Wu

Studies on persistence and enhancement of targeted gene expression 457
G.Y.Wu

The generation of tumor vaccines by adenovirus-enhanced transferrinfection of cytokine genes into tumor cells 467
G.Maass, K.Zatloukal, W.Schmidt, M.Berger, M.Cotten, M.Buschle, E.Wagner and M.L.Birnstiel

Subject Index 481

LIST OF PARTICIPANTS

M.R. ALISON
Dept. of Histopathology, RPMS, Hammersmith Hospital, London W12 ONN, UK
J.J. AMARAL-MENDES
Univ. of Evora, c/o Av.D.Carlos 1126-1o-2o, 1200 Lisboa, PORTUGAL
O. ANDRISANI
Sch.of Veterinary Medicine, Purdue University, Dept. of Vet.Physiology and Pharmacology, 1246 Lynn Hall, West Lafayette IN 47907-1246, USA
A. ANTIPA
Ippokration General Hospital, II Dept.Internal Medicine,
Athens, 115 21 Greece.
M.O. AZAMBUJA da FONSECA
Univ.of Coimbra, c/o Av.D.Carlos 1126-1o-2o, 1200 Lisboa, PORTUGAL
P. S. BHATHAL
Department of Anatomic Pathology, The Royal Melbourne
Hospital, Victoria Crattan street, Parkville, 3050, AUSTRALIA
M. BILODEAU
Unite' INSERM U-49, Hopital de Pontchaillou,
Rennes 35033, FRANCE
A. BITSCH
Institute of Pharmacology and Toxicology, University of
Wurzburg, Versbacherstrasse 9, 97078 Wurzburg, GERMANY
C. BOCACCIO
Dip.di Scienze Biomediche, e Oncologia Umana, Universita' di
Torino, C.so M.D' Azeglio 52, 10126 Torino, ITALY
O. BOJAN
Dept.of Biochemistry, Oncological Institute, "Prof.dr.I.Chriricuta"
34-36 Republicii str., 3400 Cluj-Napoca, ROMANIA
M.P. BRALET
INSERM U.370, Laboratoire de Biologie Cellulaire, Faculte de
Medecine, 156 Rue de Vaugirard, 75730 PARIS Cedex 15, FRANCE
T. BURGE
Dept.of Toxicology/B 881/608, SANDOZ AGRO Ltd., CH-4132 Muttenz 1,
SWITZERLAND

S. CARIOU
Unite' INSERM U-49, Hopital de Pontchaillou,
Rennes 35033, FRANCE
O.M.S. CARVALHO
Univ.of Evora, c/o Av.D.Carlos 1126-1o-2o, 1200 Lisboa, PORTUGAL
T. CHRISTOFFERSEN
Dept.of Pharmacology, Faculty of Medicine, University of Oslo,
PO Box 1057, Blindern, N-0316, Oslo, NORWAY
P. COMOGLIO
Dept.of Biomedical Sciences, University of Turin, School of Medicine,
Corso d' Azeglio 52, Torino, ITALY
C. COUET
Ministry of Agriculture, Fisheries and Food, Food Science Laboratory,
Colney Lane, Norwich NR4 7UQ, UK
L. COVA
INSERM Unite' 271, 151 Cours Albert Thomas, 69424 LYON,
Cedex 03, FRANCE
D. DAICOVICIU
Dept.of Biochemistry, Oncological Institute, "Prof.dr.I.Chriricuta"
34-36 Republicii street, 3400 Cluj-Napoca, ROMANIA
O.F. DAJANI
Dept. of Pharmacology, Faculty of Medicine, University of Oslo,
PO Box 1057, Blindern, N-0316, Oslo, NORWAY
J.J. DIAZ GIL
Lab. de Hepatologia Experimental, Serv. de Bioquimica Experimental,
Clinica Puerto de Hierro, San Martin de Porres 4, 28035 Madrid, SPAIN
R.M. DIXON
Dept.of Biochemistry, University of Oxford, South Parks Road, Oxford
OX1 3QU, UK
C.DRAKOULIS
II Dept. of Internal Medicine, Nikeas Hospital, P.Rali-
Fanarioton 3, Nikea 184 54 ,Athens, GREECE
T.L. ERAISER
Lab.of Immunochemistry, Institute of Carcinogenesis, Cancer Research
Center, 24 Kashirskoyo shosse, 115478 Moscow, RUSSIA

N.E. GIANNOULIS
Dept. Laboratory Medicine, Hellenic Cancer Institute,
St. Savas Hospital, 171 Alexandras Ave.
115 21 Athens, GREECE
C. GUGUEN-GUILLOUZO
Unite' des Recherches hepatologiques, INSERM U49, Hopital Pontchaillou,
Rennes, 35033, FRANCE
C. HADJIAGAPIOU
Dig. Dis. Research Center, The University of Chicago, 5841 S. Maryland
Avenue, Chicago IL 60637, USA
T. HAKVOORT
Dept.Anatomy and Embryology, Univ. of Amsterdam, Academic Medical
Center, Meibergdreef 15, 1105 AZ Amsterdam, THE NETHERLANDS
D. HATZOPOULOS
I Dept.of Internal Medicine, Nikeas Hospital, P.Rali-
Fanarioton 3, Nikea 184 54 ,Athens, GREECE
M. IATRIDES
Distomo Health Center, Fokis district, GREECE
A. IONITA
Dept.of Gastroenterology and Hepatology, Fundeni Clinical Hospital,
258 Fundeni Avenue, Bucharest 72437, ROMANIA
T.W. JORDAN
Dept. of Biochemistry, School of Biological Sciences, Victoria
University of Wellington, PO Box.600, Wellington, NEW ZEALAND
V.A. KOBLIAKOV
Lab.of Carcinogenic Agents, Institute of Carcinogenesis, Cancer Research
Center, 24 Kashirskoyo shosse, 115478 Moscow, RUSSIA
E.I. KUDRJAVTSEVA
Inst.of Carcinogenesis, Cancer Research Center, 24 Kashirskoyo shosse,
115478 Moscow, RUSSIA
N.I. KUPRINA
Lab. of Immunochemistry, Institute of Carcinogenesis, Cancer Research
Center, 24 Kashirskoyo shosse, 115478 Moscow, RUSSIA
L. LAMBOTTE
Laboratoire de Chirurgie Experimental, Universite
Catholique de Louvain, Avenue Hippocrate 55,
Brussels B-1200, BELGIUM

M. McMENAMIN
Harwell Laboratory, 313 Woodstock Road, OX2 7NY, Oxford,UK
N. MAIRESSE
I.R.B.H.N., Univ.Libre de Bruxelles, 808 route de Lennik B-1070, Brussels, Belgium.
J.G. MAJ
Dept.of Clinical Biochemistry and Environmental Toxicology, School of Medicine, Lublin, Jaczewskiego 8, 20-950 Lublin, POLAND
S.H. MARINHO
Dep.de Quimica, Fac.de Ciencias da Universidade de Lisboa,
R.Ernesto de Vasconcelos, Bloco C1, Piso 5, 1700 Lisboa, PORTUGAL
K. MATSUMOTO
Division of Biochemistry, Osaka University Medical School, Biomedical Research Center, Suita, Osaka 565, JAPAN
G.K. MICHALOPOULOS
Dept.of Pathology, University of Pittsburgh, School of Medicine,
Pittsburgh, PA 15261, USA
S-A. MIHAI
Univ. of Medicine, "Carol Davila", Hospital St. John, Medical clinic,
13 Vitan Birzesti street, Bucharest, ROMANIA
L. MIQUEROL
Institut Cochin de Genetique Moleculaire, INSERM U129,
24 rue du faubourg St. Jaques, 75014 Paris, FRANCE
J.N. PAPADOPOULOS
Laboratory of Haematology, Athens Naval Hospital, 71
Dinokratous street, 115 21, Athens, GREECE
B.-C. PARK
Dept.of Internal Medicine, Kosin University School of Medicine,
34 Amana-dong Suh Ku, 602-030 Pusan, KOREA
V.S. POLTORANINA
Laboratory of Immunochemistry, Institute of Carcinogenesis, Cancer Research Center, 24 Kashirskoyo shosse, 115478 Moscow, RUSSIA
I-D. POSTESCU
Dept.of Biochemistry, Oncological Institute, "Prof.dr.I.Chriricuta",
34-36 Republicii street, 3400 Cluj-Napoca, ROMANIA

C.J. POWELL
D.H. Dept. of Toxicology, St.Bartholomew's Hospital Medical School, Dominion House, 59 Bartholomew Close, London EC1 7ED, UK
S.P. RADAEVA
Deutsches Krebsforschungszentrum, Abteilung fur Cytopathologie, 69009 Heidelberg, Im Neuenheimer Feld 280, 69120 Heidelberg, GERMANY
B. RESULI
University Clinic of Gastroenterology, University Hospital Center, Tirana, ALBANIA
M.R. RISCA
Oncological Institute, "I. Chiricuta", Republicii St. 34-36, Cluj-Napoca, ROMANIA
V. ROGIERS
Dept.of Toxicology, Frije Universiteit Brussel, Laarbeklaan 103, B-1090 Brussels, BELGIUM
S.D. RYDER
Institute of Liver Studies, King's College Hospital, Bessemer Road, London SE5 9PJ, UK
C.E. SARRAF
Dept. of Histopathology, RPMS, Hammersmith Hospital, London W12 ONN, UK
R. SCHULTE-HERMANN
Institute for Tumor Biology, Cancer Research, Borschkegasse 8a, A-1090 Vienna, AUSTRIA
D.A. SHAFRITZ
Marion Bessin Liver Research Center, Albert Einstein College of Medicine, Bronx, New York, USA
S-H. SHEN
Biotechnology Research Institute, National Research Council Canada, 6100 Royalmoun Avenue, Montreal Quebec H4P 2R2, CANADA
N. SHIMA
Snow Brand Milk Products Co Ltd., Research Institute of Life Science, 519, Shimoishibashi, Ishibashi-machi, Shimotsuga-gun, Tochigi 329-05, JAPAN

A. SIDDIQUI
Dept. of Microbiology and Immunology, University of Colorado School of Medicine, 4200 E. 9th. Ave. B175, BRB 302, Denver, CO 80262, USA
A.C. SIMONA
Dep.of Cancer Biology, Oncological Institute of Bucharest ,Sos.Fundeni nr.252 sector 2,Bucharest, ROMANIA
P. SKETT
Dept. of Pharmacology, University of Glasgow, Glasgow G12 8QQ,UK
G.G. SKOUTERIS
Deutsches Krebsforschungszentrum, Dept.Appl.Tumor Virology, Lab.of Virus Host-cell interactions, Im Neuenheimer Feld 242, 69120 Heidelberg, GERMANY
A.J. STRAIN
Liver Unit, Queen Elisabeth Hospital, Birmingham B15 2TH, UK
S. STROM
Dept.of Pathology, University of Pittsburgh Medical Center, Pittsburgh, PA 15261, USA
S.S. THORGEIRSSON
Laboratory of Exp. Carcinogenesis, Chemical and Physical Carcinogenesis Program, Div. Cancer Etiology, NIH, 9000 Rockville Pike, Bldg. 37 Rm: 3C28 Bethesda, MD 20892, USA
C.M. TSIAPALIS
Dept.of Biochemistry, Hellenic Cancer Institute, Papanikolaou Research Center of Oncology, 171 Alexandras Avenue,
115 21 Athens, GREECE
C. TZORTZOPOULOS
Alexandras Hospital, Kifissias Avenue, 115 21 Athens, GREECE
S.K. ULGEN
Marmara University, Faculty of Medicine, Istanbul, TURKEY
A.C. VANTARAKIS
University of Patras, 103 Thessalonikis street,
26441 Patras, GREECE
L. WEIR
Medical Research Council, Hodgkin Building, PO Box 138, Lancaster road, Leicester, LE19HN, UK

J.R. WILLIAMSON
Dept. of Biochemistry and Biophysics, University of Pennsylvania, Sch.of Medicine, Philadelphia, PA 19104-6089, USA
G.Y. WU
Div.of Gastroenterology-Hepatology, University of Connecticut, School of Medicine, 263 Farmington Avenue Room AM-044, Farmington, CT 06030-1845, USA
K. ZATLOUKAL
Institute of Pathology, University of Graz,
Auenbruggerplatz 25, A-8036 Graz, AUSTRIA

MOLECULAR PATHWAYS IN HEPATOCYTE PROLIFERATION

George G. Skouteris
Deutsches Krebsforschungszentrum
Department of Applied Tumor Virology
Laboratory of Virus-Host Cell interactions
Im Neuenheimer Feld 242
D-69120 Heidelberg, Germany

In the liver, most of the hepatocyte population resides in the G_0 compartment of the cell cycle and exit into the G_1 phase is achieved after an insult or injury stimulus is given (Michalopoulos, 1990). It has recently been suggested that the cellular populations which play the major role in the replacement of the functional liver mass after insult or chemical injury are mature hepatocytes, the oval cells with characteristics of both bile duct cells and fetal hepatocytes, and the Ito (fat storing) cells (Evarts et al., 1987, Evarts et al., 1990, Schirmacher et al., 1992).

Shortly after partial hepatectomy (4 hours), the transcripts of Hepatocyte Growth Factor (HGF) and those of Transforming Growth Factor-alpha (TGF-α) are significantly increased (Evarts et al., 1993). HGF (produced by Ito cells) is coupled with its receptor (*c-met*) which shows a wide distribution in non-hepatic tissues also (lung, skin, uterus, kidney), exerts its effects on mature hepatocytes and oval cells (Naldini et al., 1991, Hu et al., 1993, Giordano et al., 1989). TGF-α and HGF transcripts detected at 4 hours after partial hepatectomy were found in Ito cells (HGF, TGF-α) and in the oval cells (TGF-α) (Evarts et al.,1992, Ramadori et al.,1992). HGF and TGF-α stimulate DNA synthesis in both mature hepatocytes and hepatic stem cells

NATO ASI Series, Vol. H 88
Liver Carcinogenesis
Edited by G. G. Skouteris

(Mead and Fausto, 1989, Nakamura et al., 1986, Strain et al., 1991, Evarts et al., 1993). Significant DNA synthesis is observed in cells of the ductular structure, in oval cells and in cells of mesenchymal origin, within four hours after the partial hepatectomy of animals treated with acetyaminofluorene (Evarts et al., 1993). This effect may be due to the circulating HGF because of limited hepatic clearance following the insult (Liu et al., 1992).

It is known that the transforming growth factor beta (TGF-β) plays an inhibitory role in hepatocyte proliferation both in vitro and *in vivo* (Oberhammer et al., 1991). In the liver, the stellate cells produce TGF-β1 and its transcripts begin to increase in parallel with the transcripts of HGF and TGF-α (Nakatsukasa et al., 1990, Evarts et al., 1990, Thorgeirsson et al. this book). TGF-β1 inhibition of hepatocyte progress through the cell cycle seems to be located just before the G1/S traverse (Thoresen et al., 1992). The stellate cells produce matrix components which promote the differentiation of all cell populations within the liver tissue (Nakatsukasa et al., 1990 . At later stages of the regenerating process, combining immunocytochemistry and *in situ* hybridization, aFGF transcripts appeared to be present in mature hepatocytes, in oval cells, as well as in perisinusoidal stellate cells (Marsden et al., 1992). The transcripts of aFGF were increased at approximately 24 hours after the regenerating stimulus, thus implying that this factor may play a role other than growth promoting in the newly developed hepatic structure. Such roles may be cell migration, angiogenesis and differentiation (Goldfarb 1990, Montesano et al., 1986).

The disruption of the tight contacts of hepatocytes after partial hepatectomy or injury is accompanied by the liberation of membrane and other substances whose identity and functions still remain unknown (Yee and Revel, 1978, Meyer et al., 1981, Traub et al., 1983). Recently a post-injury liberated substance called "injurin" has been isolated and partially characterized (Matsumoto et al., 1992). It is also shown that injurin is present in the sera of animals with organ injuries and at the same time this substance cnhances the production of HGF in the non-injured organs (Matsumoto et al. op. cit). Among other potential primary regulators of hepatocyte proliferation may be the "cell surface modulator" (CSM) which was also shown to modulate hepatocyte differentiated functions (Nakamura et al., 1984). Therefore, the changes in the liver macroenvironment such as the liberation of substances stimulating the production of liver-specific growth factors (HGF) after partial hepatectomy may contribute to a sequence of events which prepares the exit of the remaining mature hepatocytes to the G_1 phase of the cell cycle. The exit of the hepatocytes to the G1 phase, is accompanied by changes in the expression of several genes such as *c-myc* and *c-fos* whose products act by modulating the transcription of other cell cycle-related genes (Kost et al., 1990, Skouteris and McMenamin, 1992, Kruijer et al., 1986). It has been shown that c-myc is expressed throughout the *in vitro* and the *in vivo* (after resection) cell cycle (Etienne et al., 1988, Skouteris and Kaser, 1992). TGF-β1 and the persistent expression of *c-myc* may be utilized as signals regulating the percentage of hepatocytes undergoing programmed cell death (apoptosis) (Bursch et al., 1992). MYC protein regulates apoptosis in fibroblasts (Evan et al., 1992) and TGF-β1 also induces apoptosis in various cells of epithelial origin such as hepatocytes (Oberhammer et al., 1991). Transfection of primary hepatocytes with a *c-myc* construct resulted in significant cell detachment six days post-transfection (Skouteris and Kaser, 1992). However, as similar studies have suggested, primary cells transfected with *c-myc* often undergo extensive cell death before the appearance of transformed cells (Bissonnette et al., 1992). *C-myc* expression in proliferating hepatocytes under *in vivo* or *in vitro* conditions, is more likely to be due to the action of positive growth regulators (TGF-α, HGF) rather to TGF-β1. However, a rapid increase in *c-myc* expression in response to TGF-β1 has been shown for a variety of cells such as fibroblasts and vascular smooth muscle cells (Janat and Liau, 1992). Therefore, activation of *c-myc* may constitute a non-specific signal which drives the cells to either progress through the cell cycle or to apoptosis. A second signal may activate another pathway which preferentially "delays" or "inhibits" the apoptotic or proliferating activities of *c-myc* (Bissonnette et al., 1992).

Intercellular signaling in hepatocyte proliferation

TGF-α with 30-40% sequence homology with EGF binds to EGF receptor (Stromblad and Anderson, 1993). EGF receptors are decreased at about 50% 24 hours after partial hepatectomy and this may be attributed to the early increase in TGF-α mRNA, rather than to EGF (Stromblad and Anderson op. cit.). EGF and TGF-α induced tyrosine kinase and phospholipase A2 activation (Skouteris and McMenamin, 1992). Prostaglandins were shown to mediate the activities of both EGF and TGF-α and when used individually to stimulate hepatocyte DNA synthesis (Skouteris et al., 1988, Skouteris and McMenamin, 1992). Prostaglandins E_2 and F2α caused an increase of the *c-*

myc mRNA levels shortly after their individual addition in culture. Thus, the elevation of *c-myc* mRNA at the tome of the G_0/G_1 transition is a common event for EGF, TGF-α, prostaglandins and HGF (Skouteris and Kaser, 1991, Fabregat et al., 1992). Blockage of calcium channels by verapamil, affected both the EGF-induced hepatocyte DNA synthesis and the EGF-induced increase in *c-myc* mRNA levels (Skouteris and Kaser, 1991).

The elevation of the *c-myc* persisted for longer times in culture than in other cellular systems (for HGF: 8 hrs) (Fabregat et al. op. cit). In this way, the operation of an autocrine regulation of hepatocyte growth has been suggested in response to the aforementioned two factors (TGF-α, EGF) (Skouteris and McMenamin, 1992). The data on the activation of the phospholipase C pathway following the ligand activation of the EGF receptor seems to be conflicting (Okano et al., 1993). Prostaglandins E_2 and $F_{2\alpha}$ were shown to interact via the Gs and the Gi proteins with their receptors to regulate cAMP (Melien et al., 1988) and also to stimulate the production of inositol 1,4,5-trisphosphate and calcium (Athari and Jugermann, 1989, Mine et al., 1990). The increased levels of cAMP and calcium immediately after the G0-G1 transition in the hepatocyte is believed to positively regulate its DNA synthesis (Thoresen et. al., 1990). Therefore, the signals after receptor coupling of EGF and probably of TGF-α are transduced via activation of both the cAMP and the PLC-dependent pathways. Additional data on the participation of both pathways (cAMP/Phospholipase C-dependent) in the transduction of signals in the hepatocyte arises from the effects of adrenergic agonists on the hepatocyte through the a1 and the β2 adrenoreceptors (Cruise et al., 1985, Refsnes et al., 1992). The elevation of norepinephrine levels in the plasma after the regenerating stimulus is probably due to its decreased degradation through the liver monoamine oxidase system (Michalopoulos and Zarnegar, 1992). Norepinephrine was shown to amplify the effects of growth factors such as that of EGF and HGF on hepatocyte DNA synthesis (Lindroos et al., 1991). The activities of adrenergic agonists through the a1 receptor are mediated through phospholipase C, although a response is also elicited via the β2 adrenoreceptors (Refsnes et al., 1992). The latter effect implies an involvement of the cAMP-dependent pathway (Refsnes et al., op. cit.), considering the dramatic elevation in β-adrenoreceptors during culture of primary hepatocytes (Nakamura et al., 1983). The elevated cAMP levels observed in late G1 were shown to inhibit the G1/S transition of the hepatocyte. Such an effect was also identified (i.e. elevated cAMP levels in the G1/S border) after glucagon administration (Thoresen et al., 1990). HGF had no effect on both the adenylate cyclase activity and cAMP levels in the hepatocyte and, at the same time, co-administration of HGF with cAMP-increasing agents (dibutyryl cAMP, IBMX) throughout the culture time resulted in inhibition of DNA synthesis (Marker et al., 1992). The HGF-stimulated production of inositol 1,4,5, trisphosphate in the hepatocyte is likely to be mediated through activation of PLC-γ. The -SH2 domains of PLC-γ interact with receptor tyrosyl kinases via specific consensus sequences which are also present in the cytosolic domain of the HGF receptor (c-met protein) (Cantley et al., 1991, Graziani et al., 1991). Genistein, an inhibitor of protein tyrosine kinase, when present in HGF-stimulated hepatocyte cultures, abolished both the 1,4,5 P3 production and the increase in cytosolic calcium through mobilization from the extracellular stores (Baffy et al., 1992).

Therefore, PLC-γ is a molecule with a central role in the transduction of signals generated by hepatocyte-specific growth factors.

REFERENCES

Athari, A. and Jungermann, K. (1989) Direct activation by prostaglandin F2a but not thromboxane A2 of glycogenolysis via an increase in inositol 1,4,5-trisphosphate in rat hepatocytes. Biochem. Biophys. Res. Commun. 163: 1235-1242.

Baffy G., Yang L., Michalopoulos, G.K. and Williamson, J.R. (1992) HGF induces calcium mobilization and inositol phosphate production in rat hepatocytes. J. Cell Physiol. 153: 332-339.

Bissonnette, R.P., Echeverri, F., Mahboubi, A. and Green D.R. (1992) Apoptotic cell death induced by c-myc is inhibited bcl-2. Nature 359: 552-554.

Bursch, W., Oberhammer, F. and Schulte-Hermann R. (1992) Cell death by apoptosis and its protective role against disease. Tr. Pharmacol. Sci. 13: 245-251.

Cantley, L.C., Auger, K., Carpenter, C., Duckworth, B., Graziani, A., Kapeller, R., and Soltoff, S. (1991) Oncogenes and signal transduction. Cell 64: 281-302.

Cruise, J.L., Houck, K.A. and Michalopoulos, G.K. (1985) Induction of DNA synthesis in cultured rat hepatocytes through stimulation of alpha-1 adrenoreceptor by norepinephrine. Science 227: 749-751.

Etienne, P.L., Baffet, G., Desvergene, B., Boisnard-Rissel, M., Glaise, D., and Guguen-Guillouzo, C. (1988) Transient expression of c-fos and constant expression of c-myc in freshly isolated and cultured normal adult rat hepatocytes. Oncogene Res. 3: 255-262.

Evan, G.I., Wyllie, A.H., Gilbert, C.S., Littlewood, T.D., Land, H., Brooks, M., Waters, C.M., Penn, L.Z., and Hancoc, D.C. (1992) Induction of apoptosis in fibroblasts by c-myc protein. Cell 69: 119-128.

Evarts, R.P., Nagy, P., Marsden, E., and Thorgeirsson, S.S. (1987) In situ hybridization studies on expression of albumin and α-FP during the early stage of neoplastic transformation in rat liver. Cancer Res. 47: 5469-5475.

Evarts, R.P., Nakatsukasa, H., Marsden, E.R., Hsia, C-C., Dunsford, H.A., and Thorgeirsson, S.S. (1990) Cellular and molecular changes in the early stages of chemical hepatocarcinogenesis. Cancer Res. 50: 3439-3444.

Evarts, R.P., Nakatsukasa, H., Marsden, E.R., Hu, Z., and Thorgeirsson, S.S. (1992) Expression of TGF-α in regenerating liver and during hepatic differentiation. Mol. Carcinog. 5: 25-31.

Evarts, R.P., Hu, Z., Fujio, K., Marsden, E.R., and Thorgeirsson, S.S. (1993) Activation of hepatic stem cell compartment in the rat: role of TGF-α, HGF and acidic FGF in early proliferation. Cell Growth & Diff. 4: 555-561.

Fabregat, I., de Juan, C., Nakamura, T., and Benito, M. (1992) Growth stimulation of rat fetal hepatocytes in response to HGF; modulation of c-myc and c-fos expression. Biochem. Biophys. Res. Commun. 189: 684-690.

Fouad Janat, M. and Liau G. (1992) TGF-β1 is a powerful modulator of PDGF action in vascular smooth muscle cells. J. Cell Physiol. 150: 232-242.

Giordano, S., Ponzetto, C., DiRenzo, M.F., Cooper, C.S., and Comoglio, P.M. (1989) Nature 339: 155-156.

Goldfarb, M. (1990) The fibroblast growth factor family. Cell Growth Differ. 1: 439-445.

Hu, Z., Evarts, R.P., Fujio, K., Marsden, E.R., and Thorgeirsson, S.S. (1993) Expression of HGF and c-met gene during hepatic differentiation and liver development in the rat. Am. J. Pathol. 142: 1823-1830.

Kost, D.P. and Michalopoulos, G.K. (1990) Effect of EGF on the expression of protooncogenes c-myc and c-Ha-ras in short-term primary hepatocyte culture. J. Cell Physiol. 144: 122-127.

Kruijer, W., Skelly, H., Botteri, F., van der Putten, H., Barber, J.R., Verma, I.M., and Leffert, H.L. (1986) Protooncogene expression in regenerating liver is stimulated in cultures of primary adult rat hepatocytes. J. Biol. Chem. 7929-7933.

Lindroos, P.M., Zarnegar, R. and Michalopoulos, G.K. (1991) Hepatic growth factor (hepatopoietin A) rapidly increases in plasma before DNA synthesis and liver regeneration stimulated by partial hepatectomy and carbon tetrachloride administration. Hepatology 13: 743-750.

Liu, K-X., Kato, Y., Narukawa, M., Kim, D-C., Hanano, M., Higuchi, O., Nakamura T., and Sugiyama, Y. (1992) Importance of the liver in plasma clearance of hepatocyte growth factor in rats. Am. J. Physiol. 26: G642-G649.

Marker, A.J., Galloway, E., Palmer, S., Nakamura, T., Gould, G.W., McSween, R.N.M., and Bushfield, M. (1992) Role of the adenylate cyclase, phosphoinositidase C and receptor tyrosyl kinase systems in the control of hepatocyte proliferation by hepatocyte growth factor. Biochem. Pharmacol. 44: 1037-1043.

Marsden, E.R., Hu, Z., Fujio, K., Nakatsukasa, H., Thorgeirsson, S.S. and Evarts, R.P. (1992) Expression of acidic FGF in regenerating liver and during hepatic differentiation. Lab. Invest. 67: 427-433.

Matsumoto, K., Tajima, H., Hamanoue, M., Kohno, S., Kinoshita, T., and Nakamura, T. (1992) Identification and characterization of "injurin" an inducer of expression of the gene for hepatocyte growth factor. Proc. Natl. Acad. Sci. U.S.A. 89: 3800-3804.

Mead, J.E. and Fausto, N. (1989) TGF-α may be a physiological regulator of liver regeneration by means of an autocrine mechanism. Proc. Natl. Acad. Sci. U.S.A. 86: 1558-1562.

Melien, O., Wisnes, R., Refsnes, M., Gladhaug, I.P., and Christoffersen, T. (1988) Pertussis toxin abolishes the inhibitory effect of prostaglandins E1,E2,I2 and F2a on hormone-induced cAMP accumulation in cultured hepatocytes. Eur. J. Biochem. 172: 293-297.

Meyer, D.J., Yancy, M. and Revel, J.-P. (1981) Intercellular communication in normal and regenerating rat liver: a quantitative analysis. J. Cell Biol. 91: 505-523.

Michalopoulos, G.K. (1990) Liver regeneration: Molecular mechanisms of growth control. FASEB J. 4: 240-249.

Michalopoulos, G.K. and Zarnegar, R. (1992) Hepatocyte Growth Factor. Hepatology 15: 149-155.

Mine, T., Kojima, I. and Ogata E. (1990) Mechanisms of prostaglandin E2-induced glucose production in rat hepatocytes. Endocrinol. 126: 2831-2836.

Montesano, R., Vassali, J-D., Baird, A., Guillemin, R., and Orci, L. (1986) Basic FGF induces angiogenesis in vitro. Proc. Natl. Acad. Sci. U.S.A. 83: 7297-7301.

Nakamura, T., Tomomura, A., Noda, C., Shimoji, M., and Ichihara, A. (1983) Acquisition of a β-adrenergic response by adult rat hepatocytes during primary culture. J. Biol. Chem. 258: 9283-9289.

Nakamura, T., Nakayama, Y. and Ichihara, A. (1984) Reciprocal modulation of growth and liver functions of mature rat hepatocytes in primary culture by an extract of hepatic plasma membranes. J. Biol. Chem. 259: 8056-8058.

Nakamura, T., Teramoto, H. and Ichihara, A. (1986) Purification and characterization of a growth factor from rat platelets for mature parenchymal hepatocytes in primary cultures. Proc. Natl. Acad. Sci. U.S.A. 83: 6489-6493.
Nakatsukasa, H., Evarts, R.P., Hsia, C-C., and Thorgeirsson, S.S. (1990) TGF-β1 and type I procollagen transcripts during regeneration and early fibrosis of rat liver. Lab. Invest. 63: 171-180.
Naldini, L., Vigna, E., Narshiman, R.P., Gaudino, G., Zarnegar, R., Michalopoulos, G.K., and Comoglio, P.M. (1991) HGF stimulates the tyrosine kinase activity of the receptor encoded by the proto-oncogene c-met. Oncogene 6: 501-504.
Oberhammer, F., Bursch, W., Parzefall, W., Breit, P., Erber, E., Stadler, M., and Schulte-Hermann, R. (1991) Effect of TGF-β on cell death of cultured rat hepatocytes. Cancer Res. 51: 2478-2485.
Okano, Y., Mizuno, K., Nakamura, T., and Nozawa, Y. (1993) Tyrosine phosphorylation of PLCγ in c-met/HGF receptor-stimulated hepatocytes: comparison with HepG2 hepatocarcinoma cells.Biochem. Biophys. Res. Commun. 190: 842-848.
Ramadori, G., Neubauer, K., Odenthal, M., Nakamura, T., Knittel, T., Schwogler, K.-H., and Meyer zum Buschenfelde (1992) The gene of HGF is expressed in fat-storing cells of rat liver and is downregulated during cell growth and by TGF-β. Biochem. Biophys. Res. Commun. 183: 739-742.
Refsnes, M., Thoresen, G.H., Sandnes, D., Dajani, O.F., Dajani, L., and Christoffersen, T. (1992) Stimulatory and inhibitory effects of catecholamines on DNA synthesis in primary rat hepatocyte cultures: Role of alpha1- and beta-adrenergic mechanisms. J. Cell Physiol. 151: 164-171.
Schrimacher, P., Geerts, A., Pietrangelo, A., Dienes, H.P., and Rogler, C.E. (1992) HGF/hepatopoietin A is expressed in fat storing cells from rat liver but not myofibroblast-like cells derived from fat-storing cells. Hepatology 15: 5-11.
Skouteris, G.G., Ord, M.G. and Stocken, L.A. (1988) Regulation of the proliferation of primary rat hepatocytes by eicosanoids. J. Cell Physiol. 135: 516-520.
Skouteris, G.G. and Kaser M.R. (1991) Prostaglandins E2 and F2a mediate the increase in c-myc expression induced by EGF in primary rat hepatocyte cultures. Biochem. Biophys. Res. Commun. 178: 1240-1246.
Skouteris, G.G. and McMenamin, M. (1992) TGF-α-induced DNA synthesis and c-myc expression in primary rat hepatocyte cultures is modulated by

indomethacin. Biochem. J. 281: 729-733.
Skouteris, G.G. and Kaser M.R. (1992) Expression of exogenous c-myc oncogene does not initiate DNA synthesis in primary rat hepatocyte cultures. J. Cell Physiol. 150: 353-359.
Strain, A.J., Ismail, T., Tsubouchi, H., Arakaki, N., Hishida, T., Kitamura, N., Daikuhara, Y., and McMaster, P. (1991) Native and recombinant hHGF are highly potent promoters of DNA synthesis in both human and rat hepatocytes. J. Clin. Invest. 87: 1853-1857.
Stromblad, S. and Andersson, G. (1993) The coupling between TGF-α and the EGF receptor during rat liver regeneration. Exp. Cell Res. 204: 321-328.
Thoresen, G.H., Sand, T.E., Refsnes, M., Dajani, O.F., Curen, T.K., Gladhaug, I.P., Killi, A., and Christoffersen, T. (1990) Dual effects of glucagon and cAMP on DNA synthesis in cultured rat hepatocytes: Stimulatory regulation in early G1 and inhibition shortly before the S phase entry. J. Cell Physiol. 144: 371-382.
Traub, O., Druge, P.M. and Willecke, K. (1983) Degradation and resynthesis of gap junction protein in plasma membranes of regenerating liver after partial hepatectomy or cholestasis. Proc. Natl. Acad. Sci. U.S.A. 80: 755-759.
Yee, A.G. and Revel, J.-P. (1978) Loss and reappearance of gap junctions in regenerating liver. J. Cell Biol. 78: 554-564.

HGF AND LIVER GROWTH REGULATION

George K. Michalopoulos
Department of Pathology
Univ. of Pittsburgh Medical Center
Pittsburgh, PA 15261
USA

INTRODUCTION.

Liver growth has fascinated scientists for many years, perhaps even centuries. Liver has a strong regenerative potential in response to toxic agents causing loss of hepatic parenchyma. Many chemical agents tested for carcinogenicity lead to formation of hepatic tumors. Other xenobiotic agents (e.g. anti-epileptics, hypolipidemic agents, barbiturates etc.) lead to a controlled expansion of hepatic mass above normal. Liver mass is gradually restored to normal upon discontinuation of the chemical, in a process accompanied by massive apoptosis. These and other phenomena have intensified interest in molecular mechanisms that control liver growth. The introduction of 2/3 partial hepatectomy as a systematic model for the study of liver regeneration in rats as shown by Higgins and Anderson (1931) led to a definition of phenomena related to liver regeneration which were well reproduced and became the guiding lines for research in the years that followed. These phenomena were as follows:

1.Regeneration of the liver in situ elicits the creation of a blood borne stimulus which transmits a mitogenic signal for hepatocytes anywhere in the body engrafted sites of liver tissue shown by Grisham et al. (1964) or isolated

NATO ASI Series, Vol. H 88
Liver Carcinogenesis
Edited by G. G. Skouteris

hepatocytes shown by Jirtle and Michalopoulos (1982), rat livers in parabiotic circulation by Fisher (1971) etc.

2.The stimulus for regeneration was enhanced by pancreatic secretions, especially insulin. Liver regeneration, however, could start in the absence of any pancreatic or visceral stimuli shown by Bucher (1982).

3.Regeneration is carried out by mature adult parenchymal hepatocytes and not by stem cells. Hepatocytes enter into synchronized waves of DNA synthesis which start from the periportal area and eventually extend to the pericentral area of the lobule shown by Grisham (1962) and Rabes et al. (1976).

4. Liver regeneration starts very early after partial hepatectomy. Membrane hyperpolarization of plasma membrane in hepatocytes starts within 10 minutes after partial hepatectomy as shown by de Hemptinne et al. (1985). Recent studies have shown that within 30 minutes following partial hepatectomy there is dramatically enhanced expression of more than thirty genes as shown by DuBois (1990).

Essential conclusions from the above phenomena were that the stimulus or stimuli for liver regeneration start very early, almost immediately after partial hepatectomy and that they are blood borne. Despite a small increase in DNA synthesis in exocrine pancreas shown by Rao and Subbarao (1986) and enhanced expression of c-fos and c-myc in kidney cortex as shown by Roesel et al. (1989), partial hepatectomy leads to regeneration in a specific manner for liver only. There is no massive DNA synthesis or enlargement of any other organs.

The advent of hepatocyte primary cultures revealed by Bissell et al. (1973)

allowed development of bioassays in which potential growth stimuli could be screened for activity as hepatocyte mitogens. These studies were instrumental in defining essentially all the growth factors known to date to be involved in controlling mitogenesis in hepatocytes. Studies with serum free cultures in chemically defined media allowed discovery of not only mitogens per se but also of other substances that, though not mitogenic themselves, they enhance the activity of other mitogens. Thus two classes of mitogens could be defined:

1.Primary (or complete, or direct) mitogens are substances which by themselves in serum free conditions can stimulate proliferation in hepatocyte cultures.

2.Secondary (or incomplete or co-mitogens) mitogens are substances which by themselves cannot stimulate DNA synthesis in serum free primary hepatocyte cultures but they enhance the effect of primary mitogens or partially inhibit the effect of mitogenic inhibitors.

3.Mito-inhibitors are substances which inhibit the effect of primary mitogens without being toxic to hepatocytes.

The substances which were defined as belonging into these classes are as follows:

1.Primary mitogens:

a.EGF and TGFa (sharing the same receptor).

b.HGF

c.acidic FGF

d.Hepatopoietin B (a glycolipid of as yet unidentified structure).

2.Secondary mitogens

a.Insulin

b.Glucagon

c.Norepinephrine

d.Vasopressin

e.Angiotensin II

f.estrogens

g.HSS (active in vivo. In culture, co-mitogenic with EGF (54-56))

3.Mitoinhibitors.

a.$TGF\beta_{1.}$

Several reviews covering these growth factors have appeared in recent publications. This presentation relates to the role of HGF in different aspects of liver growth control and discusses some of the recent findings related to the earliest stages of liver regeneration after partial hepatectomy.

I. Liver Regeneration after Partial Hepatectomy

The very early findings showing that HGF is the only protein mitogen for hepatocytes present in the plasma revealed by Michalopoulos et al. (1984) and Lindroos et al. (1991) make HGF the most likely candidate for the cross-circulating mitogen stimulating proliferation of hepatocytes in non-hepatectomized parabiotic circulation partners of hepatectomized rats and in grafted hepatic tissue and transplanted isolated hepatocytes in extrahepatic sites shown in multiple studies by Grisham et al., (1964); Jirtle and

Michalopoulos (1982); Fisher et al., (1971); Bucher (1982) and Grisham (1962). More recent studies by Kinoshita et al. (1991) and Zarnegar et al. (1991) have measured the changes in HGF levels in plasma after 2/3 partial hepatectomy in rats or after carbon tetrachloride intoxication and have reinforced the crucial role for HGF in liver regeneration. These findings are as follows:

1.HGF levels rise in the plasma 15-17 fold within 1-2 hours after 2/3 partial hepatectomy shown by Kinoshita et al. (1991) and Zarnegar et al. (1991). The levels return to lower (but persistently higher above background) values within 24 hours. DNA synthesis ensues with a peak at 24 hours as demonstrated by Grisham (1962).

2. HGF levels also rise very shortly after carbon tetrachloride intoxication and remain elevated for 24-36 hours. DNA synthesis ensues with a peak at 48 hours after intoxication revealed by Kinoshita et al (1991).

3. HGF mRNA in rat liver increases within 3-6 hours after 2/3 partial hepatectomy, peaks at 20 hours and returns to basal levels by 72 hours revealed by Kinoshita et al (1991) and Zarnegar et al. (1991).

4. Injection of HGF directly into the portal vein led to DNA synthesis in the liver of dogs demonstrated by Francavilla et al. (1991), 30% partially hepatectomized rats as shown by Matsumoto and Nakamura (1993) and normal rats demonstrated by Liu and Michalopoulos, in press.

It should be also noted that dramatic elevations in HGF were also noted in humans after hepatic surgery demonstrated by Tomiya et al. (1992). The same study however showed that operations unrelated to liver also affected HGF levels, though much less than the hepatic resections.

The early steep increase of HGF levels in the plasma following 2/3 partial hepatectomy and preceding by many hours the initiation of DNA synthesis, makes HGF the best candidate for triggering the multiple effects described following 2/3 partial hepatectomy. Its rise within 1 hour is consistent with the early changes in gene expression in liver seen within 30 minutes after 2/3 partial hepatectomy. The mechanism for the rise is not clear at this point. The elevation of HGF protein in the plasma prior to the elevation of HGF mRNA in the liver clearly suggests that HGF increase in the plasma is not due to increased synthesis from hepatic sources. Though platelets contain large amounts of HGF, induction of severe thrombocytopenia in rats did not interfere with liver regeneration, suggesting that platelets are not involved with the generation of any relevant mitogenic signals for this process shown by Kuwashima et al. (1990). Given the wide tissue distribution of HGF revealed by Wolf et al (1991) , it is conceivable that HGF is released in the plasma from a variety of sources. Two recent studies shown by Michalopoulos and Appasamy (1993) and Liu et al. (1992) have shown that circulating HGF is normally cleared primarily by the liver , with a minor contribution by the kidneys. Time lapse autoradiographs have shown that HGF is also taken up more avidly by regenerating liver Michalopoulos and Appasamy (1993). Acute reduction in liver mass (following 2/3 partial hepatectomy or loss of the capacity to clear HGF (following toxins) may explain the acute rise of HGF prior to DNA synthesis. Such acute elevation of plasma values is also known to occur for norepinephrine.

Though the clearance hypothesis is the simplest one to explain the acute rise of HGF in the peripheral blood, several puzzles remain. The mathematics of the loss in hepatic mass and rise of HGF levels are not in straightforward correlation and perhaps can be best reconciled when the pathways of uptake and degradation of HGF are better understood. Recent studies, however, (19) have raised the possibility that the existence of other mechanisms may be involved, both in causing the elevations of plasma HGF as well as rendering hepatocytes responsive to mitogens. We have found that prolonged infusions of HGF through indwelling catheters in the portal vein, though leading to DNA synthesis of the normal liver, they do so only in periportal areas (zone 1 of liver acinus). This was also shown with TGFa. When, however, the HGF infusion was preceded by infusion of collagenase at a dose of approximately 1/1000 of that required for liver dissociation, the effects of HGF were dramatically amplified and DNA synthesis in approximately 60-70% of the hepatocytes was noted after 24 hours of HGF infusion. This is the first evidence that HGF can stimulate DNA synthesis up to levels comparable to those of liver regeneration in livers of unoperated rats. This finding, however, raises the overall question of the role of matrix degrading enzymes in rendering hepatocytes responsive to growth factors. We have observed from cell cultures that when hepatocytes are kept in collagen gels they cease being responsive to growth factors,. When these cells are treated with collagenase, they resume responsiveness to both HGF and EGF revealed by Michalopoulos et al. (1993).

If the earliest stage of liver regeneration is associated with a quick event of matrix degradation, this would potentially lead to the following events within one hour after partial hepatectomy:

1.Release of HGF in the peripheral blood. This has actually been observed, as mentioned above. The large amounts of HGF stored in liver matrix revealed by Masumoto and Yamamoto (1991) may account for the very sharp rise in HGF in the plasma, beyond the stoichiometric expectations of removing 2/3 of liver mass.

2.Hepatic matrix components should be released in the plasma. We have actually verified this by direct measurements of plasma levels of hyaluronic acid revealed by Michalopoulos et al. (unpublished data).

3.TGFβ1, also bound to hepatic matrix, should increase in the plasma. This has also been found in recent collaborative studies with Dr. Jirtle by Michalopoulos et al. (unpublished data).

These phenomena are best explained by an hypothesis that would imply that soon after partial hepatectomy, components of hepatic matrix are being rapidly degraded, causing release of HGF and TGFβ1 in the plasma. TGFβ1 in the plasma is rapidly inactivated by alpha-2- macroglobulin Crookston et al. (1993). HGF on the other hand would be available to stimulate hepatocyte proliferation. This hypothesis also addresses the specificity of liver regeneration, vis-a-vis HGF. Though HGF is a mitogen for multiple cells and tissues, the sharp rise after partial hepatectomy stimulated regeneration only in the liver because the other organs have an intact matrix surrounding the plasma membrane with mito-inhibitory components (including TGFβ1), thus preventing their proliferation in response to HGF.

The mechanisms leading to rapid degradation of the hepatic biomatrix are not clear. We have recently found, however, that urokinase activity is also rapidly increasing after partial hepatectomy Michalopoulos et al. (unpublished data). Urokinase is related to HGF activation. It has been conclusively shown that single chain HGF (the native form by which HGF is found in tissues) is inactive Lokker et al., (1992). The enzymes involved in conversion of HGF from the single chain to the heterodimer form should be able to cleave at the AVV site where the cleavage occurs. Several such enzymes may exist shown by several investigations, Mars, et al. (1993), Mizuno et al., (1992) and Naldini et al. (1992) and Naka et al. (1992). In our laboratory we have conclusive evidence that urokinase and tissue plasminogen activator are capable of converting single chain HGF to two-chain HGF in a specific fashion shown by Mars et al., (1993). The effect was demonstrated by enzymatic assays of molecular conversion which occur only with native HGF and did not occur with an HGF mutant at the AVV site. Moreover, urokinase activation of HGF led to a greatly enhanced biological activity in hepatocyte cultures. Urokinase, however, is also associated with direct Keski-Oja et al., (1993) or indirect, through plasmin shown by Rifkin (1992) activation of metalloproteinases associated with matrix breakdown. The activation of urokinase activity very soon after partial hepatectomy may lead to the following two events:

1. Activation of HGF.
2. Activation of enzymatic pathways leading to matrix degradation.

Equally puzzling is the increase in HGF mRNA at 6-24 hours after partial hepatectomy. Obviously a stimulus reaches the cells of Ito (the sites of most of HGF production in the liver Schirmacher et al. (1992) inducing an increase in HGF levels. This seems to also occur in other tissues with similar cells. A recent hypothesis was advanced by Nakamura et al. suggesting the existence of a specific protein termed "injurin" shown by Matsumoto et al. (1992) which would be a specific stimulus for release of HGF. The nature of this protein would be of great interest. Of known factors, Interleukin 1 is associated with stimulation of increase in HGF mRNA shown by Matsumoto et al. (1992) though not clearly implicated in liver regeneration until now. In recent studies, the base pair sequence of thc flanking region upstream to the 5' position of the HGF gene was elucidated by Liu et al. (1994). Several DNA responding elements were identified, including Interleukin 1, Interleukin 6, p53, glucocorticoid response element, TGFβ1, etc. This suggests that the regulation of the HGF molecule is bound to be a complex one, with several factors involved. The role of HGF produced by the Ito cells following increase in HGF mRNA remains to be determined. Mutually non-exclusive hypothescs would be that the *de novo* produced HGF stimulates adjacent hepatocytes as they enter into the second wave of DNA synthesis or that it (also) stimulates DNA synthesis in endothelial cells, for which HGF was recently shown to be a mitogen.

II.Augmentative Hepatomegaly induced by Xenobiotics

As mentioned above, several lipophilic substances used in medicine or as pesticides, chemical industry agents etc. stimulate a dramatic increase in liver weight. Most of the xenobiotic liver tumor promoters induce liver enlargement. This enlargement is mediated by a transient increase in hepatocyte DNA synthesis and hepatocyte enlargement. The transient wave of cell proliferation lasts for 5-7 days. This series of phenomena is seen with phenobarbital, peroxisome proliferators, estradiol, etc. shown by Michalopoulos et al. (1987). These findings have led to the notion that liver tumor promotion is associated with cell proliferation. Recent studies, however, have provided a different outlook toward this concept. We and others have shown that following the early phase of DNA synthesis, hepatocytes of the enlarged liver become unresponsive to the mitogenic effects of all growth factors. This was shown to be true for HGF, as well as the other hepatocyte mitogens, EGF and aFGF Tsai and Michalopoulos (1993), Tsai et al. (1991), Lindroos and Michalopoulos (1993). In the case of EGF it has been shown that this relative unresponsiveness is associated with decrease in the numbers of EGF receptors shown by Meyer and Jirtle (1989). These phenomena were found to be the case for hepatocytes isolated from rats fed phenobarbital, peroxisome proliferators, estradiol 451) etc. In view of these findings, it is highly likely that the fundamental effect of liver tumor promoters is not to stimulate cell proliferation indiscriminately but rather to suppress proliferation of the normal hepatocytes. This would lead to a compensatory increase in cell proliferation of the initiated hepatocytes resistant to the mito-inhibitory effects of the tumor promoter. This scheme is compatible with the observed data from the above in vitro studies and also some recent in vivo shown by Wanda et al. (1992), and Marsman et al. (1992) studies.

During the transient stimulation of DNA synthesis by xenobiotics in liver there is increase in circulating HGF levels in the plasma shown by Lindroos et al. (1992). The levels of HGF eventually decline after DNA synthesis ceases. The origin of the increased HGF in plasma is not clear. HGF clearance did not change in direct measurements performed in rats treated with xenobiotics. There is no increase in HGF mRNA in the liver or in other tissues. Phenomena such as matrix degradation have not been studied in relation to this phenomenon. Regardless of the mechanism leading to rise of the circulating HGF, the rise of this hepatocyte mitogen in the plasma may provide the stimulus for DNA synthesis induced by the liver tumor promoters. The mechanism by which these substances would stimulate increase in plasma levels of HGF is not clear at this point.

Conclusions and Speculations

Despite many gaps that still persist, much more is known today about regulation of hepatocyte growth than even five years ago. Most researchers in the field agree that polypeptide growth factors HGF, TGFa, EGF and acidic

FGF have a serious role to play in regulation of hepatocyte growth. The precise role by which these factors inter-relate in regulation of the every step involved still needs to be further understood. From the balance of the evidence it would appear that HGF is involved in mobilizing hepatocytes from G_0 to G_1. Once this occurs, G_1 hepatocytes become mobilized and start producing TGFα and acidic FGF, to further the proliferation of themselves and/or that of the adjacent endothelial and Ito cells. The latter cells themselves produce an array of growth factors, especially HGF, in response to not precisely defined signals from the proliferating hepatocytes. While this is an emerging hierarchy of events, the precise nature of the earliest events remains unknown. The above described matrix degradation hypothesis may account for several of these events, including rise of HGF, elimination of TGFβ1, induction of hepatocyte response to mitogens (referred as priming demonstrated by Webber et al. (1994).

A cartoon view of the overall scheme of these events, combining hypotheses and facts discussed above is shown in Figure 1, below.

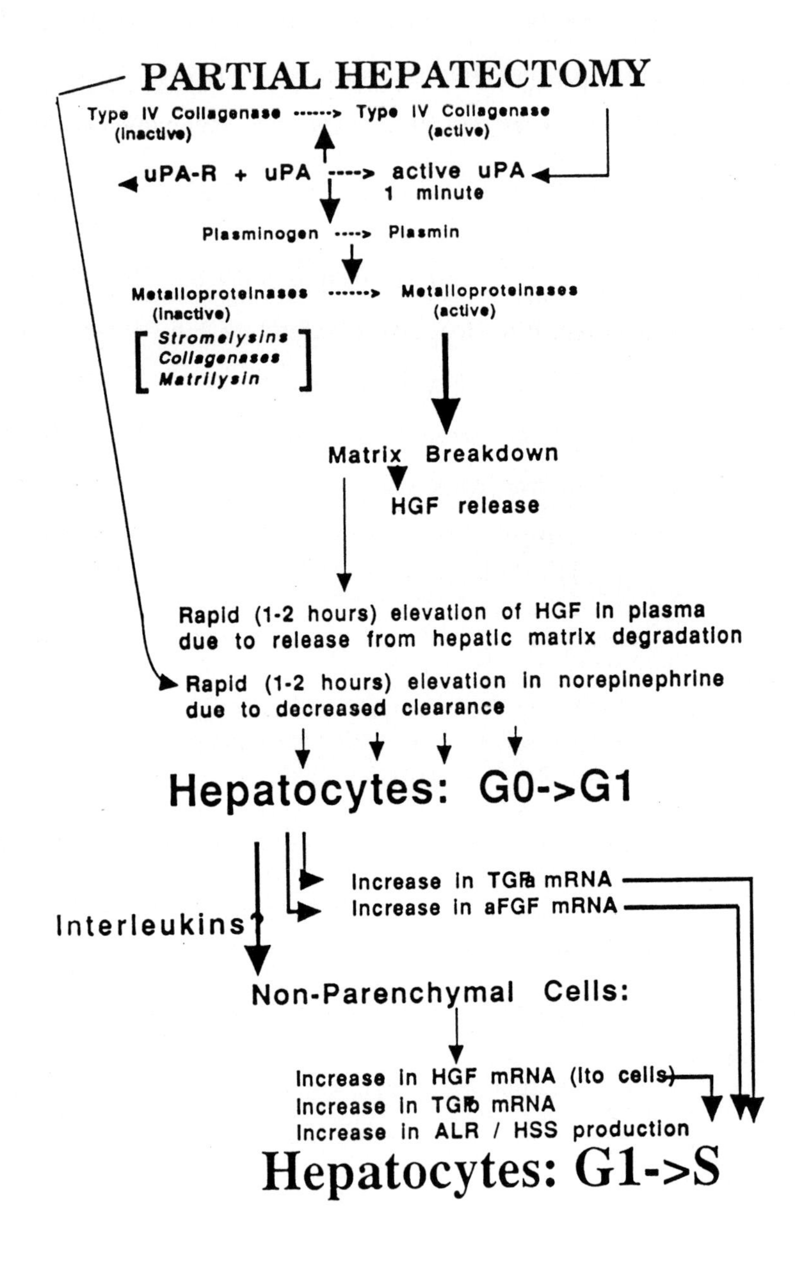

REFERENCES

Bissell DM, Hammaker LE, Meyer UA. (1973) Parenchymal cells from adult rat liver in nonproliferating monolayer culture. I. Functional studies. J Cell Biol 59:722-734

Bucher NLR (1982) Thirty Years of Liver Regeneration: A distillate. In Cold Spring Harbor Conferences on Cell Proliferation, vol. 9 (Growth of cell in chemically defined media), p15-26

Crookston KP, Webb DJ, Lamarre J, Gonias SL.(1993).Binding of platelet-derived growth factor-BB and transforming growth factor-beta 1 to alpha 2-macroglobulin in vitro and in vivo: comparison of receptor-recognized and non-recognized alpha 2-macroglobulin conformations. Biochem J 293:443-450.

de Hemptinne, B Lorge, F Kestens PJ and Lambotte L (1985) Hepatocellular hyperpolarizing factors and regeneration after partial hepatectomy in the rat. Acta Gastroenterol Belg 48: 424-431

DuBois RD (1990) Early changes in gene expression during liver regeneration: what do they mean? Hepatology 6:1079-1082

Fisher B, Szuch P, Levine M, Fisher ER (1971) A portal blood factor as the humoral agent in liver regeneration. Science 171: 575-577

Francavilla A, Starzl TE, Porter K, Foglieni CS, Michalopoulos GK, Carrieri G, Trejo Azzarone A, Barone M, Zeng QH (1991) Screening for candidate hepatic growth factors by selective portal infusion after canine Eck's fistula. Hepatology 14:665-70.

Grisham JW (1962) A morphologic study of deoxyribonucleic acid synthesis and cell proliferation in regenerating liver; autoradiography with thymidine-H3. Cancer Res 22: 842-849

Grisham JW, Leong GF, Hole BV (1964) Heterotopic partial autotransplantation of rat liver. Technique and demonstration of structure and function of the graft. Cancer Res 24: 1474-1482

Higgins GM, and Anderson RM (1931) Experimental pathology of the liver: I. Restoration of the liver of the white rat following partial surgical removal. Arch Pathol 12: 186-202

Jirtle RL, Michalopoulos G (1982) Effects of partial hepatectomy on transplanted hepatocytes. Cancer Res 42: 3000-3004

Keski-Oja J, Lohi J, Tuuttila A, Tryggvason K, Vartio T (1993) Proteolytic processing of the 72,000-Da type IV collagenase by urokinase plasminogen activator. Exp Cell Res 202:471-476

Kinoshita T, Hirao S, Matsumoto K, Nakamura T (1991) Possible endocrine control by hepatocyte growth factor of liver regeneration after partial hepatectomy. Biochem Biophys Res Commun 177:330-335.

Kuwashima Y, Aoki K, Kohyyama K, Ishikawa T. (1990) Hepatocyte Regeneration after partial hepatectomy occurs even under several thrombocytopenic conditions in the rat. Jpn. J. Cancer Res 81:607-612.

Lindroos P, Michalopoulos GK (1993) Response of phenobarbital- and ciprofibrate-exposed hepatocytes to growth factors in type I collagen gels. Carcinogenesis, 14:731-735

Lindroos P, Tsai WH, Zarnegar R, Michalopoulos GK (1992) Plasma levels of HGF in rats treated with tumor promoters. Carcinogenesis 13:139-141

Lindroos PM, Zarnegar R, Michalopoulos GK. (1991) Hepatic Growth Factor (Hepatopoietin A) Rapidly Increases in Plasma Prior to DNA Synthesis and Liver Regeneration Stimulated by Partial Hepatectomy and CCl4 Administration. Hepatology 13:743-750.

Liu KX, Kato Y, Narukawa M, Kim DC, Hanano M, Higuchi O, Nakamura T, Sugiyama Y. (1992) Importance of the liver in plasma clearance of hepatocyte growth factors in rats. Am J Physiol 263:G642-9.

Liu M, Michalopoulos G (1994) Ehancement of DNA synthesis stimulated by HGF in normal livers after infusion of Collagenase. Hepatology, In Press.

Liu Y, Michalopoulos GK, Zarnegar R. (1994) Structural and functional characterization of the mouse hepatocyte growth factor gene promoter. J Biol Chem 269:4152-4160

Lokker NA, Mark MR, Luis EA, Bennett GL, Robbins KA, Baker JB, Godowski PJ (1992) Structure-function analysis of hepatocyte growth factor: identification of variants that lack mitogenic activity yet retain high affinity receptor binding. EMBO J 11:2503-2510.

Mars W, Zarnegar R, Michalopoulos G. (1993) Activation of Hepatocyte Growth Factor by the plasminogen activators uPA and tPA. Amer J Path 143:949-958.

Marsman DS, Goldsworthy TL, Popp JA (1992) Contrasting hepatocytic peroxisome proliferation, lipofuscin accumulation and cell turnover for the hepatocarcinogens Wy-14,643 and clofibric acid.Carcinogenesis 13:1011-1017

Masumoto A, Yamamoto N (1991) Sequestration of a hepatocyte growth factor in extracellular matrix in normal adult rat liver. Biochem Biophys Res Commun 174:90-95

Matsumoto K, Nakamura T (1993) Roles of HGF as a pleiotropic factor in organ regeneration. Experientia Supplementa 65:225-249

Matsumoto K, Okazaki H, Nakamura T (1992) Up-regulation of hepatocyte growth factor gene expression by interleukin-1 in human skin fibroblasts. Biochem Biophys Res Commun 188:235-243

Matsumoto K, Tajima H, Hamanoue M, Kohno S, Kinoshita T, Nakamura T (1992) Identification and characterization of "injurin," an inducer of expression of the gene for hepatocyte growth factor. Proc Natl Acad Sci USA 89:3800-3804.

Meyer SA, Jirtle RL (1989) Phenobarbital decreases hepatocyte EGF receptor expression independent of protein kinase C activation. Biochem Biophys Res Commun 158:652-659

Michalopoulos G, Houck KA, Dolan ML, Luetteke NC (1984) Control of hepatocyte proliferation by two serum factors. Cancer Res 44:4414-4419

Michalopoulos GK, Appasamy R. (1993) Metabolism of HGF-SF and its role in liver regeneration. Experientia Supplementa 65:275-83.

Michalopoulos G, Bowen, W, Becich M, Howard TA (1993) Comparative analysis of mitogenic and morphogenic effects of HGF and EGF on rat and human hepatocytes maintained in collagen gels. J Cell Physiol 156:443-452.

Michalopoulos G, Bowen W, Mars W, Jirtle RL, Unpublished observations.

Michalopoulos GK, Eckl PM, Cruise JL, Novicki DL, Jirtle RL (1987) Mechanisms of rodent liver carcinogenesis. Toxicol Ind Health 3:119-128

Mizuno K, Takehara T, Nakamura T (1992) Proteolytic activation of a single-chain precursor of hepatocyte growth factor by extracellular serine-protease. Biochem Biophys Res Commun 189:1631-1638

Naka D, Ishii T, Yoshiyama Y, Miyazawa K, Hara H, Hishida T, Kidamura N (1992) Activation of hepatocyte growth factor by proteolytic conversion of a single chain form to a heterodimer. J Biol Chem 267:20114-20119

Naldini L, Tamagnone L, Vigna E, Sachs M, Hartmann G, Birchmeier W, Daikuhara Y, Tsubouchi H, Blasi F, Comoglio PM (1992) Extracellular proteolytic cleavage by urokinase is required for activation of hepatocyte growth factor/scatter factor. EMBO J 11:4825-4833.

Rabes HM, Wirsching R., Tuczek HV, Iseler G (1976) Analysis of cell cycle compartments of hepatocytes after partial hepatectomy. Cell Tissue Kinet 6: 517-532

Rao MS Subbarao V (1986) DNA synthesis in exocrine and endocrine pancreas after partial hepatectomy in Syrian golden hamsters. Experientia 42:833-834

Rifkin DB (1992) Plasminogen activator expression and matrix degradation. Matrix Suppl 1:20-22.

Roesel J, Rigsby D, Bailey A, Alvarez R, Sanchez JD, Campbell K, Shrestha K, Miller DM. (1989) Stimulation of protooncogene expression by partial hepatectomy is not tissue specific. Oncogene Research 5:129-136

Schirmacher P, Geerts A, Pietrangelo A, Dienes HP, Rogler CE. (1992) Hepatocyte growth factor/hepatopoietin A is expressed in fat-storing cells from rat liver but not myofibroblast-like cells derived from fat-storing cells. Hepatology 15:5-11.

Tomiya T, Tani M, Yamada S, Hayashi S, Umeda N, Fujiwara K. (1992) Serum hepatocyte growth factor levels in hepatectomized and nonhepatectomized surgical patients. Gastroenterology 103:1621-1624.

Tsai WH, Michalopoulos GK (1991) Responsiveness of hepatocytes to epidermal growth factor, transforming growth factor-beta and norepinephrine during treatment with xenobiotic hepatic tumor promoters. Cancer Lett 57:83-90

Tsai WH, Zarnegar R, Michalopoulos GK (1991) Long-term treatment with hepatic tumor promoters inhibits mitogenic responses of hepatocytes to acidic fibroblast growth factor and hepatocyte growth factor. Cancer Lett 59:103-108

Wada N, Marsman DS, Popp JA (1992) Dose-related effects of the hepatocarcinogen, Wy-14,643, on peroxisomes and cell replication. Fundam Appl Toxicol 18:149-154

Webber EM , Godowski PJ , Fausto N (1994) In vivo response of hepatocytes to growth factors requires an initial priming stimulus. Hepatology 489-497

Wolf HK, Zarnegar R, Michalopoulos GK (1991) Localization of Hepatocyte Growth Factor in human and rat tissues: An immunohistochemical study. Hepatology 14:488-494.

Yager JD, Zurlo J, Ni N (1991) Sex hormones and tumor promotion in liver.

Proc Soc Exp Biol Med 198:667-674
Zarnegar R, DeFrances MC, Kost DP, Lindroos P, Michalopous GK (1991) Expression of hepatocyte growth factor mRNA in regenerating rat liver after partial hepatectomy. Biochem Biophys Res Commun 177:559-565.

STRUCTURE, PLEIOTROPIC ACTIONS, AND ORGANOTROPHIC ROLES OF HEPATOCYTE GROWTH FACTOR

Kunio Matsumoto and Toshikazu Nakamura
Division of Biochemistry,
Biomedical Research Center,
Osaka University Medical School,
Suita, Osaka 565, Japan

INTRODUCTION

Mature parenchymal hepatocytes in the normal intact liver are in the quiescent G0 phase and their proliferative potential is highly suppressed. However, once the liver is subjected to insult, such as partial hepatectomy or hepatitis, hepatocytes actively proliferate and liver regeneration occurs. Proliferation of hepatocyte in liver regeneration is regulated by both the cytosocial environment through cell-cell and cell-matrix interactions and by the action of humoral factor. Hepatocyte growth factor (HGF) was originally identified as a potent mitogen for mature hepatocytes in primary culture in serum of partially hepatectomized rat and rat platelets (Nakamura et al.,1984; Russel et al., 1984). HGF was first purified to homogeneity from rat platelets (Nakamura et al., 1986; Nakamura et al., 1987), thereafter from human plasma (Gohda et al.,1988), rabbit plasma (Zarnegar and Michalopoulos, 1989), and livers of CCl_4-treated rats (Asami et al.,1991). In 1989, both human and rat HGF cDNAs were cloned and this factor was proved to be a novel one distinct from other known factors (Nakamura et al.,1989; Miyazawa et al., 1989;Tashiro et al.,1990).

NATO ASI Series, Vol. H 88
Liver Carcinogenesis
Edited by G. G. Skouteris

Although HGF was initially considered to have a narrow target cell specificity and to acto only as mitogen, it is now known to be a pleiotropic factor. In 1991, independent approaches used to define the chemical structure of bioactive polypeptides revealed that three polypeptide growth factors, tumor cytotoxic factor (Shima et al.,1991), scatter factor (Wedner et al., 1991;Konishi et al.,1991), and fibroblast-derived epithelial growth factor (Rubin et al.,1991), are identical molecule to HGF. Tumor cytotoxic factor was originally identified as a potent cytotoxic factor for a certain tumor cell line (Higashio et al.,1990), while scatter factor was identified and characterized as a factor which induces marked dissociation of epithelial cells, inducing "scattering" of cells in monolayer culture (Gherardi et al.,1989). Moreover, a unique biological activity of HGF as epithelial morphogen to induce blanching morphogenesis of epithelial cells was first demonstrated in 1991 (Montesano et al.,1991). On the other hand, HGF receptor was identified as *c-met* protooncogene product of transmembrane tyrosine kinase (Bottaro et al.,1991; Naldini et al.,1991;Higuchi et al.,1992).

Several lines of studies *in vitro* as *in vivo* have provided convincing evidences for the notion that HGF plays important roles as"organotrophic factor" for regeneration of organs, including the liver, kidney, and lung (Nakamura 1991; Matsumoto and Nakamura, 1993). Furthermore, HGF is also considered to be a mediator of epithelial mesenchymal interactions, critical tissue interactions required for the inductive and morphogenic tissue organization during embryogenesis and organogenesis (Montesano et al.,1991a; Matsumoto and Nakamura,1993). In the present paper, we present our works on structural characteristics ,pleiotropic activities, and the mechanisms of organ regeneration by HGF.

Molecular Characteristics of HGF:

(1) Chemical Structure

The purified rat HGF gene gave a single band corresponding to a Mr of 82 kDa, under non-reducing condition on SDS-PAGE, while in the reducing condition it split into two bands with Mr of 69 kDa and 34 kDa, respectively (Nakamura et al.,1987). The larger subunit (69 kDa) is the α-subunit and the lower subunit (34 kDa) is β-subunit. Molecular cloning of human and rat HGF revealed that the amino acid sequence of human HGF is>90% identical with that of rat HGF, and that HGF is synthesized as a single chain precursor of 728 amino acid residues, which is proteolytically processed to two chain mature HGF by cleavage at Arg^{494}-Val^{495} site (Fig.1)(Nakamura et al.,1989; Tashiro et al.,1990; Seki et al.,1990; Mizuno and Nakamura, 1993a). Single chain HGF has no biological activities and the activation of HGF is coupled with the conversion from a single-chain to a two chain HGF by extracellular serine protease, HGF-converting enzyme (Mizuno et al., 1993b; Mizuno et al., 1994a).

The α-chain of HGF has an N-terminal hairpin structure and four kringle domains and the β-chain has a serine-protease like domain. Thus HGF has a 38% homology with plasminogen, whereas neither HGF has serine-protease activity nor plasminogen or its active form plasmin has HGF activity (Nakamura et al.,1989; Tashiro et al., 1990; Mizuno and Nakamura, 1993a).

(2) HGF gene

We cloned and characterized human HGF gene. HGF gene is composed of 18 exons interrupted by 17 introns (Fig.1)(Seki et al.,1991). Structural domains of the HGF molecule, such as signal sequence, hairpin loop, four kringles, serine protease-like domain, are encoded by separate exons. The HGF gene is located in the long arm of chromosome 7 (7q 11.2-21) in humans.

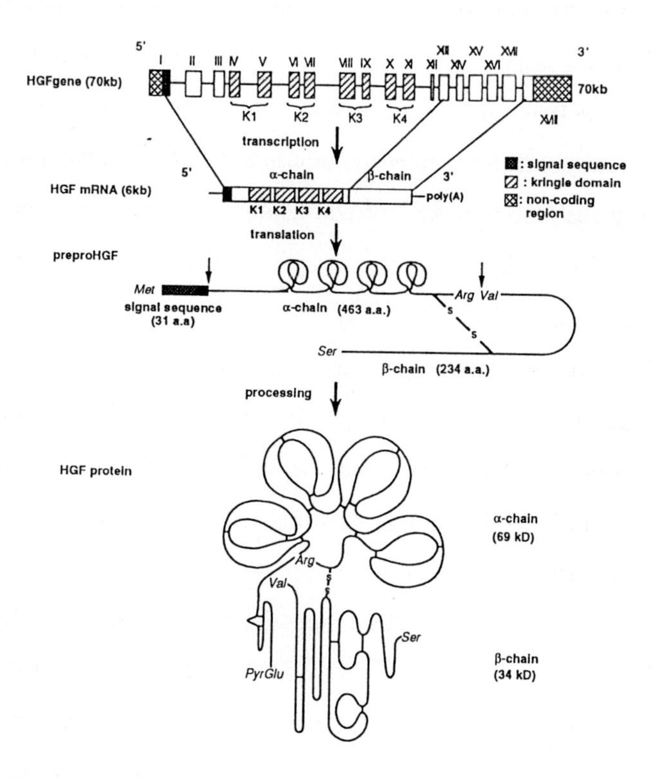

Fig. 1. Schematic structures of HGF gene, HGF mRNA, prepro-HGF, and mature HGF.

(3) Structure-function relationship

Among various growth factor, HGF is unique in possessing a structure of four kringle domains and a serine protease-like domain. To determine which domain is essential for biological activities of HGF, we prepared various deletion or substitution mutant HGF's and examined biological activities (Matsumoto et al., 1991a). Deletion of the first or second kringle domain led to an almost complete inactivation of HGF activity to stimulate DNA synthesis of hepatocytes in primary culture. Likewise, deletion of N-terminal hairpin structure almost completely inactivated biological activities of HGF. On the contrary, the mutant HGF deleted with the third or fourth kringle domain resulted in marked decrease of HGF activities, but the remaining HGF activity was significant. These results seem to be inconsistent with the finding that N-terminal hairpin structure, the first and second kringle domains in HGF molecule are responsible for the binding to its signal-transducing cellular receptor, *c-met* (Chan et al., 1991; Hartmann et al., 1992; Lokker et al., 1992).

HGF has significant affinity to heparin and the specific affinity to heparin has been used for the purification of HGF> Although biological significance of the specific affinity to heparin has not yet fully been understood, heparin/heparan sulfate proteoglycans in cell surface of extracellular matrix are likely to function as reservoir for HGF. We also determined domain structure of HGF responsible for the binding to heparin/heparan sulfate using various deletion mutants of HGF. N-terminal hairpin structure and the second kringle domain are essential to bind heparin/heparan sulfate (Mizuno et al., 1994b).

Pleiotropic Actions of HGF:

(1) Mitogenic activity and anti-tumor activity

Although HGF was initially characterized as a potent mitogen for hepatocytes, HGF exerts mitogenic activity to various types of epithelial cells and other cells. We found that HGF acts a potent mitogen for renal tubular epithelial cells (Igawa et al., 1991), epidermal keratinocytes and melanocytes (Matsumoto et al., 1991b; Matsumoto et al., 1991c), gastric mucosal epithelial cells (Takahashi et al., 1993), alveolar type II epithelial cells, bronchial epithelial cells (mason et al., 1994), Articular chondrocytes, and other epithelial cell lines and hematopoietic stem cell line (Tajima et al., 1992a; Mizuno et al., 1993c).

The first strong evidence the HGF acts as growth inhibitor for tumor cells was based on the finding that tumor cytotoxic factor was identical molecule with HGF (Shima et al., 1991). We found that HGF inhibits growth of several tumor cells, such as Hep G2 human hepatoma cells, KB human squamous cell carcinoma cells, and B6/F1 murine melanoma cells (Tajima et al., 1991). Likewise, the growth of various hepatoma cells was also inhibited by the addition of HGF. Moreover, it is noteworthy that the growth inhibitory effect of HGF is functional also *in vivo*. When Fao rat hepatoma cells were stably transfected with HGF cDNA, their growth was remarkably retarded compared to that of parental cells. Moreover, when these cells were implanted into nude mice, both tumorigenicity and tumor size were remarkably decreased (Shiota et al., 1992).

(2) Motogenic activity

As fibroblast-derived motogenic factor, scatter factor was shown to be identical to HGF, HGF enhanced motility of various types of cells in monolayer culture, including MDCK canine renal epithelial cells (Konishi et all, 1991), Hep G2 hepatoma cells, PAM212 murine keratinocytes, SBC-5 and Lu99 lung carcinoma cells normal human keratinocytes (Tajima et al., 1992a; Matsumoto et al., 1991b). Activation of rho p21 small GTP-binding protein is involved in motogenic activity of HGF (Takaishi et al., 1994).

(3) Morphogenic activity

Montesano et al. established an unique experimental system to search for a factor which induces epithelial morphogenesis *in vitro*. When MDCK renal epithelial cells were grown in collagen gel matrix in the presence of conditioned medium of fibroblasts, they specifically formed blanching tubular structure, resembling renal tubules (Montesano et al., 1991b). The result suggested that fibroblasts produce a factor(s) which induces blanching tubulogenesis of MDCK cells. Notably, the fibroblast-derived factor which induces blanching morphogenesis was identified HGF (Montesano et al., 1991a). It is noteworthy that various well-characterized growth factors did not exert such a unique epithelial morphogenic activity, and that HGF exerted similar morphogenic activity of hepatic epithelial cells (Johnson et al, 1993).

Organotrophic Role of HGF:

(1) Induction of HGF and down-regulation of HGF receptor

From the initial characterization of HGF, HGF has been considered as a long-sought hepatotrophic factor for liver regeneration. When various types of hepatic injuries (CCL_4-induced hepatitis, ischemia, physical liver crush, *etc.*) were induced in rats, HGF mRNA expression was rapidly and markedly induced in the liver (Kinoshita et al, 1989; Hamanoue et al., 1992). Fig. 2 shows summarized changes in HGF mRNA, the number of cell surface HGF receptor, and DNA synthesis in the liver after hepatic injuries. In a reciprocal relationship to induction in HGF mRNA and following HGF activity, the cell surface HGF receptor was markedly down-regulated, presumably due to the internalization after binding of HGF to the receptor (Higuchi et al., 1991). About 24 hours after the down-regulation of the receptor, hepatocyte actively underwent DNA synthesis (Fig. 2).

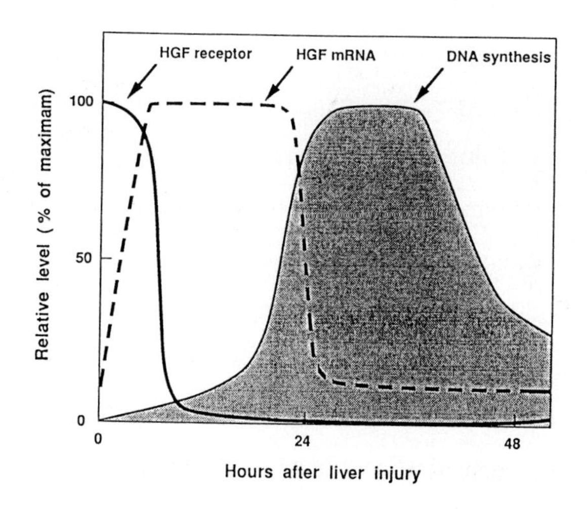

Fig. 2. Changes of HGF mRNA, HGF receptor, and regenerative DNA synthesis after liver injury in rat.

(2) Paracrine and endocrine mechanisms

We identified which types of cells produce HGF during liver regeneration, using *in situ* hybridization, immunohistochemical analysis, and isolated cells. HGF mRNA was expressed in non-parenchymal liver cells, including Kupffer cells, sinusoidal endothelial cells, and Ito (fat-storing) cells (Noji et al.,1991; Ramadori et al., 1992), but not in parenchymal hepatocytes. In consistent with the localization of HGF mRNA, HGF protein was also specifically localized in sinusoidal non-parenchymal liver cells as detected by immunohistochemical analysis (Fig.3). Therefore, the proliferation of hepatocyte after liver injuries may be triggered by HGF produced in non-parenchymal liver cells by a paracrine mechanism.

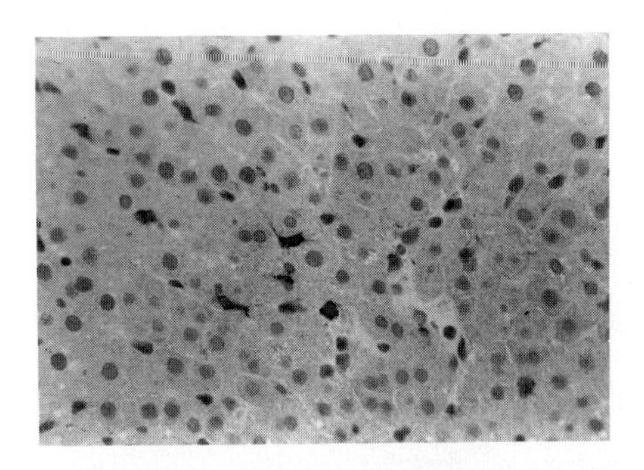

Fig. 3. Distribution of HGF in rat liver as detected by immunohistochemical analysis. HGF protein is located in non-parenchymal liver cells.

On the other hand, we thereafter found that HGF mRNA expression in intact distant organs such as the lung and spleen was rapidly induced after hepatic injuries and plasma HGF level also increased rapidly (Kinoshita et al., 1991; Yanagita et al.,1992). These results implicate the presence of humoral factor which induces HGF expression in distant intact organs in response to hepatic injuries and that HGF may trigger liver regeneration by an endocrine mechanism, in addition to a paracrine mechanism functioning in local injured sites.

Based on these findings, we prepared a large amount of human recombinant HGF and injected into experimental animals with various types of hepatic injuries. Recombinant HGF strongly accelerated liver regeneration and exerted anti-hepatitis action *in vivo* (Ishiki et al.,1992; see also our another paper in this book), and therefore, HGF is a long-sought hepatotrophic factor for liver regeneration.

(3) Regulation of receptor function by cell-cell contact

It is worth noting that the down-regulation of HGF receptor specifically occurred in the liver, but not in other intact organs, even though HGF expression increased in intact organs such as the lung, kidney and spleen, and circulating HGF level was elevated (Tajima et al.,1992b). To know the mechanisms governing specific response to HGF in the injured liver but not in intact organs, we analyzed expression and function of HGF receptor on plasma membranes of cultured hepatocytes (Mizuno et al.,1993d). DNA synthesis of hepatocytes cultured at a low cell density was highly stimulated by HGF, but this stimulatory effect was not so obvious in hepatocytes cultured at high cell density. In close parallel to the potency of DNA synthesis, the number of HGF receptor on hepatocytes cultured at low cell density was much greater than in high cell density. Moreover, the rate of HGF-induced down-regulation of the receptor on hepatocytes cultured at low cell density was much faster than in case of high cell density. Because tight cell-cell contact is rapidly loosened following onset of liver injuries, these results implicate that regulation of expression and activation of HGF receptor through cell-cell contact may be responsible as underlying mechanism for injured-organ specific activation of HGF receptor.

(4) HGF is renotrophic and pulmonotropic factor

Unilateral nephrectomy and acute renal injury in human and animals induces proliferation and tissue reconstruction of the kidney as known renal regeneration. In rats, the peak of DNA synthesis in the remaining or injured kidney occurred 48 hours after unilateral nephrectomy or treatment with renal toxins or renal ischemia. HGF mRNA in the remaining kideny increased markedly, as early as 6 hours after the operation, and HGF activity in the kidney reached a peak at 12 hours (Nagaike et al.,1991). Similarly, HGF mRNA expression in the kidney reached a peak 6-12 hours after nephrotoxic and ischemic treatments which caused severe acute renal failure in rats (Fig.4) (Igawa et al.,1993). Following the induction of HGF expression in the kidney, HGF receptor on plasma membranes of the kidney was down-regulated, in injured organ specific manner (Tajima et al.,1992b). More importantly, injection of human recombinant HGF into rats with acute renal failure caused by nephrotoxic compounds such as $HgCl_2$ and cisplatin, a most widely used anti-tumor drug, markedly enhanced renal regeneration and prevented onset of acute renal failure (Kawaida et al., 1994; see also our another paper in this book). These results are strong evidence for that HGF functions as a renotropic factor for renal regeneration.

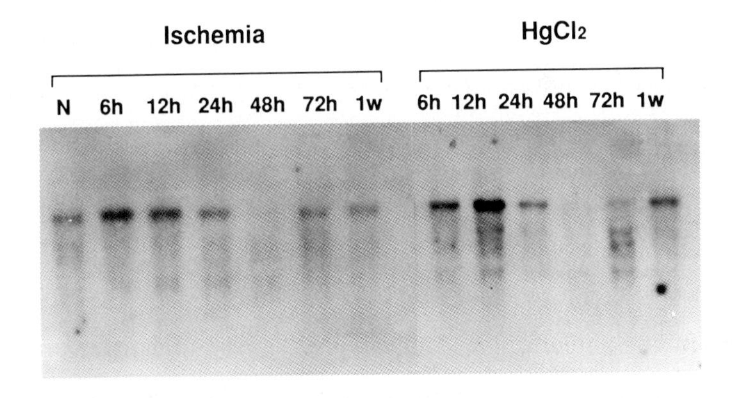

Fig. 4. Induction of HGF mRNA in rat kidney following the onset of acute renal failure caused by ischemia or $HgCl_4$-administration.

On the basis of the finding that HGF is a potent mitogen for bronchial epithelial cells and alveolar type II cells in the lung (Mason et al.,1994), we recently examined whether HGF plays a role as pulmonotrophic factor for lung regeneration (Yanagita et al., 1993). We have obtained evidence that HGF functions as a potent pulmonotrophic factor when lung injury was caused by intratracheal hydrochloride injection; expression of HGF mRNA and HGF activity in the lung was markedly increased and the down-regulation of HGF receptor specifically occurred in the lung before regenerative DNA synthesis of alveolar and bronchial epithelial cells. HGF concentration in sera of patients with various lung diseases was much higher than that in healthy donors and intravenously injected HGF accelerated lung regeneration *in vivo*, following acute lung injury (not shown).

(5)Regulatory factors for expression of HGF

Based on the finding that HGF mRNA was rapidly induced in non-injured organs such as the lung and spleen after hepatic and renal injuries, the presence of a humoral mediator which induces expression of HGF was postulated. We obtained evidence for the presence of an inducer of HGF expression in the blood of rats after hepatic or renal injuries: we partially purified it and named it "injurin", from its putative physiological function to mediate organ injuries (Matsumoto et al.,1992a). Chemical characterization of the partially purified injurin showed it is a proteinous factor with Mr of 10-30 kD, determined by molecular sieve chromatography.

In addition to injurin, we further identified several well-characterized factors have "injurin activity" to induce expression and production of HGF. Interleukin-1 (Matsumoto et al.,1992b), tumor necrosis factor-α (TNF-α), prostaglandin-E_1 and prostaglandin-I_2 are potent inducers of HGF through transcriptional activation of HGF gene, while tissue-derived heparan sulfate and heparin induces HGF production by post-transcriptional mechanisms (Matsumoto et al.,1993b).

Transforming growth factor-β1 (TGF-β1) is also a multipotent growth factor that regulates cell growth bidirectionally in a cell type-dependent manner (Roberts and Sporn, 1993). Although it is the most potent growth inhibitor for epithelial cells, including mature hepatocytes (Nakamura et al.,1985), TGF-β1 strongly suppressed expression of HGF gene in stroma-derived cells (Matsumoto et al., 1992c). Taken together, TGF-β1 seems to be involved in the restraint or termination of organ regeneration through its potential to suppress expression of HGF gene, as well as to directly inhibit epithelial cell proliferation.

CONCLUSION

Fig.5 illustrates possible molecular mechanism for the initiation of liver regeneration. Following the onset of liver injuries (hepatitis, partial hepatectomy, ischemia, etc.), the tight cell-cell contact between hepatocytes is rapidly loosened, the cell cycle stage changes from quiescent Go to the competent G_1 stage, and putative nonfunctional HGF receptor in Go stage may modified into those capable of transducing the growth-promoting signal of HGF. On the other hand, concomitant with the cell cycle transition of hepatocytes, injurin IL-1, TNF-α, prostaglandins, and heparin/heparan sulfate, inducers of HGF expression may be rapidly produced or secreted. In response

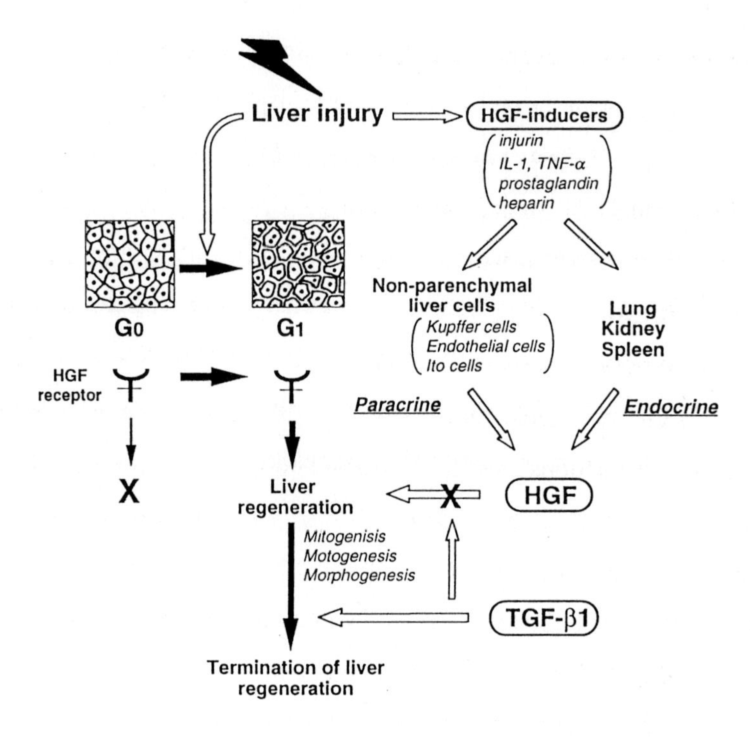

Fig. 5. Possible mechanisms of liver regeneration.

to these inducers, expression of HGF is rapidly induced in non-parenchymal liver cells (Kupffer cells, sinusoidal endothelial cells, Ito cells) in the liver and also in intact organs such as the lung, kidney, and spleen. HGF acts as hepatotropic factor for the initiation of liver regeneration trough both a paracrine and an endocrine mechanisms. At the later stage of liver regeneration, TGF-β1 functions as a growth inhibitor for hepatocytes, and also as a potent suppressor of HGF expression for stromal non-parenchymal liver cells. We suppose that renotropic and pulmonotrophic functions of HGF may be exerted through the similar mechanisms for liver regeneration.

HGF is a mesenchymal (or stromal)-derived pleiotropic growth factor which target predominantly a wide variety of epithelial cells and several other cells. During embryogenesis and organogenesis, HGF functions as a humoral mediator of morphogenic and inductive tissue interaction, "epithelial-mesenchymal interactions". While in adult tissues, HGF plays a crucial role as "organotrophic factor", eliciting hepatotropic, renotropic and pulmonotrophic functions to enable regeneration of injured organs such as the liver, kidney and lung. Thus, on the basis of its pleiotropic functions as mitogen, motogen, and morphogen, HGF is considered to be a key molecule to construct normal tissue architecture during embryogenesis, organogenesis, as well as organ regeneration.

REFERENCES

Asami O., Ihara I.,Shimidzu N.,Tomita Y., Ichihara A. and Nakamura T. (1991) Purification and characterization of hepatocyte growth factor from injured liver of carbon tetrachloride-treated rats. J.Biochem. 109:8-13.

Bottaro DP., Rubin JS.,Faletto DL., Chan AM-I, Kmiecik TE., Vande Wounde GF. and Aaronson SA. (1991) Identification of the hepatocyte growth factor receptor as the c-met proto-oncogene product. Science 251:802-804.

Chan AM-L., Rubin JS., Bottaro DP., Hirschfield DW., Chedid M. and Aaronson SA. (1991) Identification of a competitive antagonist encoded by an alternative transcript. Science 254:1382-1385.

Gherardi E., Gray J., Stoker M.,Perryman M. and Furlong R. (1989) Purification of scatter factor, a fibroblast-derived basic protein that modulates

epithelial interactions and movement. Proc.Natl.Acad.Sci.U.S.A. 86:5844-5848.
Gohda E., Tsubuchi H., Nakayama H., Hirono S., Sakiyama O.,Takahashi K., Miyazaki H., Hashimoto S. and Daikuhara Y. (1988) Purification and partial characterization of hepatocyte growth factor from plasma of a patient with hepatic failure. J.Clin.Invest. 81:414-419.
Hamanoue M., Kawaida K., Takao S., Shimazu H., Noji S., Matsumoto K. and Nakamura T. (1992) Rapid and marked induction of hepatocyte growth factor during liver regeneration after ischemic or crush injury. Hepatology 16:1485-1492.
Hartmann G., Naldini L., Weidner KM., Sachs M., Vigna E., Comoglio PM., Bircheimer W. (1992) A functional domain in the heavy chain of scatter factor/hepatocyte growth factor binds the *c-Met* receptor and induces cell dissociation but not mitogenesis. Proc.Natl.Acad.Sci.U.S.A. 89:11574-11578.
Higashio K., Shima N., Goto M., Itagaki Y.,Nagao M., Yasuda H., Morinaga T. (1990) identity of a tumor cytotoxic factor from human fibroblasts and hepatocyte growth factor. Biochem.Biophys.Res.Comm. 170:397-404.
Higuchi O. and Nakamura T. (1991) Identification and change in the receptor for hepatocyte growth factor in rat liver after partial hepatectomy or induced hepatitis. Biochem.Biophys.Res.Commun. 176:599-607.
Higuchi O., Mizuno K.,Vande Wounde GF and Nakamura T.(1992) Expression of *c-met* oncogene in cos cells induces signal transducing high-affinity receptor for hepatocyte growth factor. FEBS Lett. 301:282-286.
Igawa T., Kanda S., Kanetake H., Saitoh Y.,Ichihara A., Tomita Y. and Nakamura T.(1991) Hepatocyte growth factor is a potent mitogen for cultured rabbit renal tubular epithelial cells. Biochem.Biophys.Res.Commun. 174:831-838.
Igawa T., Matsumoto K., Kanda S., Saito Y. and Nakamura T. (1993) hepatocyte growth factor may function as a renotropic factor for regeneration in rats with acute renal injury. Am.J.Physiol. 265:F61-69.
Ishiki Y., Ohnishi H., Muto Y., Matsumoto K. and nakamura T. (1992) Direct evidence that hepatocyte growth factor is a hepatotrophic factor for livr regeneration and has potent antihepatitis effects *in vivo*. Hepatology 16:1227-1235.
Johnson M., Koukoulis G., Matsumoto K., Nakamura T. and Iyer A.(1993) hepatocyte growth factor induces proliferation and morphogenesis in nonparenchymal epithelial liver cells. Hepatology 17:1052-1061.
Kawaida T.,matsumoto K., Shimazu H. and Nakamura T. (1994) Hepatocyte

growth factor prevents acute renal failure and accelerates renal regeneration. Proc.Natl.Acad.Sci.U.S.A. ,in press.

Kinoshita T., Tashiro K. and Nakamura T. (1989) Marked increase of HGF mRNA in non-parenchymal liver cells of rats treated with hepatotoxins. Biochem.Biophys.Res.Commun. 165:1229-1234.

Kinoshita T.,Hirao S., Matsumoto K. and Nakamura T. (1991) Possible endocrine control by hepatocyte growth factor of liver regeneration after partial hepatectomy. Biochem.Biophys.Res. Commun. 177:330-335.

Konishi T., Takehara T., Tsuji T., Ohsato K., Matsumoto K. and Nakamura T. (1991) Scatter factor from human embryonic lung fibroblasts is probably identical to hepatocyte growth factor. Biochem.Biophys.Res.Commun. 180:765-773.

Lokker NA., Mark MR., Luis EA.,Bennet GL., Robbins KA., Baker JB. and Godowski PJ. (1992) Structure-function analysis of hepatocyte growth factor: identification of variants that lack mitogenic activity yet retain high affinity receptor binding. EMBO J. 11:2503-2510.

Mason RJ., Leslie CC., McCormic-Shannon K., Deterrding R., Nakamura T., Rubin JS, Aaronson SA. and Shannon JM. (1994) hepatocyte growth factor is a growth factor for rat alveolar type II cells. Am. J. Resp.Cell & Mol.Biol., in press.

Matsumoto K., Takehara T., Inoue H.,Hagiya M., Shimizu S. and Nakamura T. (1991a) Deletion of kringle domains or the N-terminal hairpin structure in hepatocyte growth factor results in marked decreases in related biological activities. Biochem. Biophys.Res.Commun. 181:691-699.

Matsumoto K., Hashimoto K., Yoshikawa K. and Nakamura T. (1991b) Marked stimulation of growth and motility of human keratinocytes by hepatocyte growth factor. Exp.Cell Res. 196:114-120.

Matsumoto K., Tajima H. and Nakamura T. (1991c) hepatocyte growth factor is a potent stimulator of human melanocyte DNA synthesis and growth. Biochem.Biophys.Res. Commun. 176:45-51.

Matsumoto K., Tajima H., Hamanoue M., Kono S., Kinoshita T. and Nakamura T. (1992a) Identification and characterization of "injurin", an inducer of the gene expression of hepatocyte growth factor. Proc.Natl.Acad. Sci.U.S.A. 89:3800-3804.

Matsumoto K., Okazaki H. and Nakamura T. (1992b) Up-regulation of hepatocyte growth factor gene expression by interleukin-1 in human Skin fibroblasts. Biochem.Biophys.Res. Commun. 188:235-243.

Matsumoto K., Tajima H., Okazaki H. and Nakamura T. (1992c) Negative regulation of hepatocyte growth factor gene expression in human lung fibroblasts and leukemic cells by transforming growth factor-β1 and glucocorticoids. J.Biol.Chem. 267:24917-24920.

Matsumoto K. and Nakamura T.(1993a) Roles of HGF as a pleiotropic factor in organ regeneration :"*Hepatocyte growth factor-scatter factor (HGF-SF) and the c-met receptor*" eds. Goldberg ID. and Rosen EM. ,pp.225-249. Birkauser Verlag, Basel, Switzerland.

Matsumoto K., Okazaki H. and Nakamura T. (1993b) Heparin as an inducer of hepatocyte growth factor. J.Biochem. 114:820-826.

Miyazawa K., Tsubouchi H., Naka D., Takahashi K., Okigaki M., Gohda E., Daikuhara Y. and Kitamura N. (1989) Molecular cloning and sequence analysis of cDNA for human hepatocyte growth factor. Biochem.Biophys. Res.Commun. 163:967-973.

Mizuno K. and Nakamura T. (1993a) Roles of HGF as a pleiotropic factor in organ regeneration :"*Hepatocyte growth factor-scatter factor (HGF-SF) and the c-met receptor*" eds. Goldberg ID. and Rosen EM. ,pp.1-29. Birkauser Verlag, Basel, Switzerland.

Mizuno K., Takehara T. and nakamura T. (1993b) Proteolytic activation of a single-chain precursor of hepatocyte growth factor by extracellular serine-protease. Biochem.Biophys.Res.Commun. 189:1631-1638.

Mizuno K., Higuchi O.,Ihle JN. and Nakamura T. (1993c) hepatocyte growth factor stimulates growth of hematopoietic progenitor cells. Biochem.Biophys. Res.Commun. 194:178-186.

Mizuno K., Higuchi O., Tajima H., Yonemasu T. and Nakamura T. (1993d) Cell density-dependent regulation of hepatocyte growth factor receptor on adult rat hepatocytes in primary culture. J.Biochem. 114:96-102.

Mizuno K., Tanoue Y., Okano I., Herano T., Takada K. and Nakamura T. (1994a) Purification and characterization of hepatocyte growth factor (HGF)-converting enzyme: activation of pro-HGF. Biochem.Biophys.Res.Commun. 198:1161-1169.

Mizuno K., Inoue H., Hagiya M., Shimizu S., Nose T., Shimohigashi Y. and Nakamura T. (1994b) Hairpin loop and second kringle domain are essential sites for heparin binding and biological activity of hepatocyte growth factor. J.Biol.Chem. 269:1131-1136.

Montesano R., Matsumoto K., Nakamura T. and Orci L. (1991a) Identification of a fibroblast-derived epithelial morphogen as hepatocyte growth factor. Cell

67:901-908.

Montesano R., Schaller G. and Orci L. (1991b) Induction of epithelial tubular morphogenesis in vitro by fibroblast-derived soluble factors. Cell 66:697-711.

Nagaike M., Hirao S., Tajima H., Noji S., Taniguchi S., Matsumoto K. and Nakamura T. (1991) renotropic functions of hepatocyte growth factor in renal regeneration after unilateral nephrectomy. J.Biol.Chem. 266:22781-22784.

Nakamura T., Nawa K. and Ichihara A. (2984) Partial purification and characterization of hepatocyte growth factor from serum of hepatectomized rats. Biochem.Biophys.Res.Commun. 122:1450-1459.

Nakamura T., Tomita Y., Hirai R., Yamaoka K., Kaji K. and Ichihara A. (1985) Inhibitory effect of transforming growth factor-β1 on DNA synthesis of adult rat hepatocytes in primary culture. Biochem.Biophys.Res.Commun. 133:1042-1050.

Nakamura T., Teramoto H. and Ichihara A. (1986) Purification and characterization of hepatocyte growth factor from rat platelets for mature parenchymal hepatocytes in primary cultures. Proc.Natl.Acad.Sci.U.S.A. 83:6489-6493.

Nakamura T., Nawa K., Ichihara A., Kaise N. and Nishino T. (1987) Purification and subunit structure of hepatocyte growth factor from rat platelets. FEBS Lett. 224:311-318.

Nakamura T., Nishizawa T., Hagiya M., Seki T., Shimonishi M., Sugimura A., Tashiro K. and Shimizu S. (1989) Molecular cloning and expression of human hepatocyte growth factor. Nature 342:440-443.

Nakamura T. (1991) Structure and function of hepatocyte growth factor. Prog.Growth Factor Res. 3:67-86.

Naldini L., Vigna E., Narchimham RP., Gaudino G., Zarnegar R., Michalopoulos GK. and Comoglio PM. (1991) Hepatocyte growth factor stimulates the tyrosine kinase activity of the receptor encoded by the proto-oncogene *c-Met*. Oncogene 6:501-504.

Noji S., Tashiro K., Koyama E., Nohno T., Ohyama K., Taniguchi S.,and Nakamura T. (1990) Expression of hepatocyte growth factor gene in endothelial and Kupffer cells of damaged rat liver, as revealed by in situ hybridization. Biochem.Biophys.Res.Commun. 173:42-47.

Ramadori G., Neubauer K., Odenthal M., Nakamura T., Knittel T., Schwogler K-H. and Meyer zum Buschenfelde (1992) The gene for hepatocyte growth factor is expressed in fat-storing cells of rat liver and is down-regulated during cell growth by transforming growth factor-b. Biochem.Biophys.Res. Commun.

183:739-742.
Roberts AB. and Sporn MB. (1993) Physiological action and clinical application of transforming growth factor-beta. Growth Factors 8:1-9.
Rubin JS., Chan AM-L., Bottaro DP., Burges WH., Taylor WG., Cech AC., Hirschfiled DW., Wong J., Miki T., Finch PW. and Aaronson SA. (1991) A broad-spectrum human lung fibroblast-derived mitogen is a variant of hepatocyte growth factor. Proc.Natl.Acad.Sci.U.S.A. 88:415-419.
Russell WE., McGowan JA. and Bucher NLR. (1984) Partial characterization of hepatocyte growth factor from rat platelets. J.Cell Physiol. 119:183-192.
Seki T., Ihara I., Sugimura A., Shimonishi M., Nishizawa T., Asami O., Hagiya M., Nakamura T. and Shimizu S. (1990) Isolation and expression of cDNA for different forms of hepatocyte growth factor from human leukocytes. Biochem.Biophys.Res.Commun. 172:321-327.
Seki T., Hagiya M., Shimonishi M., Nakamura T. and Shimizu S. (1991) Organization of the human hepatocyte growth factor-encoding gene. Gene 102:213-219.
Shima N., Nagao M., Ogaki F., Tsuda E., Murakami A. and Higashio K. (1991) Tumor cytotoxic factor-hepatocyte growth factor from human fibroblasts: cloning of its cDNA, purification and characterization of recombinant protein. Biochem.Biophys. Res.Commun. 180:1151-1158.
Shiota G., Rhoads DB., Wang TC., Nakamura T. and Schmidt EV. (1992) hepatocyte growth factor inhibits growth of hepatocellular carcinoma cells. Proc.Natl.Acad.Sci.U.S.A. 89:373-377.
Tajima H., Matsumoto K. and Nakamura T. (1991) Hepatocyte growth factor has a potent anti-proliferative activity for various tumor cell lines. FEBS Lett 291:229-232.
Tajima H., Matsumoto K. and nakamura T. (1992a) Regulation of cell growth and motility by hepatocyte growth factor and receptor expression in various cell species. Exp.Cell Res. 202:423-431.
Tajima H., Higuchi O., Mizuno K. and Nakamura T. (1992b) Tissue distribution and hepatocyte growth factor receptor and its exclusive down-regulation in regenerating organ after injury. J. Biochem. 111:401-406.
Takahashi M., Ota S., Terano A., Yoshiura K., Matsumura M., Niwa Y., Kawabe T., nakamura T. and Omata M. (1993) hepatocyte growth factor induces mitogenic reaction to the rabbit gastric epithelial cells in primary culture. Biochem.Biophys.Res.Commun. 191:528-534.
Takaishi K., Sasaki T.,. kato M.,Yamori W., Kuroda S., Nakamura T., Takeichi

M. and Takai Y. (1994) Involvement of *rho* p21 small GTP-binding protein and its regulatory protein in the HGF-induced cell motility. Oncogene 9:273-279.

Tashiro K., Hagiya M., Nishizawa T., Seki T., Shimonishi M., Shimizu S. and Nakamura T. (1990) Deduced primary structure of hepatocyte growth factor and expression of the mRNA in rat tissues. Proc.Natl.Acad.Sci.U.S.A. 87:3200-3204.

Weidner KM., Arakaki N., Hartmann G., Vanderkerkhove J., Weingart S., Rieder H., Fonatsch C., Tsubouchi H., Hishida T., Daikuhara Y. and Bircheimer W. (1991) Evidence for the identity of human scatter factor and human hepatocyte growth factor. Proc.Natl.Acad.Sci.U.S.A. 88:7001-7005.

Yanagita K., Nagaike M., Ishibashi H., Niho Y., Matsumoto K. and Nakamura T. (1992) Lung may have an endocrine function producing hepatocyte growth factor in response to injury of distal organs. Biochem. Biophys.Res. Commun. 182:802-809.

Yanagita K., Matsumoto K., Sekiguchi K., Ishibashi H., Niho Y., Nakamura T. (1993) Hepatocyte growth factor may act as a pulmotrophic factor on lung regeneration after acute lung injury. J.Biol.Chem. 268:21212-21217.

Zarnegar R.and Michalopoulos GK. (1989) Purification and biological characterization of human hepatopoietin A, a polypeptide growth factor for hepatocytes. Cancer Res. 49:3314-3320.

HEPATOCYTE GROWTH FACTOR AND ITS VARIANT WITH A DELETION OF FIVE AMINO ACIDS ARE DISTINGUISHABLE IN BIOLOGICAL, PHYSICOCHEMICAL, AND IMMUNOCHEMICAL PROPERTIES.

Nobuyuki Shima and Kanji Higashio *
Research Institute of Life Science
Snow Brand Milk Products, Co., Ltd.
519 Ishibashi-machi, Shimotsuga-gun, Tochigi
Japan

INTRODUCTION

Hepatocyte growth factor (HGF) (Nakamura et al.,1987; Gohda et al., 1988), also designated as scatter factor (Gherardi et al., 1989) or fibroblast-derived tumor cytotoxic factor (Higashio et al., 1990), is a heparin-binding basic protein with an approximate molecular mass of 80 kD. HGF is a disulfide-linked heterodimer composed of an α-chain of 52-56 kD and a β-chain of 30-34 kD. Recent studies have revealed that HGF not only functions as a mitogen for hepatocytes, it also stimulates the growth of various epithelial and endothelial cells (Rubin et al., 1991); it inhibits the growth of tumor cells (Higashio et al., 1990; Shima et al., 1991a); and it stimulates the mobility of epithelial (Gherardi et al., 1989) and endothelial cells (Rosen et al., 1990).

Recently, Bottaro et al. (1991) identified the c-Met protooncogene product, a membrane-spanning tyrosine kinase, as a functional receptor for HGF. The c-Met receptor has been detected on various epithelial cells (Bottaro et al., 1991) and endothelial cells (Grant et al., 1993),whose growth is stimulated by HGF. The receptor has also been found on Meth A mouse sarcoma cells,whose

*To whom correspondence should be addressed.

NATO ASI Series, Vol. H 88
Liver Carcinogenesis
Edited by G. G. Skouteris

growth is inhibited by HGF (Komada et al., 1992), and on some carcinoma cell lines that are scattered by HGF (Naldini et al., 1991a). HGF stimulates autophosphorylation of the c-Met tyrosine kinase in these target cells. These observations indicated that the biological signals to the various types of target cells seem to be mediated by a common receptor, c-Met.

The nucleotide sequence analysis of cDNA clones for human HGF (Nakamura et al., 1989; Miyazawa et al., 1989) predicted that HGF consists of 6 major domains: a hairpin and four kringles in the α-chain, and a serine protease-like domain in the β-chain. In addition to the originally reported cDNA, another major variant, which lacks 15 nucleotides encoding a 5-amino acid residue (the FLPSS sequence) in the first kringle domain, has been isolated (Seki et al., 1990). The significance of the existence of the deleted form of HGF (dHGF), however, remains to be established. Structure-function studies with partially truncated HGF molecules (Matsumoto et al.; 1991, Okigaki et al., 1992) have demonstrated that N-terminal three domains in the α-chain of HGF (the hairpin and the first two kringles) are essential for the binding of HGF to the c-Met receptor. This implies that the deletion of 5 amino acids in the first kringle might affect the biological activity of HGF. Our preliminary result indicated that dHGF had higher mitogenic activity than HGF for rat hepatocytes (Shima et al., 1991b), but another study reported that there was no obvious difference between the biological activities of the two HGFs (Seki et al., 1990). To elucidate whether the roles of these two forms of HGF are the same, we made detailed analyses of the physicochemical, immunochemical, and biological properties of HGF and dHGF using highly purified recombinant proteins.

MATERIALS AND METHODS

Expression of HGF and dHGF cDNAs: Plasmid to express HGF or dHGF was constructed by inserting a 2.3-kb fragment of HGF or dHGF cDNA (Shima et al., 1991b) into pcDNAI (Invitrogen) under cytomegarovirus promoter, and a

2.4-kb fragment of a mouse dihydrofolate reductase (DHFR) transcription unit from pAdD26SVA (Kaufman and Sharp, 1982) into the unique NheI site of pcDNAI. The expression plasmid (10μg) and pSV2 neo (1μg) (Southern and Berg, 1982) were co-transfected into Namalwa cells, CHO cells, and C-127 cells by the published method (Felgner and Holm, 1989). The transformed cells were selected with G418 and subsequently gene-amplified with methotrexate. Clones were screened for high HGF or dHGF production by an ELISA using monoclonal antibodies that recognize both HGFs equally (Shima et al., 1991c).

Target cells: Adult rat hepatocytes were prepared by the method of Seglen (1976). Human umbilical vein endothelial cells, HUVEC, and human aorta smooth muscle cells, AOSMC, were purchased from Kurabo Co. (Japan). Mouse sarcoma 180 cells and Meth A sarcoma cells were obtained from National Cancer Center Research Institute (Japan). LLC-PK1 (pig kidney epithelial cells) and OK (American opossum kidney epithelial cells) were purchased from American Type Culture Collection (ATCC, USA).

Preparation of highly purified HGF and dHGF: Transformed cells producing HGF or dHGF were cultured in DMEM (Gibco) with 5% calf serum. HGFs in the conditioned media were purified by a combination of CM sepharose, Heparin CL-6B, Mono S-HPLC, and Heparin 5PW-HPLC, as described previously (Shima et al., 1991b). The protein concentration was determined by the method of Lowry using bovine serum albumin as a standard protein. For biological assays, the HGFs were diluted in PBS containing 0.25% human serum albumin and 0.001% Tween 80, and were sterilized by filtration through a 0.22-μm membrane filter. For physicochemical studies, the HGFs were diluted in PBS containing 0.01% Tween 20. The concentration of each HGF in the solution was reconfirmed by the ELISA as described above.

Solubility test: The purified HGFs were dialyzed against water and lyophilized. The lyophilized HGF and dHGF were suspended in PBS containing 0.01% Tween 20 at 37°C and were agitated for 30 minutes. After centrifugation at 30,000 x g for 30 minutes at 5°C, the concentration of protein in the supernatant was determined by the method of Lowry.

Cell culture and assay for cell growth: Sarcoma 180 cells were cultured in

DMEM with 10% fetal bovine serum (FBS). Meth A cells were cultured in RPMI 1640 with 10% FBS. The cells were seeded in 96-well plates at 1 X 10^3 cells/50 µl/well. To each test well, 50 µl of culture medium containing serially diluted HGF or dHGF was added. The plates were incubated at 37°C for four days. Viable cells were counted with a hemocytometer. Adult rat hepatocytes suspended in William's E medium (Gibco) with 10% fetal bovine serum (FBS) and 10 nM dexamethasone were inoculated at an initial cell density of 10^4 cells/50 µl/well in 96-well plates and were incubated at 37°C for 24 hours. HUVE cells were grown in E-GM UV medium (Kurabo). The trypsinized cells were suspended in Medium 199 (Gibco) with 10% FBS and were inoculated at a cell density of 5 x 10^3 cells/50 µl/well. AOSM cells were grown in S-GM (Kurabo) . The trypsinized cells suspended in S-BM (Kurabo) with 5% FBS were inoculated at a cell density of 10^4 cells/50 µl/well and were incubated at 37°C for 48 hours. LLC-PK1 cells were maintained in DMEM with 10% FBS. The cells suspended in serum-free DMEM were inoculated at a cell density of 10^4 cells/50 µl/well and were incubated at 37°C for 48 hours. A serially diluted HGF or dHGF in the respective medium was added to each well of the cell cultures (50 µl/well). The plates were then incubated at 37°C for 24 hours. Subsequently, 1µCi/10µl of [methyl-^{3}H]thymidine (Amersham, UK; 85Ci/mmol for rat hepatocytes and 5Ci/mmol for the other cells) was added to each well and the plates were further incubated for 2 hours. The culture plates for anchorage-dependent cells were washed with PBS (200 µl/well) and trypsinized. The radioactivity incorporated into the cells was determined using a Direct Beta Counter MATRIX 96 (Packard, USA). Data are presented as the mean ± SD of triplicate cultures. Each experiment was repeated by using at least two separate preparations of the HGFs.

Selection of monoclonal antibodies against HGF or dHGF. Monoclonal antibodies were prepared by the published method (Kohler and Milstein, 1975) with our modifications (Shima et al., 1991c). Briefly, BALB/c mice were immunized intraperitoneally with 100 µg of purified dHGF. Splenocytes from the immunized mice were fused with P3X63-Ag8.653 mouse myeloma cells (ATCC, CRL1580). The specificity of each monoclonal antibody was determined by using a solid phase ELISA as follows. A 96-well plate

(MaxiSorp, Nunc, Denmark) was coated with HGF, dHGF, or reduced dHGF (1μg/well). Subsequently, the plate was filled with 50% Block Ace (Snow Brand Milk Products, Japan) and incubated for 1 hour at room temperature. After the plate was washed 3 times with PBS containing 0.1% Tween 20, test samples (hybridoma culture medium or purified monoclonal IgG solution) diluted in 0.2 M Tris HCl, pH 7.3, containing 50% Block Ace and 0.1% Tween 20, were applied to the plate, followed by incubation for 3 hours at 37°C. After washing, the amount of the bound antibody was detected with peroxidase-linked goat anti-mouse IgG (Cappel, Belgium).

RESULTS

Biological activities of the two forms of HGF.

To clarify the difference between the two HGFs, we compared the biological activity of each HGF. The purified HGFs obtained from CHO cells were tested for cytotoxic activity on tumor cells. Both HGFs showed almost the same dose-responses for cytotoxicity on Sarcoma 180 or Meth A sarcoma cells (Fig. 1).

The purified HGFs obtained from Namalwa cells, CHO cells, and C-127 cells were respectively tested for rat hepatocyte assay. Dose-response curves for the stimulation of DNA synthesis in rat hepatocytes by dHGF and HGF were very similar up to about 10 ng/ml, but differed significantly at higher concentrations (Fig. 2A, B, and C). Although the cDNAs were expressed from the three different host cells, the same phenomena were observed. HGF markedly decreased its activity in a tested dose range of 10 to 100 ng/ml, while dHGF gave the maximal activity in the same dose range. Specific activity of dHGF was maximally 1.5 to 1.9-fold higher than that of HGF in this dose range.

The purified HGFs obtained from Namalwa cells were tested for the stimulation of DNA synthesis in other target cells. HGF was more potent than dHGF in the stimulation of DNA synthesis in mesenchymal cells, such as HUVEC and AOSMC (Fig. 3A and B). In contrast, dHGF was more potent

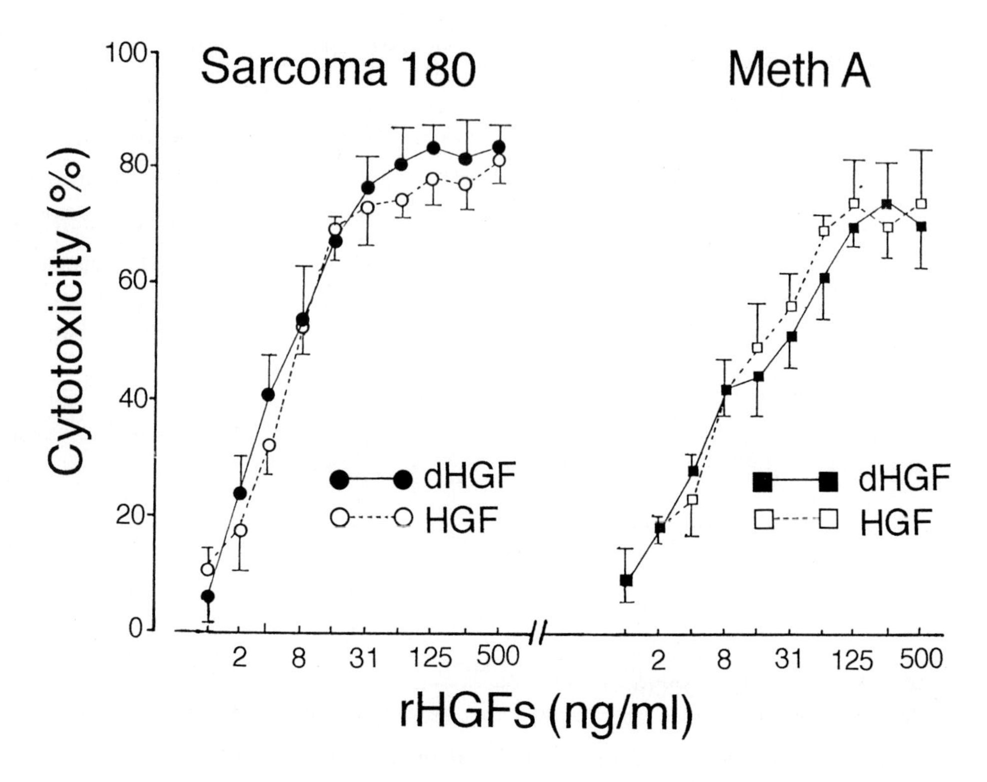

Fig.1. Cytotoxic activity of HGF or dHGF against sarcoma 180 (the left panel) and Meth A sarcoma cells (the right panel). Results are presented as mean ± SD of triplicate culture.

than HGF in the stimulation of DNA synthesis in epithelial cells, such as LLC-PK1 and OK (Fig. 3C and D). To evaluate the difference in the growth-stimulating potency of the two HGFs on various types of cells, we compared a half-maximal dose of HGF and a dose of dHGF that gave the same level of activity as the half-maximal activity of HGF. HGF was respectively about 20- and 10-fold more potent than dHGF in the growth stimulation of HUVEC and AOSMC, and dHGF was respectively about 3- and 2-fold more potent than HGF in LLC-PK1 and OK cells.

Physicochemical properties of the two forms of HGF

The purified HGFs obtained from Namalwa cells were used for

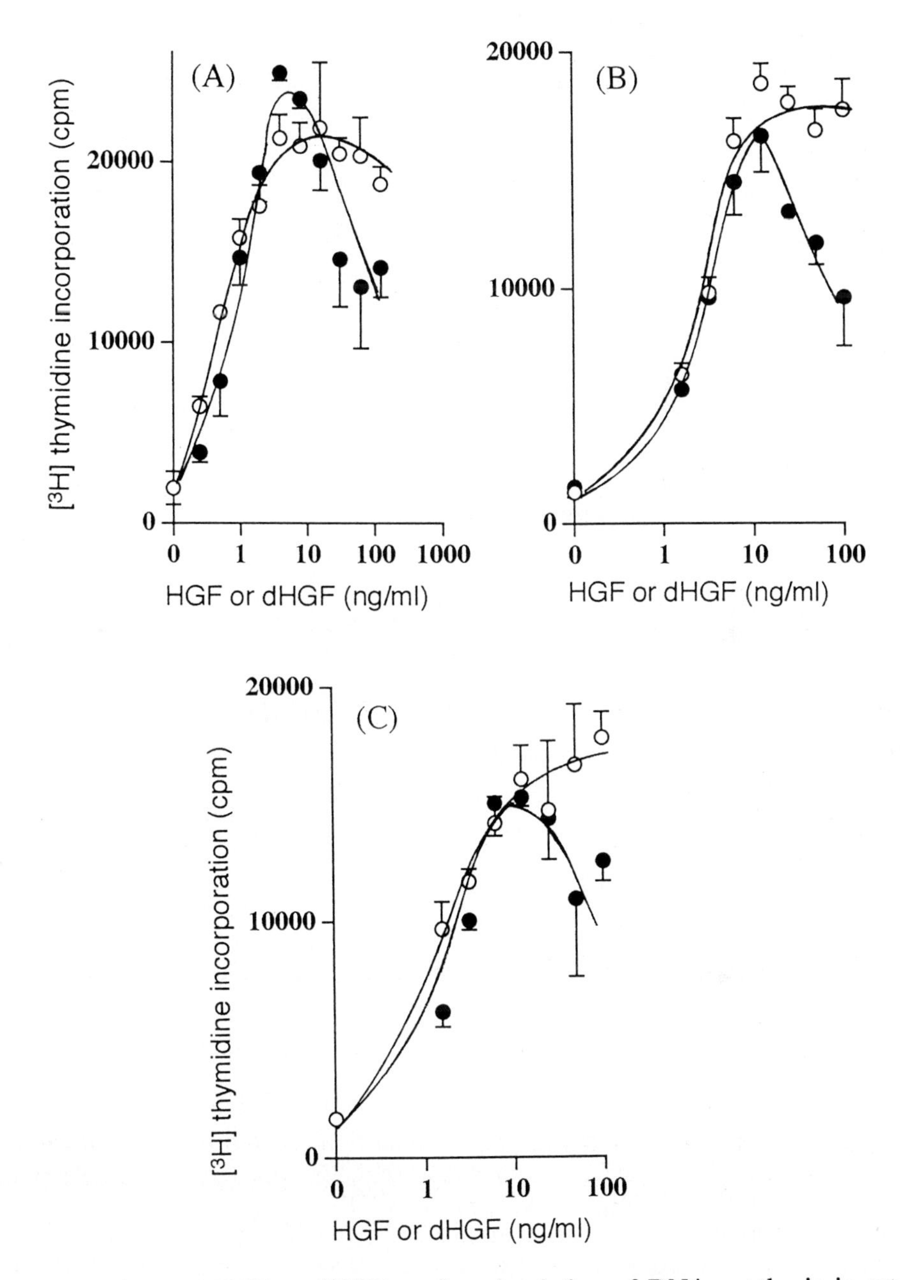

Fig. 2. Effect of HGF or dHGF on the stimulation of DNA synthesis in rat hepatocytes. Hepatocytes were cultured with HGF (closed circles) or dHGF (open circles). The purified recombinant HGFs obtained from Namalwa cells (A), CHO cells (B), and C-127 cells (C) were used for the assay. Results are presented as the mean ± SD of triplicate cultures.

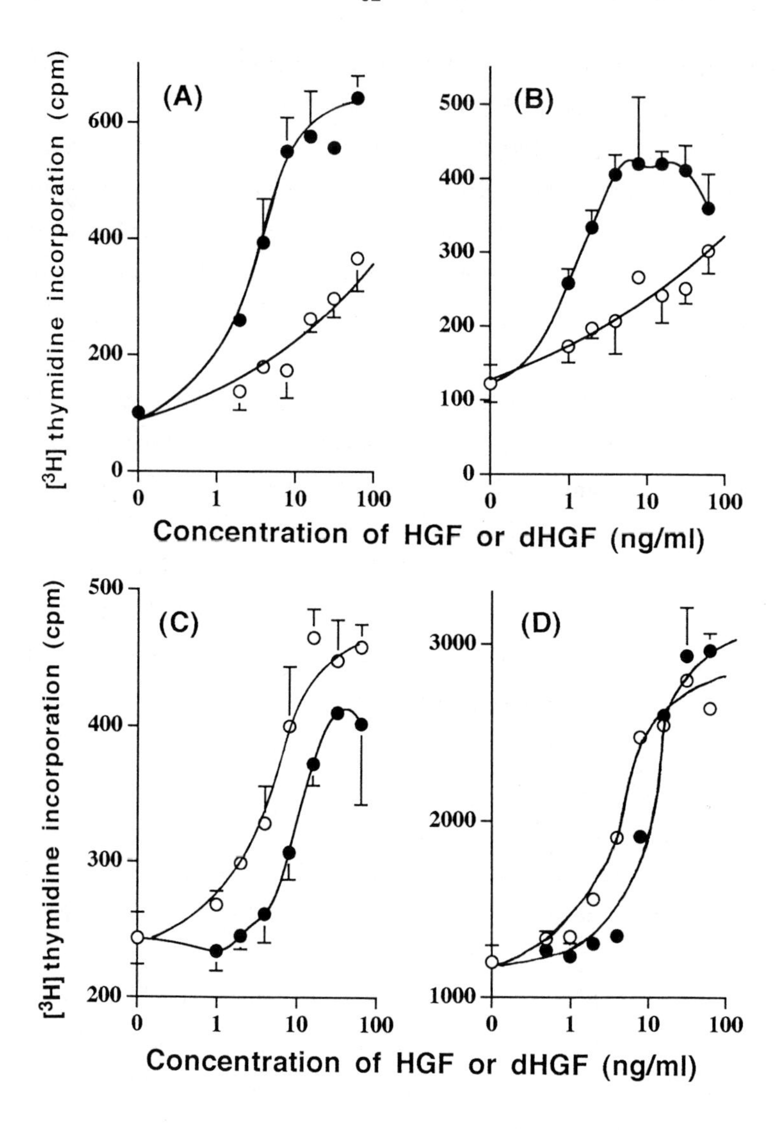

Fig. 3. Effect of HGF or dHGF on the stimulation of DNA synthesis in various kinds of mesenchymal cells (A and B) and epithelial cells (C and D). Cells were cultured with HGF (closed circles) or dHGF (open circles). Results are presented as the mean ± SD of triplicate cultures. (A) human umbilical vein endothelial cells; HUVEC, (B) human aorta smooth muscle cells; AOSMC, (C) LLC-PK1 (pig kidney epithelial cells), (D) OK (American opossum kidney epithelial cells).

physicochemical study. For initial characterization of the two HGFs, we analyzed the purity and molecular weight of the purified HGFs. The two HGFs were subjected to SDS-PAGE under reducing and non-reducing conditions. Each HGF showed homogeneous electrophoresis bands with purity of over 99%, and the band sizes were as predicted (data not shown). The two HGFs were next applied to a high performance gel chromatography column (Superose 12/30, Pharmacia) equilibrated with PBS containing 0.01% Tween 20. Each HGF showed the same profile with a single peak, and the molecular mass of each HGF was calculated to be 74 kD, indicating that each HGF dissolved as a monomer in the aqueous solution.

There was no significant difference between the two HGFs in sialic acid contents, N-linked sugar structure, and heat- and pH-stability (data not shown). A marked difference in solubility, however, was observed between the two HGFs. dHGF was soluble up to 1.34 mg/ml. In contrast, HGF was soluble over 90 mg/ml under the same condition. Since our preliminary experiment indicated that the presence of NaCl (greater than 150 mM) affects the solubility of the HGFs, the concentration of sodium ion in the lyophilized HGFs was determined by an atomic absorption analysis. The calculated NaCl concentrations in the HGF and dHGF solutions at a concentration of 1 mg protein per ml of water were 0.55 and 0.94 μM, respectively, indicating the absence of any contribution of NaCl from the lyophilized materials in the solubility test.

Screening of dHGF specific monoclonal antibody: To demonstrate that the deletion causes a conformational change, we screened dHGF specific monoclonal antibodies by a solid phase ELISA. The recognition of antigens varied depending on the antibodies used. P1C8, P2D6, and B4A2 recognized HGF and dHGF equally, but they did not recognize dHGF reduced with 2-mercaptoethanol. D3B3 recognized both HGF and dHGF, but it recognized HGF to a lesser extent. A2G9 and H9E3 recognized dHGF but not HGF or reduced dHGF (Table 1).

Table 1. Antigen specificity of monoclonal antibodies

Antibodies	Maximal antibody binding (OD 492 nm)		
	Antigen		
	dHGF	reduced-dHGF	HGF
P1C8	1.21	0.03	1.19
P2D6	1.23	0.03	1.28
B4A2	1.02	0.03	1.04
D3B3	1.31	0.07	0.56
A2G9	1.42	0.02	0.02
H9E3	1.21	0.09	0.09

Blank value was 0.03 OD at 492 nm.
Results are presented as the mean of duplicate experiments.

DISCUSSION

The comparison of specific activities of HGF and dHGF revealed that the deletion of five amino acids did not affect cytotoxic activity on tumor cells, but it did affect growth stimulating activity on various types of normal cells. This implies that different mechanisms mediate the effect on growth inhibition and stimulation in different types of cells. Dose-response curves for the stimulation of DNA synthesis in rat hepatocytes by dHGF and HGF were very similar up to about 10 ng/ml, but differed significantly at higher concentrations. We observed the same phenomena even when we expressed the cDNAs using three different host cells: Namalwa, CHO, and C-127 cells (Fig. 2A, B, and C). Matsumoto et al.(1991) also showed the same result in the dose response curves between the two forms of HGF, which are expressed transiently in COS cells. These results indicate that the difference does not depend on expression systems but is due to a structural change caused by a deletion of five amino acids (FPLSS) in the first kringle of HGF. Seki et al. (1990), however, reported that specific activities of HGF and dHGF transiently expressed in COS cells were almost the same. It is likely that the dosages of both materials in that experiment were in the dose range (< 10 ng/ml) in which both materials show almost the same specific activities. The difference in the specific activity at higher concentrations implies that mechanisms of stimulation of DNA synthesis are probably different between the two HGFs. Ligand induced oligomerization of tyrosine kinase receptors has been shown and is required for signal transduction (Schlessinger, 1988). A mechanism of c-Met oligomerization in the presence of an excess amount of ligand may be different between the two HGFs.

In addition to the difference between the biological activities of the two HGFs observed in the hepatocyte bioassay, marked differences were observed in other target cells, especially in the target cell specificity between the two HGFs (Fig. 3). This may account for discrepancies in the results of past experiments using HGF. It was reported that HGF acts only on epithelial cells and not mesenchymal cells (Kan et al., 1991). On the other hand, other studies have shown that HGF acts as a mitogen for various mesenchymal cells, such as melanocytes, endothelial cells, and myeloblastic cells, NFS-60 (Kmiecik et al., 1992). In addition to these observations, we have found that HGF acts as a mitogen for human smooth muscle cells, AOSMC, although no mitogenic effect of HGF on a rat AOSM cell line had been previously reported (Harris et al., 1993). Our results therefore suggest that HGF and dHGF should be distinguished from each other in order to avoid confusion caused by their different biological actions. dHGF may be more specific for the growth of epithelial cells of ectodermal origin than HGF and the two HGFs may play different roles in physiological actions.

A marked difference in solubility was observed between the two HGFs. This suggests that the deletion in the first kringle affects the tertiary structure of HGF. The conformational change was demonstrated by the existence of dHGF specific monoclonal antibodies that reacted to dHGF but did not react to HGF. Moreover, we observed that these dHGF specific antibodies recognized a mutant dHGF with no N-linked oligosaccharide chains as well as the wild-type dHGF (data not shown), but they did not recognize reduced dHGF (Table 1). These findings demonstrate that dHGF specific antibodies recognize three-dimensional structures newly formed in the protein moiety by the deletion of 5 amino acids. Thus, we concluded that HGF and dHGF are different in tertiary structure, and the structural change in HGF may alter its biological activities and solubility.

In the hepatocyte bioassay, both HGFs showed very similar dose response curves up to 10 ng/ml, suggesting that the two forms share the same receptor with the same affinity. Moreover, Weidner et al. (1993) have reported that HGF and dHGF bind equally to the c-Met receptor based on inhibition assay of cross-linking of ^{125}I-HGF to the c-Met receptor. Taken together, it is likely that the structural change caused by the deletion in the first kringle does not alter the affinity of HGF for the c-Met receptor. In addition to the high-affinty receptor, c-Met , a receptor for HGF with lower-affinity has been reported (Naldini et al., 1991b). Since the lower-affinity binding is suppressed by the addition of an excess amount of heparin, the binding sites are thought to be cell surface heparan sulfate proteoglycans. To characterize the lower-affinity binding of HGF and dHGF, we compared the binding of the HGFs to heparin or heparan sulfate using a solid phase ELISA. Our results showed that there was no significant difference between the affinities of the two HGFs either for heparin or heparan sulfate (data not shown). Thus, the difference between the biological activities of the two HGFs may not be explained by the binding of HGFs to the c-Met receptor or heparan sulfate proteoglycans.

A possible explanation for the difference in the target cell specificity between the two HGFs may be different receptors from the c-Met receptor. Several isoforms of the c-Met receptor, generated by alternative splicing, have been reported (Rodrigues et al., 1991). Scatchard analysis also suggests the presence of two high-affinity binding sites for HGF (Bottaro et al., 1991). In addition, a cross-linking study has identified an additional larger molecular weight complex than the complex of HGF and the c-Met (Arakaki et al., 1992). Novel putative tyrosine kinase receptors of the met family have also been reported (Ronsin et al., 1993; Huff et al., 1993; Yee et al., 1993). Since kringle structure is well known to play a role in protein-protein interaction, the deletion in the first kringle may cause a conformational change that alter the function of binding of the HGFs to a receptor. Further study of HGF receptors, including intracellular signaling cascades, should facilitate understanding of the biological difference between the two HGFs.

ACKNOWLEDGMENTS

We thank Drs. A. Murakami and K. Yamaguchi for supporting recombinant expression; Drs. E. Tsuda, M. Goto, and K. Yano for performing thymidine incorporation assays on various types of cells; Mr. H. Hayasaka for performing solubility test; Miss F. Ogaki for excellent technical assistance; and Mr. S. Whiffen for helpful comments on the English composition.

REFERENCES

Arakaki, N., Hirono, S., Ishii, T., Kimoto, M., Kawakami, S., Nakayama, H., Tsubouchi, H., Hishida, T., and Daikuhara, Y. (1992) Identification and partial characterization of two classes of receptors for human hepatocyte growth factor on adult rat hepatocytes in primary culture. J. Biol. Chem. 267: 7101-7107.

Bottaro, D. P., Rubin, J. S., Faletto, D. L., Chan, A. M.-L., Kmiecik, T. E., Vande Woude, G. F., and Aaronson S. A. (1991) Identification of the hepatocyte growth factor receptor as the *c-met* proto-oncogene product. Science 251: 802-804.

Felgner, P.L., and Holm M. (1989) Cationic liposome-mediated transfection. focus 11: 21-37.

Gherardi, E., Gray, J., Stoker, M., Perryman, M., and Furlong, R.(1989) Purification of scatter factor, a fibroblast-derived basic protein that modulates epithelial interactions and movement. Proc. Natl.Acad. Sci. USA 86: 5844-5848.

Gohda, E., Tsubouchi, H., Nakayama, H., Hirono, S., Sakiyama, O., Takahashi, K., Miyazaki,H.,Hashimoto,S.,and Daikuhara, Y.(1988) Purification and partial characterization of hepatocyte growth factor from plasma of a patient with fulminant hepatic failure. J. Clin. Invest. 81:414-419.

Grant, D.S., Kleinman, H.K., Goldberg, I.D., Bhargava, M.M., Nickoloff, B.J., Kinsella, J.L., Polverini, P., and Rosen, E.M. (1993) Scatter factor induces blood vessel formation *in vivo.* Proc. Natl. Acad. Sci. USA 90: 1937-1941.

Harris, R.C., Burns, K.D., Alattar, M., Homma, T., and Nakamura, T. (1993) Hepatocyte growth factor stimulates phosphoinositide hydrolysis and mitogenesis in cultured renal epithelial cells. Life Science 52: 1091-1100.

Higashio, K., Shima, N., Goto, M., Itagaki, Y., Nagao, M., Yasuda, H., and Morinaga, T. (1990) Identity of a tumor cytotoxic factor from human fibroblasts and hepatocyte growth factor. Biochem. Biophys. Res. Commun. 170: 397-404.

Huff, J.L., Jelinek, M.A., Borgman, C.A., Lansing, T.J.,and Parsons, J.T. (1993) The protooncogene *c-sea* encodes a transmembrane protein-tyrosine kinase related to the Met/hepatocyte growth factor/scatter factor receptor. Proc. Natl. Acad. Sci. USA 90: 6140-6144.

Kan, M., Zhang, G., Zarnegar, R., Michalopoulos, G., Myoken, Y., McKeehan, W.L., and Stevens, J. I. (1991) Hepatocyte growth factor/hepatopoietin A stimulates the growth of rat kidney proximal tubule epithelial cells (RPTE) rat nonparenchymal liver cells, human melanoma cells, mouse keratinocytes and stimulates anchorage-independent growth of SV-40 transformed RPTE. Biochem. Biophys. Res. Commun. 174: 331-337.

Kaufman, R.J., and Sharp, P. (1982) Construction of a modular dihydrofolate reductase cDNA gene: Analysis of signals utilized for efficient expression. Mol. Cell. Biol. 2, 1304-1319.

Kmiecik, T.E., Keller,J.R., Rosen,E., and Vande Woude,G.F.(1992) Hepatocyte growth factor is a synergistic factor for the growth of hematopoietic progenitor cells. Blood 80: 2454-2457.

Kohler, K., Milstein, C. (1975) Continous cultures of fused cells secreting antibody of predifined specificity. Nature 256: 495-497.

Komada,M., Miyazawa,K., Ishii, T., and Kitamura, N. (1992) Characterization of hepatocyte growth factor receptors on Meth A cells. Eur. J. Biochem. 204: 857-864.

Matsumoto,K., Takehara,T., Inoue,H., Hagiya, M., Shimizu, S., and Nakamura, T. (1991) Deletion of kringle domains or the N-terminal hairpin structure in hepatocyte growth factor results in marked decreases in related biological activities. Biochem. Biophys. Res. Commun. 181: 691-699.

Miyazawa, K., Tsubouchi, H., Naka, D., Takahashi, K., Okigaki, M., Arakaki, N., Nakayama, H., Hirono, S., Sakiyama, O., Takahashi, K., Gohda, E., Daikuhara, Y., and Kitamura, N. (1989) Molecular cloning and sequence analysis of cDNA for human hepatocyte growth factor. Biochem. Biophys. Res. Commun. 163: 967-973.

Nakamura, T., Nawa, K., Ichihara, A., Kaise, N., and Nishino, T. (1987) Purification and subunit structure of hepatocyte growth factor from rat platelets. FEBS Lett. 224: 311-316.

Nakamura, T., Nishizawa, T., Hagiya, M., Seki, T., Shimonishi, M., Sugimura, A., Tashiro, K., and Shimizu, S. (1989) Molecular cloning and expression of human hepatocyte growth factor. Nature 342: 440-443.

Naldini, L., Vigna, E., Ferracini, R., Longati, P., Gandino, L., Prat, M., and Comoglio, P.M. (1991a) The tyrosine kinase encoded by the MET proto-oncogene is activated by autophosphorylation. Mol. Cell. Biol. 11, 1793-1803.

Naldini,.L., Weidner, M., Vigna, E., Gaudino, G., Bardelli, A., Ponzetto, C., Narsimhan, R., Hartmann, G., Zarnegar, R., Michalopoulos, G., Birchmeier, W., and Comoglio, P.M. (1991b) Scatter factor and hepatocyte growth factor are indistinguishable ligands for the Met receptor. EMBO J. 10; 2867-2878.

Okigaki, M., Komada, M., Uehara, Y., Miyazawa, K.,and Kitamura, N. (1992) Functional characterization of human hepatocyte growth factor mutants obtained by deletion of structural domains. Biochemistry 31: 9555-9561.

Rodrigues, G.A., Naujokas, M.A.,and Park, M.(1991) Alternative splicing generates isoforms of the *met* receptor tyrosine kinase which undergo differential processing. Mol.Cell.Biol. 11: 2962-2970.

Ronsin, C., Muscatelli, F., Mattei, M.G. and Breathnach, R. (1993) A novel putative receptor protein tyrosine kinase of the met family. Oncogene 8, 1195-1202.

Rosen, E.M., Carley, W., and Goldberg, I.D. (1990) Scatter factor regulates vascular endothelial cell motility. Cancer Invest. 8: 647-650.

Rubin, J. S., Chan, A. M.-L., Bottaro, D. P., Burgess, W. H., Taylor, W. G., Cech, A. C., Hirschfield, D. W., Wong, J., Miki, T., Finch, P. W., and Aaronson S.A. (1991) A broad-spectrum human lung fibroblast-derived mitogen is a variant of hepatocyte growth factor. Proc. Natl. Acad. Sci. USA 88: 415-419.

Schlessinger,J.(1988)Signal transduction by allosteric receptor oligomerization. TIBS 13: 443-447.

Seglen, P. O. (1976) Preparation of isolated rat liver cells. Methods in Cell Biology (D. M. Prescott, Ed.) Vol.13,p.29-83. Academic Press, New York.

Seki, T., Ihara, I., Sugimura, A., Shimonishi, M., Nishizawa, T., Asami, O., Hagiya, M., Nakamura, T., and Shimizu, S. (1990) Isolation and expression of cDNA for different forms of hepatocyte growth factor from human leukocyte. Biochem. Biophys. Res. Commun. 172: 321-327.

Shima, N., Itagaki, Y., Nagao, M., Yasuda, H., Morinaga, T., and Higashio, K. (1991a) A fibroblast-derived tumor cytotoxic factor/F-TCF(hepatocyte growth factor) has multiple functions in vitro. Cell Biol. Int. Rep. 15: 397-408.

Shima, N., Nagao, M., Ogaki, F., Tsuda, E., Murakami, A., and Higashio, K. (1991b) Tumor cytotoxic factor/hepatocyte growth factor from human fibroblast: Cloning of its cDNA, purification and characterization of recombinant protein. Biochem. Biophys. Res. Commun. 180, 1151-1158.

Shima, N., Higashio, K., Ogaki, F., and Okabe, K. (1991c) ELISA for F-TCF (human hepatocyte growth factor/hHGF)/fibroblast-derived tumor cytotoxic factor antigen employing monoclonal antibodies and its aplication to patients factor antigen employing monoclonal antibodies and its aplication to patients with liver diseases. Gastroenterol. Japonica 26: 477-482.

Southern, P. J., and Berg, P. (1982) Transformation of mammalian cells to antibiotic resistance with a bacterial gene under control of the SV-40 early region promotor. J. Mol. Appl. Genet. 1, 327-342.

Weidner, K.M., Hartmann, G., Sachs, M.,and Birchmeier W. (1993) Properties and functions of Scatter factor/hepatocyte growth factor and its receptor *c-Met*. Am. J. Respir. Cell Mol. Biol. 8: 229-237.

Yee, K., Bishop, T.R., Mather, C., and Zon, L.I. (1993) Isolation of a novel receptor tyrosine kinase cDNA expressed by developing erythroid progenitors. Blood 82: 1335-1343

GROWTH STIMULATION MEDIATED BY G PROTEIN COUPLED RECEPTORS IN HEPATOCYTES: SYNERGISM WITH EPIDERMAL GROWTH FACTOR AND MECHANISMS OF SIGNAL TRANSDUCTION

Thoralf Christoffersen, G. Hege Thoresen, Olav F. Dajani, Dagny Sandnes & Magne Refsnes.

Department of Pharmacology,
Faculty of Medicine,
University of Oslo,
P.O.Box 1057 Blindern,
N-0316 Oslo, Norway.

INTRODUCTION

Hepatocyte proliferation is stimulated not only by mitogenic growth factors through tyrosine kinase receptors, but also by a number of other hormonal agents, most of which exert their effects via G protein-coupled receptors. Although factors in the latter category usually seem to act permissively rather than as complete, independent, hepatic mitogens, the magnitude of their effects may under optimal conditions be quite substantial. A more precise understanding of how their signals are mediated and integrated with the actions of mitogenic growth factors in hepatocytes may yield insights into mechanisms of both hepatic and general growth regulation.

The family of G proteins, i.e. heterotrimeric guanine nucleotide-binding regulatory proteins with the subunits α,β,γ (Gilman 1987), couple a large number of receptors to their effector pathways. These receptors have a characteristic basic structure, with an extracellular N-terminal chain, seven hydrophobic, apparently plasma membrane-spanning, segments, and an intracellular C-terminal part (Dohlman et al. 1991). The effector molecules that are activated by the G proteins comprise various messenger-generating

NATO ASI Series, Vol. H 88
Liver Carcinogenesis
Edited by G. G. Skouteris

enzymes and ion channels. Important examples of specific coupling by G protein subtypes are Gs → stimulation of adenylyl cyclase and formation of cyclic AMP (cAMP), Gi → inhibition of adenylyl cyclase, and Gq → stimulation of (phosphoinositide-specific) phospholipase C-β with generation of inositol 1,4,5-trisphosphate ($InsP_3$) and diacylglycerol (DAG). These are the mechanisms that are primarily focused on in the discussion below, although other effectors might be involved, and both the complexity of the coupling process (Clapham and Neer 1993) and the multifunctional interactions in the signalling (Milligan 1993) that are now being recognized, are likely to have implications for in growth regulation.

Among the agonists that act on liver cells through G protein-linked receptors are several polypeptide hormones (e.g. glucagon, vasopressin, angiotensin II), catecholamines, and various prostaglandins and other eicosanoids. Many of these agents affect hepatocyte proliferation (reviews: Michalopoulos 1990, Bucher and Strain 1992). In the studies discussed below we have used glucagon, adrenergic agents, and prostaglandins as examples to examine growth effects mediated by G protein-linked receptors, with cultured hepatocytes as the experimental model. These agents synergized with the proliferogenic effect of EGF (or TGFα), which served as the mitogen in these experiments. Then, using their ability to enhance the proliferative response in EGF-treated hepatocytes as the growth parameter, we tried to explore some of the initial G protein-mediated mechanisms that might be responsible for the growth-promoting effects of these agonists.

Temporal and quantitative aspects of interacting hormonal growth effects

Most of the experiments described were done in serum-free primary monolayer cultures of adult rat hepatocytes (details in Sand and Christoffersen 1987, 1988). The cultures were mainly used as a $G_1 \rightarrow S$ model in the present studies. We wanted to know where in G_1 the different regulations are exerted. Data summarized in Fig. 1 indicated that insulin promotes hepatocyte growth by acting relatively early in the prereplicative period (G_0 or early G_1), while the cells (if pretreated with insulin) become more sensitive to EGF (and to TGFα) at later stages in G_1 (Sand and Christoffersen 1987, Sand et al. 1992, and unpublished data). Concomitantly, the high number of EGF receptors present in the freshly isolated hepatocytes (Gladhaug and Christoffersen 1988) undergoes a marked downregulation, which particularly affects the high-affinity population (Gladhaug et al. 1988, 1992). Studies of negative control showed that the hepatocytes were most sensitive to the inhibitory effect of TGFβ when exposed to the factor shortly ($\approx$ 1 h) before the S phase entry (Thoresen et al. 1992).

There are several similarities between the growth effects of glucagon, norepinephrine, and prostaglandins (Thoresen et al. 1990, Refsnes et al. 1992, 1994). First, the positive growth response is most pronounced if these agonists are administered during the first few hours of culturing (whereas EGF addition can be postponed). Second, all the agonists seem to advance the DNA synthesis in time, producing a faster recruitmentof cells to S phase, especially in the early period. And third, all the agents mentioned, having weak growth effects alone, interact with the effect of EGF in essentially a synergistic (multiplicative) manner. Typical effects, exemplified by norepinephrine/EGF, are shown in Fig. 2 A,B . Fig. 2C compares the ability of various hormones and prostaglandins to amplify the effect of EGF on DNA synthesis. The order of efficacy (in terms of maximal responses) was: norepinephrine > $PGE_2 \approx PGF_{2\alpha}$ > vasopressin ≈ angiotensin II ≈ glucagon.

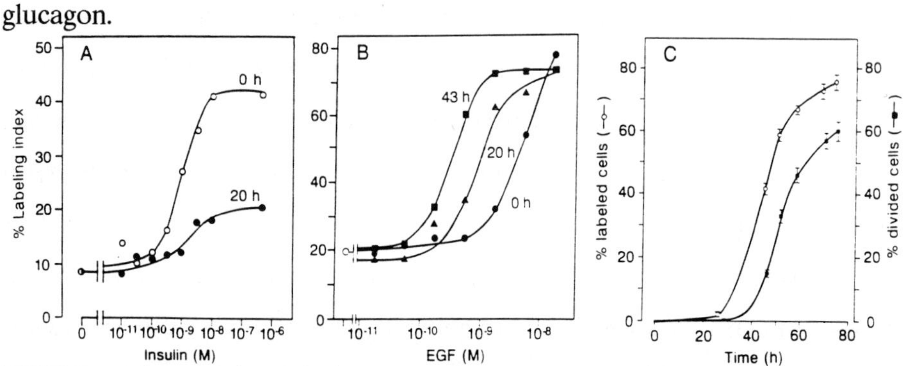

FIGURE 1: Characteristics of the growth response of hepatocytes to insulin and EGF in vitro. A and B: Differential time-dependent changes in sensitivity tothese hormones during the G_1 phase, showing that the effect of insulin is most pronounced at an early stage, while the cells increase their sensitivity to EGF with time in culture. C: Time course of the DNA synthesis and cell division in hepatocytes treated with insulin plus EGF, demonstrating that a under opti mal conditions a majority of the hepatocytes can be induced to enter S phase and divide. (With permission from Sand and Christoffersen 1987, 1988, Sand et al. 1992.)

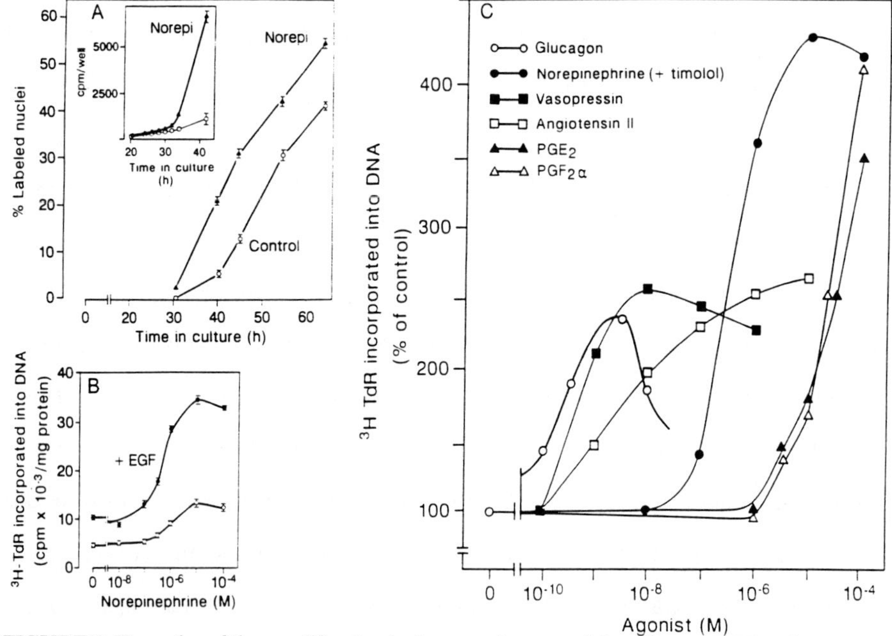

FIGURE 2: Examples of the amplification by hormonal agents of the effect of EGF on hepatocyte DNA synthesis. A: Time course of the accelerating effect of norepinephrine on S phase recruitment in cells also receiving EGF. B: Synergism between norepinephrine and EGF. C: Dose-response curves comparing the effects of various hormones and prostaglandins on the DNA synthesis in hepatocytes that also received EGF. In B and C the DNA synthesis was assessed in the period 20-50 h of culturing. (With permission from Refsnes et al. 1992, 1994, plus some previously unpublished data in C.)

Glucagon: cAMP-mediated bidirectional effect on the DNA synthesis.

The hepatocytes, each having approximately 200.000 glucagon receptors on its surface (Sonne et al. 1978), are extremely sensitive to this hormone. The responses to glucagon display all the characteristics of G protein-dependent processes and the recent molecular cloning of the glucagon receptor (Jelinek et al. 1993) has shown a structure compatible with a seven-transmembrane molecule typical of the G protein-coupled receptors. It has long been known that glucagon strongly activates adenylyl cyclase (review: Rodbell 1983), with a resultant large elevation of cAMP in the hepatocytes (Christoffersen and Berg 1974, Sonne et al. 1978, Christoffersen et al. 1984). Adenylyl cyclase/cAMP has commonly been regarded as the sole signal pathway of this hormone. However, there are data which indicate that glucagon receptors may also couple to phospholipase C and thereby both release intracellular Ca^{2+} and stimulate protein kinase C (Wakelam et al. 1986, Mine et al. 1988, Jelinek et al. 1993), although it is still not unequivocally proven that the latter effects occur in a direct, cAMP-independent, way.

Studies both in vivo and in vitro have indicated that glucagon infuences hepatic growth. However, while some investigators have reported growth-promoting effects (see e.g. Bucher and Swaffield 1975, McGowan et al. 1981, Friedman et al. 1981) others have found inhibition (e.g. Starzl et al. 1978, Vintermyr et al. 1989). It has, therefore, been hard to define the precise role of this hormone in the regulation of hepatocyte proliferation.

A clue to explaining the discrepant data may come from the finding that glucagon exerts dual, bidirectional, effects on the DNA synthesis in cultured hepatocytes (Br nstad et al. 1983); when glucagon was added early (0-6 h) after the seeding, low concentrations (1 pM - 1 nM) were stimulatory while higher concentrations inhibited. Further studies (Thoresen et al. 1990, 1992) showed that the stimulation is primarily due to a positive modulation early in G_1, whereas the inhibition results from an effect late in G_1, shortly before the S phase entry (Fig. 3). Both effects are exerted in low, physiological concentrations of glucagon if the cells are exposed at the appropriate time points in G_1.

Due to the possible involvement of cAMP-independent mechanisms, we have also compared the effects of glucagon on hepatocyte DNA synthesis with those of other agents that either activate the adenylyl cyclase (such as forskolin and cholera toxin) or are synthetic protein kinase A-activating analogs of cAMP (e.g. dibutyryl cAMP and 8-bromo cAMP).

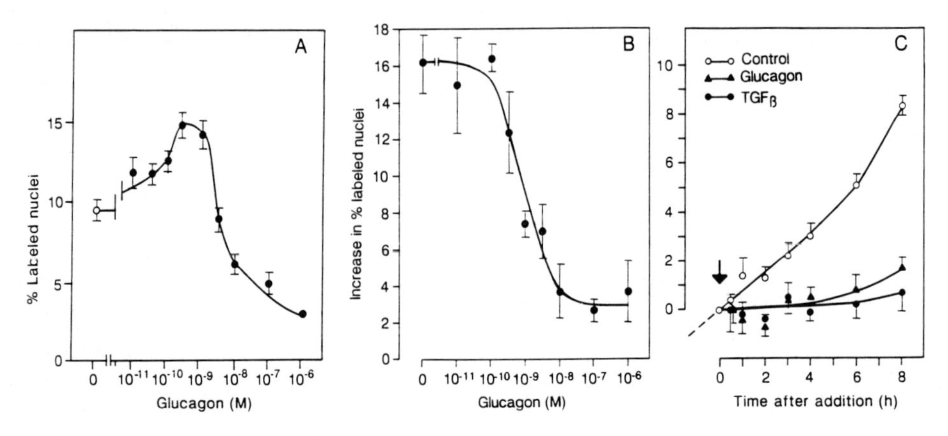

FIGURE 3: Bidirectional growth effects of glucagon in cultured hepatocytes. The cells routinely received insulin and EGF. A: Dose-response curve for the effect of glucagon added 3 h after plating, showing stimulatory effect at low concentrations. B and C: Dose-response and time curves for the inhibitory effect of glucagon administered at 43 h after plating, i.e. when hepatocytes were continuously entering S phase. Panel C also demonstrates that glucagon, like TGFβ, exerts the inhibition at a point immediately before the G_1/S border. All the cells were treated with EGF, insulin, and dexamethasone. (With permission from Thoresen et al. 1990, 1992.)

All these agents basically mimicked the bidirectional concentration- and time-dependent effects of glucagon (Bronstad et al. 1983, Thoresen et al. 1990). Taken together, the results suggest that glucagon has the ability to promote as well as inhibit hepatocyte growth and that both these effects can be accounted for by the adenylyl cyclase-dependent pathway. It has been suggested that this dual growth control may be related to cAMP-dependent regulation of both the level and degree of phosphorylation of the Rb protein (Okamoto et al. 1993).

Adrenergic agents: Role of α_1- and β_2-adrenergic signal mechanisms in stimulatory and inhibitory growth effects in hepatocytes

The three main types (α_1, α_2, and β) of adrenoceptors, with subtypes, have been cloned (Lefkowitz and Caron 1988). In hepatocytes, the major part of adrenergic effects seem to be mediated by α_1- and β-adrenoceptors, although a small number of α2-receptors are present and might contribute to certain functions. (In rat liver the most abundant subtypes are α_{1B} and β2.) For each of the three main types of the adrenoceptors one signal pathway has been demonstrated beyond doubt (Lefkowitz and Caron 1988, Exton 1994), namely: activation of adenylyl cyclase via Gs proteins for the β-receptors, activation of phosphoinositide-specific phospholipase C mediated by the Gq/G_{11} subfamily for the α_1-receptors, and inhibitory modulation of the adenylyl cyclase through Gi proteins for the α_2-receptors. However, there is increasing evidence for the existence of additional signal mechanisms (recent discussion: Garcia-Sainz 1993).

Adrenergic mechanisms have long been implicated in the regulation of hepatic growth. Studies using various adrenergic blockers in partially hepatectomized rats suggested involvement of both α_1- and β-type receptors in the regenerative response (MacManus et al. 1973, Ashrif et al. 1974, Morley and Royse 1981, Cruise et al. 1987). In culture, DNA synthesis could be enhanced by activation of β-adrenoceptors in both regenerating and normal hepatocytes (Br nstad and Christoffersen 1980, Friedman et al. 1981). However, it was later demonstrated very clearly that α_1-

adrenoceptors mediate marked growth stimulation in vitro (Cruise et al. 1985, 1986, Cruise and Michalopoulos 1985, Takai et al. 1988), and a model implicating α_1-mechanisms in the initiation of hepatocyte proliferation has been proposed (Michalopoulos 1990).

We investigated further the mechanisms mediating the proliferative responses to adrenergic stimuli, and tried to sort out the contributions by both α_1- and β-mechanisms. One reason why it was of interest to explore the effects of β-adrenergic activation, was the striking observation in a number of previous studies that there is a many-fold increase in the abundance of hepatocyte β-adrenoceptors and activity of β-receptor-linked adenylyl cyclase in various conditions associated with proliferation of these cells.

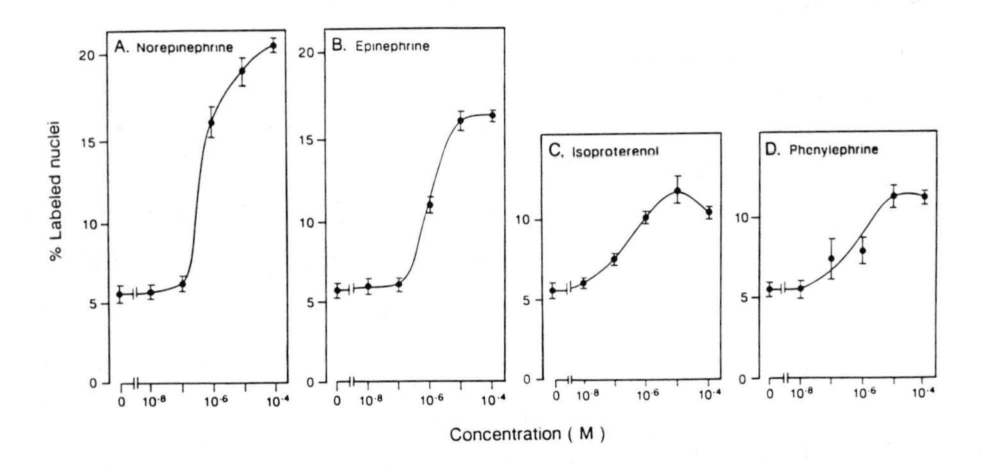

FIGURE 4: Stimulatory effect of various adrenergic agents on DNA synthesis in cultured hepatocytes. EGF, insulin, and dexamethasone were present. (With permission from Refsnes et al 1992).

This is the case both in the perinatal period (Christoffersen et a. 1973), during liver regeneration (Br nstad and Christoffersen 1980, Aggerbeck et al. 1983, Huerta-Bahena et al. 1983, Sandnes et al. 1986), as well as in vitro upon primary culturing of hepatocytes (Nakamura et al. 1983, Refsnes et al. 1983). A concomitant, almost reciprocal, decline in α_1-adrenoceptors has been found (Schwarz et al. 1985, Sandnes et al. 1986, Refsnes et al. 1992). A very large upregulation of the β-adrenoceptor-mediated signal pathway has also been found during hepato-carcinogenesis (Christoffersen et al. 1972, Boyd et al. 1973, Christoffersen and Berg 1975, Refsnes et al. 1986).

Under experimental conditions designed to assess the early recruitment of hepatocytes to S phase, the process was enhanced by norepinephrine and epinephrine (both of which act at α- and β-receptors) and also, but to a lesser extent, by isoproterenol (a β-receptor agonist) and phenylephrine (an α-receptor agonist) (Fig 4). Experiments with the α_1-receptor blocker prazosine and the β-receptor blocker timolol (Fig. 5A,B) showed that the stimulatory effect of norepinephrine consisted of both an α_1- and a β-adrenergic component. The α_1-component was the most prominent in terms of the maximal response at high concentrations of the agonist, but the β-component contributed significantly and predominated at low concentrations of the agonist.

However, adrenergic mechanisms could also mediate inhibitory control of the DNA synthesis (Refsnes et al. 1993). When administered at later stages (24-48h) of culturing, when a large fraction of the cells were near the G_1 exit, norepinephrine strongly decreased the number of new cells entering S phase (Fig. 5C,D). This inhibition was not affected by prazosine but was blocked by timolol (Fig. 5C) and was mimicked by isoproterenol (Fig. 5D); it thus showed the characteristics of a pure β-adrenergic effect, which is in accordance with the growth-inhibitory effect of cAMP late in G_1, as discussed above. Taken together, the results show that adrenergic receptors may convey both positive and negative regulation of hepatocyte proliferation.

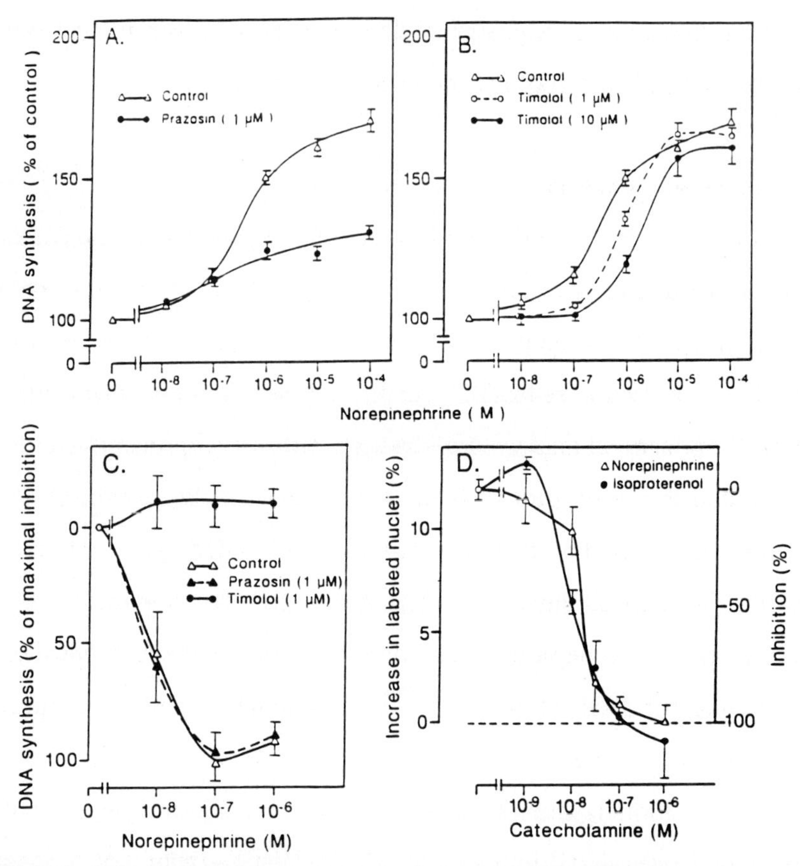

FIGURE 5: Role of α_1- and β-components in the growth effects of norepinephrine. A and B: Stimulatory effects of norepinephrine added at 3 h of culturing. C and D: Inhibitory effects after addition at 40 h. The figure illustrates that the stimulation under the conditions used is conveyed by both α_1- and β-adrenergic mechanisms, while the inhibition is a purely β-adrenoceptor-mediated (cAMP-dependent) effect. EGF, insulin and dexamethasone were present. (With permission from Refsnes et al. 1992.)

The data are compatible with the suggestion that adrenergic stimuli may play a role in the triggering of hepatocyte proliferation, and they confirm that the stimulation is mediated to a large extent through α_1-adrenoceptors, although β-adrenoceptors also contribute. In addition, the data show that adrenergic influence may inhibit hepatocyte proliferation, through a β-adrenergic, cAMP-mediated mechanism. Due to the dramatic upregulation of the β-adrenoceptors and greatly increased responsiveness of the β-receptor/adenylyl cyclase pathway in actively growing populations of hepatocytes such as in regenerating liver (see above), adrenergic control might have a significant role in limitation of the growth.

Effects of prostaglandins

Although it is clear that multiple receptors and pathways mediate the diverse biological effects of prostaglandins (Halushka et al. 1989), these are not fully elucidated. Receptors of the EP_3 subtype have been cloned (Sugimoto et al.1992), revealing the typical seven membrane-spanning domains. It appears, in analogy with the adrenergic receptors, that at least three mechanisms may be involved in prostaglandin-induced signalling: stimulation of adenylyl cyclase, inhibition of adenylyl cyclase, and stimulation of phospholipase C. All these effects can be elicited by prostaglandins also in hepatocytes, depending on the conditions (Br nstad and Christoffersen 1981, Melien et al. 1988, Garrity et al. 1989, Athari and Jungermann 1989, Mine et al. 1990, Pschel et al. 1993).

A role for prostaglandins in the induction of hepatocyte growth is suggested by in vivo experiments (MacManus and Braceland 1976, Miura et al. 1979). In culture, several prostaglandins promote the DNA synthesis in neonatal and adult hepatocytes (Andreis et al. 1981, Skouteris et al. 1988, Skouteris and McMenamin 1992, Refsnes et al. 1994) and upregulate c-myc (Skouteris and Kaser 1991). While the ability of prostaglandins to enhance growth in hepatocytes is thus well documented, relatively little information exists on the receptor mechanisms by which they mediate this effect.

We have investigated initial steps in the growth-promoting effects of prostaglandins in hepatocytes, especially the possible involvement of adenylyl cyclase and phosphoinositide-specific phospholipase C . It has been

suggested that PGE_2 enhances liver regeneration largely through a cAMP-dependent mechanism (see e.g. Tsuji et al. 1993). Our results have not supported the notion that cAMP mediates the growth effect of the prostaglandins on the hepatocytes. Although a small rise in cAMP can be produced in hepatocytes by e.g. PGE_2 if the response is amplified by inhibiting breakdown of the nucleotide and blocking
concomitant inhibitory influence on the adenylyl cyclase (Melien et al. 1988), we have not detected any significant cAMP increase when such precautions are not taken. We performed (Refsnes et al., manuscript submitted) parallel experiments with PGE_2 and glucagon, both administered to the cultures 3 h after seeding, and assessed both the acute cAMP response and the DNA synthesis (at 50 h). Glucagon elicited a large, dose-dependent, increase in cAMP, from 2.7   0.4 pmol/mg p in the controls to 15.9   2.4 at 10 nM of glucagon (i.e. the concentration that exerted the peak effect on the DNA synthesis). Under identical conditions PGE_2 did not detectably elevate cAMP (see also P schel et al. 1993). However, the DNA synthesis showed a stronger response to PGE_2 than to glucagon in these experiments (see also Fig. 2C). Further evidence against a role of cAMP in proliferative actions of the prostaglandins was obtained in experiments showing that the effects of both PGE_2, $PGF_{2\alpha}$, and PGI_2 were fully additive with those of glucagon or 8-bromo-cAMP (Refsnes et al., unpublished).

In contrast to the lack of effect on cAMP, both PGE_2 and $PGF_{2\alpha}$ moderately but consistently elevated the concentration of $InsP_3$ and very potently increased free Ca^{2+} (measured in single hepatocytes). This is compatible with, but does not prove, a role of phosphoinositide-specific phospholipase C in the growth effects of these agents in hepatocytes. To explore this question further, we compared the effects on $InsP_3$ and DAG of various agonists that promote DNA synthesis. Both norepinephrine and PGE_2 and $PGF_{2\alpha}$ increased $InsP_3$, typical responses being approximately 2-3 fold. However, vasopressin and angiotensin II, which were less effective as stimulators of the DNA synthesis (Fig. 2C), produced much larger increases in $InsP_3$ (approximately 10-fold increase) than the prostaglandins and norepinephrine (Sandnes et al. 1993). The same relative effectivity among these agents existed with respect to DAG.

The possibility of growth-promoting signalling via the Gi proteins was also considered. Prostaglandins can inhibit adenylyl cyclase, presumably through Gi. We have shown evidence for this mechanism in hepatocytes (Melien et al. 1988, see Fig. 6). It is conceivable that prostaglandin receptors that couple to Gi might also affect other Gi-dependent pathways.

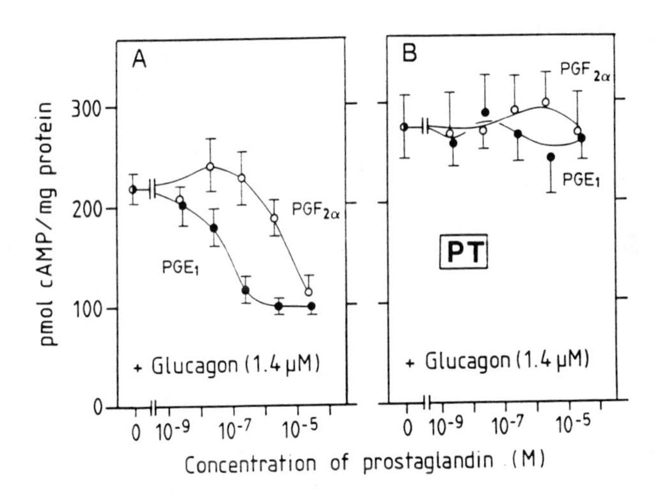

FIGURE 6: PGE_1 and $PGF_{2\alpha}$ depress the glucagon-induced cAMP formation in cultured hepatocytes (panel A) by a mechanism that is sensitive to pertussis toxin (panel B), presumably reflecting the involvement of an inhibitory effect via Gi on the adenylyl cyclase. (With permission from Melien et al. 1988)

We found that after pretreatment of hepatocytes in culture with concentrations of pertussis toxin that seemed to completely abolish the inhibitory effects on the adenylyl cyclase, evidenced as a reversal of the depressed glucagon responsiveness (typical results in Fig. 6), prostaglandins could still enhance the DNA synthesis (Refsnes et al., unpublished). This would argue against an involvement of Gi in the growth response. However, since it is hard to exclude the possibility that a small number of Gi proteins are left intact after the treatment with pertussis toxin, and that they suffice to mediate a growth effect, the results are still not conclusive. Recent reports suggesting that Gi2 is involved in linking EGF receptors to phospholipase C-γ1 in hepatocytes (Yang et al. 1991, 1993) increase the suspicion that these G proteins may have a role in integrating different growth-regulatory signal pathways.

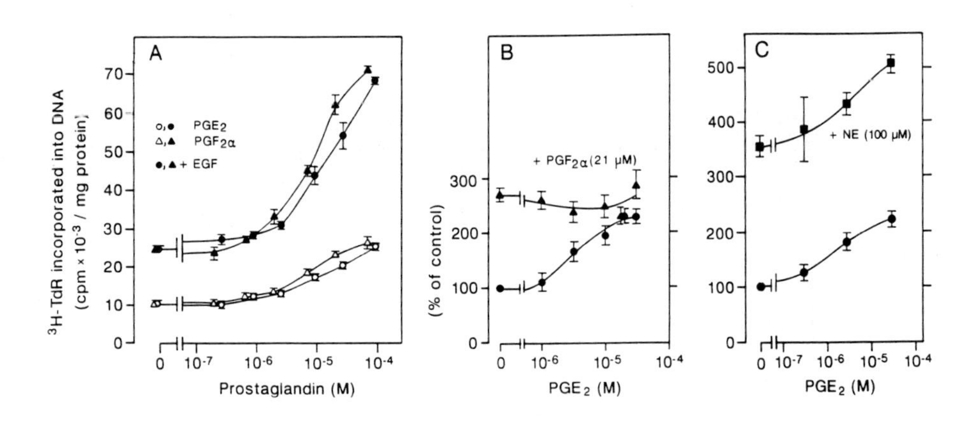

FIGURE 7: Characteristics of the enhancing effects of prostaglandins on DNA synthesis in cultured hepatocytes. A: Synergistic effect of PGE_2 or $PGF_{2\alpha}$ with EGF. B and C: Nonadditivity between PGE_2 and $PGF_{2\alpha}$ but additivity with norepinephrine. Insulin and dexamethasone were present, and EGF was added to all the cultures in B and C. Timolol was given in C, to exclude the β-adrenergic effects of norepinephrine. (With permission from Refsnes et al. 1994.)

Further data are required to finally decide if Gi proteins play a role in prostaglandin-enhanced hepatocyte growth.

Finally, we asked whether the effects of the prostaglandins and norepinephrine can substitute for each other, which would have suggested that they involve the same mechanisms. We found that whereas the effects of PGE_2 and $PGF_{2\alpha}$ on the DNA synthesis were essentially nonadditive (Fig. 7B), the combination of PGE_2 and norepinephrine, used in supramaximal concentrations, showed virtually complete additivity (Fig. C, see also Refsnes et al. 1994).

Recently we have found that this applies also to combinations of norepinephrine and prostaglandins with various other Ca^{2+}-mobilizing hormones (unpublished results). Taken together, these results indicate either that growth effects of norepinephrine and prostaglandins are conveyed via separate pathways, or, if the effects converge on a common pathway, that early agonist-specific steps (receptor, coupling) have much less capacity than, and thus do not saturate, the downstream events.

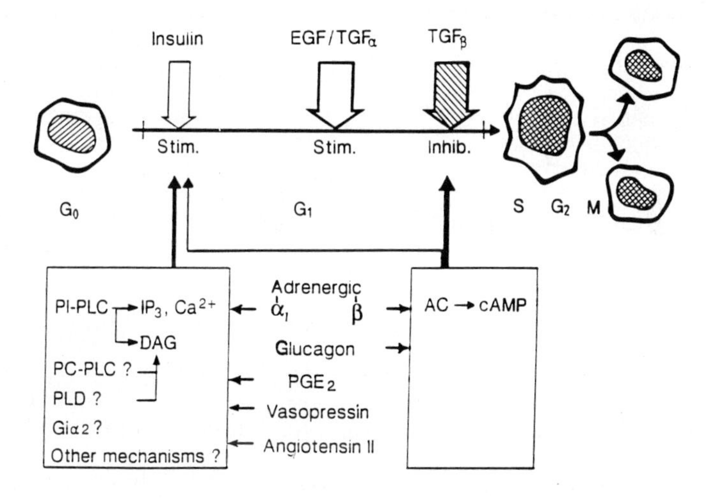

FIGURE 8: Schematic summary of some of the growth-stimulatory and -inhibitory effects hormones and growth factors on hepatocytes in vitro, suggesting where in the G_1 phase the cells are most sensitive to different agents, and possible signal mechanisms for growth-regulating effects via G protein-linked receptors. AC: adenylyl cyclase, cAMP: cyclic AMP, PI-PLC and PC-PLC: phosphatidylinositol-specific and phosphatidylcholine-specific phospholipase C, PLD: phospholipase D. IP_3: inositol-1,4,5-trisphosphate, DAG: diacylglycerol.

Conclusions

Several agonists that act through G protein-coupled receptors enhance DNA synthesis in hepatocytes and synergize with EGF/TGFα. They appear to do this largely by eliciting alterations in the hepatocytes that facilitate or accelerate the traverse through the G_1 phase, so that more cells become responsive to EGF and are stimulated to enter the S phase. Underlying mechanisms are not clarified, but certain conclusions may be drawn (Fig. 8):

Receptors linked (by Gs) to activation of the adenylyl cyclase mediate a dual regulation of hepatocyte proliferation, where cAMP exerts both growth-promoting effects early in G_1 and a strong inhibition shortly before the cells enter the S phase. This pathway conveys the bidirectional effects of glucagon and β-adrenergic activation, while no evidence has been found for its involvement in the growth effects of the prostaglandins.

Phosphoinositide-specific phospholipase C is activated by a number of Ca^{2+}-mobilizing hormonal factors that also promote growth in hepatocytes, such as norepinephrine (via α_1-adrenoceptors), prostaglandins, vasopressin and angiotensin II. However, although this activation presumably is a necessary component of the signalling, it does not seem to be the sole determinant for the magnitude of the enhancement by these agents of hepatocytes proliferation.

The large capacity for additivity of growth effects of different agonists that act on G protein-coupled receptors suggest that several pathways are operating. It is important to explore further the role of alternative mechanisms. Notable candidates include phospholipase D and phosphatidylcholine-specific phospholipase C, as well as Gi-mediated signaling.

Acknowledgements: Work from the authors' laboratory was supported by grants from The Norwgian Cancer Society, The Nordic Insulin Fund, and The Research Council of Norway.

REFERENCES

Aggerbeck M, Ferry N, Zafrani ES, Billon MC, Barouki R and Hanoune J (1983) Adrenergic regulation of glycogenolysis in rat liver after cholestasis. Modulation of the balance between $alpha_1$ and $beta_2$ receptors. J Clin Invest 71:476-4

Andreis PG, Whitfield JF and Armato U (1981) Stimulation of DNA synthesis and mitosis of hepatocytes in primary cultures of neonatal rat liver by arachidonic acid and prostaglandins. Exp Cell Res 134:265-272.

Ashrif S, Gillespie JS, and Pollock D (1974) The effects of drugs or denervation on thymidine uptake into rat regenerating liver. Eur J Pharmacol 29:324-327.

Athari A and Jungermann K (1989) Direct activation by prostaglandin $F_{2\alpha}$ but not thromboxane A_2 of glycogenolysis via an increase in inositol 1,4,5-trisphosphate in rat hepatocytes. Biochem. Biophys Res Commun 163:1253-1242.

Boyd H, Louis CJ and Martin TJ (1974) Activity and hormone responsiveness of adenyl cyclase during induction of tumours in rat liver with 3'-methyl-4-dimethylaminoazobenzene. Cancer Res 34:1720-1725.

Br nstad G and Christoffersen T (1980) Increased effect of adrenaline on cyclic AMP formation and positive β-adrenergic modulation of DNA synthesis in regenerating hepatocytes. FEBS Lett *120*:89-93.

Br nstad GO and Christoffersen T (1981) Inhibitory effect of prostaglandins on the stimulation by glucagon and adrenaline of formation of cyclic AMP in rat hepatocytes. Eur.J.Biochem. 117: 369-374.

Br nstad GO, Sand TE and Christoffersen T (1983) Bidirectional concentration-dependent effects of glucagon and dibutyryl cyclic AMP on DNA synthesis in cultured adult rat hepatocytes. Biochim Biophys Acta 763:58-63.

Bucher NLR and Strain AJ (1992) Regulatory mechanisms in hepatic regeneration. In: Wright's Liver an Biliary Disease. GH Milward-Sadler, R Wright, and MJP Arthur, eds. WB Saunders, London, Vol 3, pp. 258-274.

Bucher NLR and Swaffield MN (1975) Regulation of hepatic regeneration in rats by synergistic action of insulin and glucagon. Proc Natl Acad Sci USA 72:1157-1160.

Christoffersen T and Berg T (1974) Glucagon control of cyclic AMP accumulation in isolated intact rat liver parenchymal cells in vitro. Biochim Biophys Acta 338:408-417.

Christoffersen T and Berg T (1975) Altered hormone control of cyclic AMP formation in isolated parenchymal liver cells from rats treated with 2-acetylaminofluorene. Biochim Biophys Acta 381:72-77.

Christoffersen T, M rland J, Osnes JB and Elgjo K (1972) Hepatic adenyl cyclase: Alterations in hormone response during treatment with a chemical carcinogen. Biochim Biophys Acta 279:3663-366

Christoffersen T, M rland J, Osnes JB and ye I (1973) Development of cyclic AMP metabolism in rat liver. Biochim Biophys Acta 313:338-349.

Christoffersen T, Refsnes M, Br nstad GO, stby E, Huse J, Haffner F,

Sand TE, Hunt NH, and Sonne O (1984) Changes in hormone responsiveness and cyclic AMP metabolism in rat hepatocytes during primary culture and effects of supplementing the medium with insulin and dexamethasone. Eur J Biochem *138*:217-226.

Clapham DE and Neer EJ (1993) New roles for G-protein βγ-dimers in transmembrane signalling. Nature 365:403-406.

Cruise JL, Cotecchia S & Michalopoulos G (1986) Norepinephrine decreases EGF binding in primary rat hepatocyte cultures. J Cell Physiol 127:39-44.

Cruise JL, Houck KA and Michalopoulos GK (1985) Induction of DNA synthesis in cultured rat hepatocytes through stimulation of α_1 adrenoreceptors by norepinephrine. Science *227*:749-751.

Cruise JL, Knechtle SJ, Bollinger RR, Kuhn C and Michalopoulos GK (1987) α_1-Adrenergic effects and liver regeneration. Hepatology *7*:1189-1194.

Cruise JL and Michalopoulos GK (1985) Norepinephrine and epidermal growth factor: Dynamics of their interaction in the stimulation of hepatocyte DNA synthesis. J Cell Physiol *125*:45-50.

Dohlman HG, Thorner J, Caron MG and Lefkowitz RJ. (1991) Model systems for the study of seven-transmembrane-segment receptors. Annu Rev Biochem 60:653-688.

Exton JH (1994) Phosphoinositide phospholipases and G proteins in hormone action. Annu Rev Physiol 56:349-369.

Friedman DL, Claus TH, Pilkis SJ and Pine GE (1981) Hormonal regulation of DNA synthesis in primary cultures of adult rat hepatocytes. Exp Cell Res 135:283-290.

Garcia-Sainz JA (1993) α1-Adrenergic action: Receptor subtypes, signal transduction and regulation. Cell Signal 5:539-547.

Garrity MJ, Reed MM and Brass EP (1989) Coupling of hepatic prostaglandin receptors to adenylate cyclase through a pertussis toxin sensitive guanine nucleotide regulatory protein. J Pharmacol Exp Therap 248:979-983.

Gilman Ag (1987) G proteins: transducers of receptor-generated signals. Annu Rev Biochem. 56:615-649.

Gladhaug IP and Christoffersen T (1988) Rapid constitutive internalization and externalization of epidermal growth factor receptors in isolated rat hepatocytes. J Biol Chem 263: 12199-12203.

Gladhaug IP, Refsnes M and Christoffersen T (1992) Regulation of surface expression of high-affinity receptors for epidermal growth factor (EGF) in hepatocytes by hormones, differentiating agents, and phorbol esters. Dig Dis Sci 37:233-239

Gladhaug IP, Refsnes M, Sand TE & Christoffersen T (1988) Effects of butyrate on epidermal growth factor binding, morphology, and DNA synthesis in cultured rat hepatocytes. Cancer Res 48:6560-6564.

Halushka PV, Mais DE, Mayeux PR and Morinelli TA (1989) Thromboxane, prostaglandin and leukotriene receptors. Annu Rev Pharm Tox *29*:213-239.

Huerta-Bahena J, Villalobos-Molina R and Garcia-Sainz JA (1983) Roles of $alpha_1$- and beta-adrenergic receptors in adrenergic responsiveness of liver cells formed after partial hepatectomy. Biochim Biophys Acta 763:112-119.

Jelinek LJ, Lok S, Rosenberg GB et al. (1993) Expression cloning and signaling properties of the rat glucagon receptor. Science 259:1614-1616.

Lefkowitz RJ and Caron M (1988) Adrenergic receptors: Models for the

study of receptors coupled to guanine nucleotide regulatory proteins. J Biol Chem 263:4993-4996.
MacManus JP and Braceland BM (1976) A connection between the production of prostaglandins during liver regeneration and the DNA synthetic response. Prostaglandins *11*:609-620.
MacManus JP, Braceland BM, Youdale T and Whitfield JF (1973) Adrenergic antagonists and a possible link between the increase in cyclic adenosine 3',5'-monophosphate and DNA synthesis during liver regeneration. J Cell Physiol *82*:157-164.
McGowan JA, Strain AJ and Bucher NLR (1981) DNA synthesis in primary cultures of adult rat hepatocytes in a defined medium: Effects of epidermal growth factor, insulin, glucagon, and cyclic-AMP. J Cell Physiol 108: 353-363.
Melien , Winsnes R, Refsnes M, Gladhaug IP and Christoffersen T (1988) Pertussis toxin abolishes the inhibitory effects of prostaglandins E_1, E_2, I_2 and $F_{2\alpha}$ on hormone-induced cAMP accumulation in cultured hepatocytes. Eur J Biochem *172*:293-297.
Michalopoulos GK (1990) Liver regeneration: Molecular mechanisms of growth control. FASEB J *4*:176-187.
Milligan G (1993) Mechanisms of multifunctional signalling by G protein-linked receptors. Trends Pharmacol Sci 14:239-244.
Mine T, Kojima I and Ogata E (1988) Evidence of cyclic AMP-independent action of glucagon on calcium mobilization in rat hepatocytes. Biochim Biophys Acta 970:166-171.
Mine T, Kojima I and Ogata E (1990) Mechanisms of prostaglandin E_2-induced glucose production in rat hepatocytes. Endocrinology 126: 2831-2836.
Miura Y and Fukui N (1979) Prostaglandins as possible triggers for liver regeneration after partial hepatectomy: A review. Cell Molec Biol *25*:179-184.
Morley CGD and Royse VL (1981) Adrenergic agents as possible regulators of liver regeneration. Int J Biochem *13*:969-973.
Nakamura T, Tomomura M, Noda C, Shimoji M and Ichihara A (1983) Acquisition of a Ω-adrenergic response by adult rat hepatocytes during primary culture. J Biol Chem 258:9283- 9289.
Okamoto Y, Kikuchi E, Matsumoto M and Nakano H. (1993) Dibutyryl cyclic AMP enhances the down-regulation of RB protein during G_1 phase in the proliferating primary rat hepatocytes, but inhibits their entries into S phase and RB's phosphorylation. Biochem. Biophys. Res. Commun. 195: 84-89.
P schel GP, Kirchner C, Schr der A and Jungermann K. (1993) Glycogenolytic and antiglycogenolytic prostaglandin E_2 actions in rat hepatocytes are mediated via different signalling pathways. 218:1083-1089.
Refsnes T, Sager G, Sandnes D, Sand TE, Jacobsen S and Christoffersen T (1986) Increased number of β-adrenoceptors in hepatocytes from rats treated with 2-acetylaminofluorene. Cancer Res *46*:2285-2288.
Refsnes M, Sandnes D, Melien , Sand TE, Jacobsen S and Christoffersen T (1983) Mechanisms for the emergence of catecholamine-sensitive adenylate cyclase and β-adrenergic receptors in cultured hepatocytes: Dependence on protein and RNA synthesis and suppression by isoproterenol. FEBS Lett 164:291-298.
Refsnes M, Thoresen GH, Sandnes D, Dajani OF, Dajani L and Christoffersen T (1992) Stimulatory and inhibitory effects of catecholamines on DNA synthesis in primary rat hepatocyte cultures: Role

of alpha1-and beta-adrenergic mechanisms. J Cell Physiol *151*:164-171.
Refsnes M, Thoresen GH, Dajani OF and Christoffersen T (1994) Stimulation of hepatocyte DNA synthesis by prostaglandin E_2 and prostaglandin $F_{2\alpha}$: Additivity with the effect of norepinephrine and synergism with epidermal growth factor. J Cell Physiol *154*:35-40.
Rodbell M (1983) The actions of glucagon at its receptor: Regulation of adenylate cyclase. In: Handbook of Experimental Pharmacology. PJ Lefebvre, ed. Springer-Verlag, Berlin, Vol 66/I, pp. 263-290.
Sand TE and Christoffersen T (1987) Temporal recquirement for epidermal growth factor and insulin in the stimulation of hepatocyte DNA synthesis. J Cell Physiol *131*:141-148.
Sand TE and Christoffersen T (1988) A simple medium for the study of hepatocyte growth in culture under defined conditions. In Vitro Cell Dev Biol *24*:981-984.
Sand TE, Thoresen GH, Refsnes M and Christoffersen T (1992) Growth-regulatory effects of glucagon, insulin, and epidermal growth factor in cultured hepatocytes: Temporal aspects and evidence for bidirectional control by cyclic AMP. Dig Dis Sci *37*:84-92.
Sandnes D, Dajani OF, Refsnes M, Thoresen GF and Christoffersen T (1993) Role of phospholipase C-mediated enhancement of DNA synthesis in hepatocytes. Cell Signalling and Cancer Treatment, AACR/EORTC: B-24 (abstr).
Sandnes D, Sand TE, Sager G, Br nstad GO, Refsnes M, Gladhaug IP, Jacobsen S and Christoffersen T (1986) Elevated level of Ω-adrenergic receptors in hepatocytes from regenerating liver. Time study of $[^{125}I]$iodopindolol binding followig partial hepatectomy and its relationship to catecholamine-sensitive adenylate cyclase. Exp Cell Res 165:117-126.
Schwarz KR, Lanier SM, Carter EA, Homcy CJ and Graham RM (1985) Rapid reciprocal changes in adrenergic receptors in intact isolated hepatocytes during primary cell culture. Mol Pharmacol 27:200-209.
Skouteris GG and Kaser MRT (1991) Prostaglandin-E2 and prostaglandin-F2alpha mediate the increase in c-myc expression induced by EGF in primary rat hepatocyte ultures. Biochem Biophys Res Commun *178*:1240-1246.
Skouteris GG and McMenamin M (1992) Transforming growth factor -α -induced DNA synthesis and c-myc expression in primary rat hepatocyte cultures is modulated by indomethacin. Biochem J *281*:729-733.
Skouteris GG, Ord MG and Stocken LA (1988) Regulation of the proliferation of primary rat hepatocytes by eicosanoids. J Cell Physiol *135*:516-520.
Sonne O Berg T and Christoffersen T (1978) Binding of ^{125}I-labeled glucagon and glucagon-stimulated accumulation of adenosine 3':5'-monophosphate in isolated intact rat hepatocytes: Evidence for receptor heterogeneity. J Biol Chem 253:3203-3210.
Starzl TE, Francavilla A, Porter KA, Benichou J and Jones AF (1978) The effect of splanchnic viscera removal upon canine liver regeneration. Surg Gynecol Obstet 147:193-207.
Sugimoto Y, Namba T, Honda A, Hayashi Y, Negishi M, Ichikawa A and Narumiya S (1992) Cloning and expression of a cDNA for mouse prostaglandin E receptor EP_3 subtype. J Biol Chem 267:6463-6466.
Takai S, Nakamura T, Komi N and Ichihara A (1988) Mechanism of stimulation of DNA synthesis induced by epinephrine in primary culture of adult rat hepatocytes. J Biochem *103*:848-852.
Thoresen GH, Refsnes M and Christoffersen T (1992) Inhibition of

hepatocyte DNA synthesis by transforming growth factor β1 and cyclic AMP: Effect immediately before the G_1/S border. Cancer Res *52*: 3598-3603.

Thoresen GH, Sand TE, Refsnes M, Dajani OF, Guren TJ, Gladhaug IP, Killi A and Christoffersen T (1990) Dual effects of glucagon and cyclic AMP on DNA synthesis in cultured rat hepatocytes: Stimulatory regulation in early G1 and inhibition shortly before the S phase entry. J Cell Physiol *144*:523-530.

Tsuji H, Okamoto Y, Kikuchi E, Matsumoto M and Nakano H (1993) Prostaglandin E_2 and rat liver regeneration. Gastroenterology 105:495-499.

Vintermyr OK, Mellgren G, B e R and D skeland SO (1989) Cyclic adenosine monophosphate acts synergistically with dexamethasone to inhibit the entrance of cultured adult rat hepatocytes into S-phase: With a note on the use of nucleolar and extranucleolar [^{3}H]Thymidine labelling patterns to determine rapid changes in the rate of onset of DNA replication. J Cell Physiol 141:371-382.

Wakelam MJO, Murphy GJ, Hruby VJ and Houslay MD (1986) Activation of two signal-transduction systems in hepatocytes by glucagon. Nature, 323: 68-71.

Yang L, Baffy G, Rhee SG, Mannings D, Hansen CA & Williamson JR (1991) Pertussis toxin-sensitive Gi protein involvement in epidermal growth factor-induced activation of phospholipase C-γ in rat hepatocytes. J Biol Chem 266:22451-22458.

Yang L, Camoratto AM, Baffy G, Raj S, Manning DR & Williamson JR (1993) Epidermal growth factor-mediated signaling of Gi-protein to acti vation of phospholipases in rat cultured hepatocytes. J Biol Chem 268:3739-3746.

PROTEIN TYROSINE PHOSPHATASES IN SIGNAL TRANSDUCTION

Shi-Hsiang Shen and Denis Banville
Biotechnology Research Institute
National Research Council Canada
6100 Royalmount Avenue
Montreal, Que. H4P 2R2, Canada

INTRODUCTION

The phosphorylation of proteins at tyrosine residues is implicated in numerous cellular processes such as signal transduction, neoplastic transformation and the mitotic cycle. These processes are regulated by the activities of both protein-tyrosine kinases (PTKs) and protein tyrosine phosphatases (PTPs). PTPs represent a highly diversified and rapidly expanding family of enzymes (Fischer et al., 1991, Walton and Dixon, 1993). Indeed, since the identification of PTP1B six years ago (Tonks et al., 1988), more than 40 PTPs, excluding species homologies, have been identified. This is in striking contrast to the relatively small number of serine/threonine phosphatases identified over the past forty years. As with protein tyrosine kinases (PTK), there are two classes of PTPs: transmembrane (receptor-type) and intracellular enzymes. Most transmembrane PTPs have two conserved tandem intracellular catalytic domains with diversified receptor-like extracellular sequences. All the intracellular enzymes contain only one conserved catalytic domain with widely diversified non-catalytic amino or carboxyl termini.

NATO ASI Series, Vol. H 88
Liver Carcinogenesis
Edited by G. G. Skouteris

Current data indicate that PTPs do not act as mere negative switches of cellular processes or "house-keeping" enzymes to antagonize the action of PTKs, but actively participate in nearly every aspect of signal transduction. For instance, CD45, a receptor-type PTP, transmits a signal by activating $p56^{lck}$ via dephosphorylation of one of its tyrosine residue (Y-505) (Mustelin, 1989). Similarly, CDC25, a yeast enzyme which belongs to a subfamily of tyrosine phosphatase with homology to a vaccinia protein ,VHI, controls the cell cycle by dephosphorylating - thereby activating - $p34^{cdc2}$, thus promoting transition from G2 to M phase in eukaryotic cells (Murry,1992). Some PTPs may act as tumour suppressor or anti-oncogene, as it was originally proposed as a likely function for these enzymes. An example of this mechanism was provided by the finding that over-expression of PTP1B prevented the transformation of 3T3 cells by the *neu* oncogene product (Brown-Shimer et al., 1992). On the other hand, over-expression of certain PTPs can promote cell transformation by activating cellular oncogenes. Such neoplastic transformation was observed in rat embryonic fibroblasts where over-expression of receptor-type enzyme RPTPα activated c-*src* via tyrosine dephosphorylation (Zheng et al.,1992). Thus, the function of PTPs appears to be versatile, though in most cases, the precise mechanisms by which they regulate cellular signal transduction are yet to be revealed. In this paper, we will describe two subfamilies of PTPs to further illustrate that these enzymes can either positively or negatively regulate cellular signal transduction by interacting with specific substrates via different functional, non-catalytic, motifs present on the enzymes.

MATERIALS AND METHODS

Isolation of PTPs by the polymerase chain reaction (PCR). To clone the catalytic domains of PTPs by PCR, a set of degenerate primers was designed from two highly conserved regions within the catalytic portion of known PTPs. One primer corresponded to the amino acid sequence KC(A\D)QYWP, and the other primer was designed from the sequence VHCSAGV. PCR fragments of about 250 bp were amplified from a human cDNA library. PCRs were performed for 30 cycles with 94^{0}C denaturation and 50^{0}C annealing in standard conditions. The amplified individual bands were purified after electrophoresis on 1.6% agarose, and cloned into the pGEM vector (Promega, Madison, WI), and transformed into *E.coli.* The cloned catalytic domains of PTPs were re-amplified from individual colonies in by PCR as above. All PCR fragments were resolved by electrophoresis on 1.6% agarose and transferred on nitrocellulose membranes. Southern blots were hybridised with radioactively-labelled probes prepared from a mixture of known PTPs (the catalytic domains). The fragments which did not hybridize to these probes were used as probes to screen human cDNA libraries (4x SSC, 2x Denhardt's solution, 0.5% SDS, 0.1 M sodium phosphate buffer, pH 7.0, 100 mg/ml sodium dextran sulfate, 100 µg/ml sonicated and denatured salmon sperm DNA at 50^0 C). The inserts from the positive clones were subcloned into the Bluescript KS vector and sequenced with oligonucleotides primers or internal restriction fragments (Chretien and Shen, 1992) primers using T7-polymerase sequencing kit (Pharmacia).

Northern blot analysis. RNAs were isolated from various tissues by the standard guanidinum isothiocyanate method. The blot was pre-hybridized and hybridized in 50% formamide 2.5x Denhardt solution, 25 mM potassium phosphate buffer pH 7.4, 0.1% SDS, 10% Sodium Dextran sulfate, 5x SSPE, 250 µg/ml sonicated and denatured salmon sperm DNA, 500 µg/ml yeast tRNA. The probes were labelled with the T7 Quick Prime kit (Pharmacia). Pre-hybridization was performed for 3 hours and the hybridization was carried out overnight at 50^{0}C. The blot was washed twice with 2x SSPE, 0.1% SDS, then with 1x SSPE, 0.1% SDS and finally with 0.2x SSPE 0.5% SDS at room temperature for 15 minutes each and was exposed to Kodak X-OMAT AR film at -80^{0}C with screen.

Expression and activity assay of recombinant enzyme. The sequences encoding the PTP domains were cloned in frame into the expression vector pET-3c (Studier et al., 1990) and transformed into *E. coli* BL21 (DE3) cells. The expression of the recombinant enzyme was performed as described (Shen et al., 1991). Assays for PTP activity with low molecular weight compounds and peptides were performed at pH 5.0 and with protein substrates at pH 7.0 as described (Zhao et al., 1992).

RESULTS AND DISCUSSION

Identification of new PTPs. For the isolation of novel PTPs, we have designed degenerate primers from the conserved regions of known PTPs for PCR cloning. While this method is very powerful to clone PTPs, it was found that specific tissues always predominantly expressed certain types of PTPs which in most cases, had already been identified due to their abundance in the tissues. In order to avoid repeatedly cloning known PTPs, PTP fragments amplified by PCR were subjected to Southern blotting. The blots were hybridized with known PTP probes. Fragments which did not hybridise to the probes were further characterized by sequencing. The sequences obtained were compared with that of known PTPs, thus establishing whether a new PTP sequence had been identified. Fragments containing potential new PTPs were then used as probes to screen cDNA libraries in order to clone the full length cDNAs. Figure 1 shows the results of a Southern blotting in which #86, 88, 90 and 91 were negative to the PTP probes used. The SH2 domain-containing PTPs, PTP2C, and hPTP1E described below were identified from this blot.

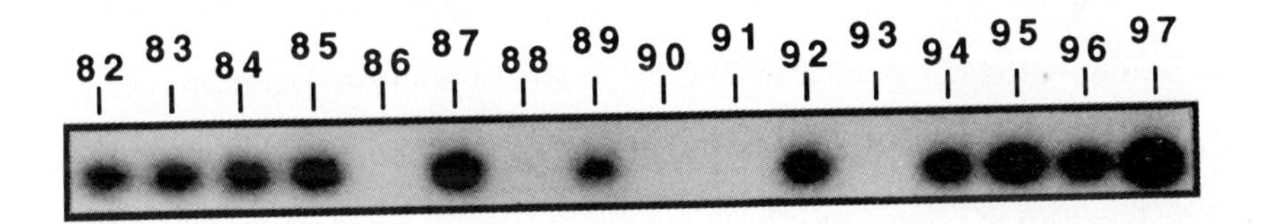

Figure 1. Southern blot of PCR fragments.

Characterization of *src* Homology 2 (SH) Containing PTPs. One subfamily of intracellular PTPs which we and others have identified contains two *src* homology 2 (SH2) domains. Three members of the SH2 domain-containing PTPs (SH-PTPs) have been cloned. They are: PTP1C (also known as SH-PTP1, HCP, or SHP) (Shen et al., 1991; Plutzky et al., 1992; Yi et al., 1992), PTP2C (also designated as SH-PTP2, Syp, or PTP1D) (Ahmad et al., 1993; Freeman et al., 1992; Feng et al., 1993; Vogel et al., 1993), and *Drosophila* corkscrew (*csw*) (Perkins et al., 1992). SH2 domains were initially identified at the N-terminal, non-catalytic region of various Src-like PTKs, and were subsequently found in a number of cytoplasmic signalling proteins. SH2 domains bind to the autophosphorylation sites of tyrosine kinase receptors and to other tyrosine-phosphorylated proteins in the signal transduction pathway (Pawson and Gish, 1992). Several external signals, such as certain growth factors and hormones, initiate their action by binding to a ligand-specific cell surface receptor which immediately autophosphorylates multiple tyrosine residues within its cytoplasmic domain. The autophosphorylated tyrosine residues create high-affinity binding sites for specific cellular proteins containing SH2 domains, including GAP, (PLC-γ), PI-3 kinase, src, SHC, GRB2/SEM5, Vav (Pawson and Gish, 1992). These SH2 domain-phosphopeptide complexes recruit other signalling molecules to the receptor where they can be phosphorylated by the receptors and/or interact with other proteins. Formation of these signal transduction complexes coordinates the multiple intracellular programs that initiate various changes in cell proliferation.

Since both PTP1C and PTP2C contain SH2 domains, it immediately raised the question as to whether these SH2 domains also interact with tyrosine phosphorylated signalling proteins, such as growth factor receptors, thus regulating downstream cellular signal transduction. The *in vitro* binding experiments showed that PTP1C indeed interacted with activated EGFR, demonstrating for the first time, the direct linkage of a phosphatase to a growth factor-receptor complex through the SH2 domain (Shen et al., 1991). Subsequently, the *in vivo* interaction of PTP1C with c-kit was demonstrated in haematopoietic cells where this enzyme is abundantly expressed (Yi and Ihle, 1993). Recently, it was further showed that PTP1C associates with tyrosine phosphorylated cytokine receptors, such as interleukin-3 (IL-3) and interleukin-5 (IL-5) receptor as well as with the erythropoietin (Epo) receptor (Yi et al., 1993). Finally, it has been demonstrated that the *motheaten*(*me*) and *viable motheaten* (me^v) phenotypes in mice are caused by the absence of the PTP1C protein or by a reduction in the amount of PTP1C respectively (Shultz et al., 1993). The associated pathological changes in homozygous mice with the *me* and me^v, include a dramatic increase in myeloid precursor cells, an increase cell in proliferation, spontaneous CSF-1 production, accumulation of the macrophage/monocyte population and severe immunodeficiencies. Thus, the phenotypes of *me* and me^v mice suggest that PTP1C may directly or indirectly negatively regulate signal transduction pathway in a number of haematopoietic lineages.

Although it is most abundantly found in haematopoietic cells, PTP1C is also expressed in numerous non-haematopoietic cells. The different level of expression of PTP1C in hematopoietic vs non-hematopoietic cells is the result of usage of different tissue specific promoters in the PTP1C gene (D.Banville, R. Stocco and S. Shen unpublished results). To study its physiological relevance in non-haematopoietic cells, we have assessed the *in vivo* interaction of PTP1C with PDGFR and EGFR both of which are widely distributed in a variety of cell type. In 293 cells, PTP1C was transiently tyrosine-phosphorylated by PDGFR upon the stimulation by its ligand, then the tyrosine-phosphorylated PTP1C underwent autodephosphorylation. In co-immunoprecipitation experiments, it was found that the ligand-activated PDGFR associated with PTP1C. During the interaction, both PTP1C and PDGFR are modulating each other's activity via phosphorylation and dephosphorylation of tyrosine residues. This association occurs through the SH2 domains because deletion of SH2 domains on PTP1C prevents formation of this complex. Similar results were found for ligand stimulated EGFR. These data further support our previous conclusion that this enzyme directly links to growth factor receptors, thus regulating signal transduction in both haematopoietic as well as in non-haematopoietic cells. The physiological consequence of the interaction of PTP1C with growth factor receptors on signal transduction in non-hematopoietic cells remains to be established.

Unlike PTP1C, PTP2C is ubiquitously expressed (Ahmad et al., 1993 and Freeman et al., 1992). In response to growth factor stimulation, PTP2C also binds to the activated EGF and PDGF receptors (Feng et al., 1993 and Vogel et al., 1993). In addition, upon insulin stimulation, PTP2C binds to tyrosine phosphorylated insulin receptors substrate 1 (IRS-1) in vitro and in vivo (Kuhne et al., 1993). Recently, we found that PTP2C binds to the insulin receptor itself in vitro. PTP2C is highly similar to the *Drosophila csw* (corkscrew) phosphatase. Corkscrew is a maternally required gene product, essential for the normal development of the anterior and posterior segments of the Drosophila embryo. Corkscrew functions in concert with l(1)polehole (D-*raf*), the counterpart of the mammalian serine/threonine kinase c-raf, to positively transduce signals generated by the activated by *torso* receptor-PTK, a PDGF receptor homologue (Perkins et al., 1992). Interestingly, following its association with the EGF and PDGF receptors, PTP2C does not dephosphorylate the receptors but is itself phosphorylated by the kinase moiety of the receptors (Vogel et al., 1993; Lechleider et al., 1993). This is in contrast with PTP1C which dephosphorylates the receptor to which it binds (Vogel et al. 1993; and this report). These properties are consistent with the notion that PTP1C negatively regulates signal transduction by dephosphorylation of receptor PTKs, whilst PTP2C, the mammalian functional homologue of *Drosophila csw*, positively regulate the downstream signal transduction.

In comparison with other intra-cellular PTPs, such as PTP1B and T-cell PTP, the purified PTP1C and PTP2C proteins display a very low specific activity towards all substrates when assayed *in vitro* (less than 0.1% of that observed with PTP1B) (Zhao et al., 1992). In order to investigated the mechanism(s) by which the phosphatase activity of this subfamily is regulated, deletions of either the N-terminal, or the C-terminal, or both non-catalytic domains were created. As shown in Figure 2, truncation of either the N-terminal SH2 domain or both SH2 domains resulted in an increase of the phosphatase activity by 10-100 fold depending on substrates assayed. Similarly, truncation of 40 amino acids on the C-termini by limited trypsin digestion increased the specific activity by more than 20 fold (Zhao et al., 1992). However, removal of both of the N-termini and C-terminal segments did not increase the activity above that achieved by the deletion of any one of these segments alone. These results indicate that the non-catalytic sequences in PTP1C and PTP2C inhibit the phosphatase activity. Deletion of these sequences may changes the conformation of the enzyme, thus activating the enzymatic activity.

An increase in the phosphatase activity of PTP2C was also observed after binding to a phosphorylated peptide suggesting that the binding of PTP2C to

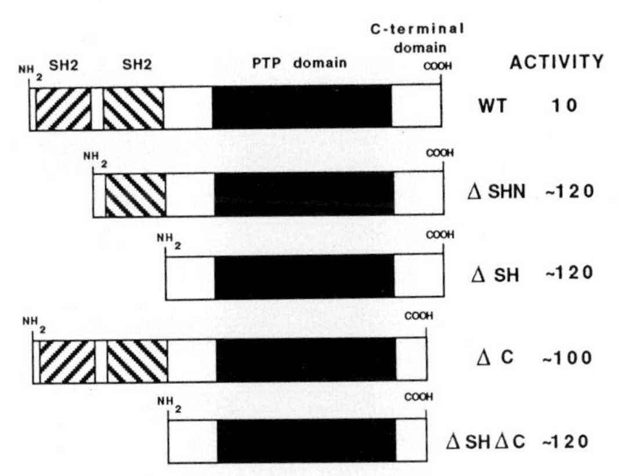

Fig. 2 Enzymatic activity of wild-type and truncated PTP1C

an activated growth factor receptor in vivo, leads to the enzymatic activation of the phosphatase (Lechleider et al.,1993).

ERM-PTP subfamily. A second type of protein tyrosine phosphatase with structural features suggesting an obvious involvement in cellular signal transduction is exemplified by hPTP1E and PTPE1, two novel PTPs which we have cloned and described here. These two enzymes possess a large N-terminal non-catalytic domain with a region of significant homology to the cytoskeletal-associated proteins ezrin, radixin, moesin and band 4.1 which will be referred here as the ERM-domain (Lankes et al., 1991). hPTP1E and PTPE1, together with the previously reported PTPH1 (Yang and Tonks, 1991) and PTPMEG (Gu et al., 1991) constitute a new class of PTPs which we termed the ERM-PTP subfamily (Figure 3).

In addition to a common structural organisation, the members of the ERM family of protein share an important functional feature. Indeed, in each case for which the subcellular localisation of these protein has been studied, it was determined that they co-localisc with the cortical actin cytoskeleton where they interact with both the cell membrane and the actin cytoskeleton. The presence of a highly similar domain in the ERM-PTPs suggest that these enzymes will be targeted to the interface between the plasma membrane and the cytoskeleton ,

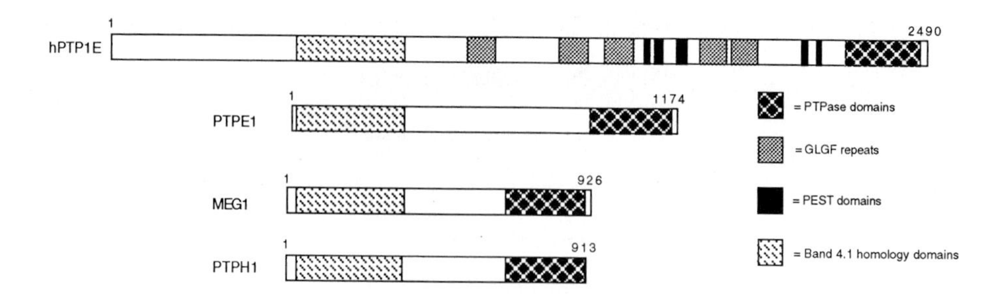

Fig.3 ERM-PTP family

a prime location for their participation in cellular signal transduction. Several members of the ERM protein family, including ezrin, talin and band 4.1, are substrates for protein tyrosine kinases. The change in the phosphorylation state of these proteins is correlated with changes in protein-protein interaction and/or their subcellular distribution. The importance of the ERM protein family has been emphasised recently with the finding that the tumour suppressor neurofibromatosis type 2 gene (*NF2*) encodes a cytoplasmic protein with striking sequence similarity to ERM family members (Tsukita et al., 1993). Although there is as yet no evidence to demonstrate their involvement, it seems reasonable to postulate a role for the ERM-PTPs in the regulation of the association of the ERM proteins with cytoskeletal elements and the plasma membrane through alteration of their phosphorylation status.

Among the members of the ERM-PTP subfamily, hPTP1E stands out because of several unique features. Its 2490 amino acid residues make it the largest of all tyrosine phosphatases identified so far (Figure 3). At its N-terminus, preceding the ERM domain, there is a large segment of approximately 560 amino acid residues which does not have homology to any other known protein. It is therefore the only member of the ERM protein family not to have the ERM-domain at its very N-terminus. Whether this feature influences its subcellular localisation or activity is not known yet. In addition to the ERM domain, hPTP1E contains five repeated domains, each of approximately 90 amino acid residues, located between the ERM domain and the PTPase domain. These sequences which have been termed GLGF repeats, were first identified within the product of the lethal discs large-1 (dlg) gene of Drosophila, a tumor suppressor gene encoding a guanylate kinase located in the septate junctions of epithelial cells (Woods and Bryant, 1991). Similar domains were later identified in three other mammalian guanylate kinases: PSD-95, from the rat brain synapses (Cho et al., 1992), p55, a human erythrocyte membrane protein (Ruff et al., 1991), and ZO-1, a component of the tight junctions of polarised epithelial cells (Willott et al., 1993). The precise function of these repeated structure within these proteins is unknown. However, they are believed to be involved in the subcellular distribution of the proteins which are all found in specialised structures of the cell membrane where cell-cell contacts occur. The presence of these repeats within hPTP1E suggest that this phosphatase may be a component of the signal transduction pathways mediating cell-cell and/or cell-substrate interaction. Several protein tyrosine kinases, including proto-oncogenic gene products such as c-yes and c-src, are present in junctional plaques. It is therefore most likely that phosphatases will also be involved in regulating the interaction of these proteins with adhesion receptors, cytoskeletal components and signal transducer proteins. hPTP1E possesses the features which make it a likely candidate for this job.

Another feature of hPTP1E is the presence of five stretches of sequences that are rich in proline (P), glutamic acid (E), serine (S) and threonine (T) residues, scattered between the GLGF repeats. These PEST sequence usually confers a short intracellular half-live to the polypeptides containing them (Dice, 1987) and may provide hPTP1E a built in feature for auto-termination of its function when required. It is not clear whether the function of PEST sequences requires other factors or if it is influenced by the subcellular localization of the protein.

Finally, the phosphatase domain, located at the extreme carboxyl end of the protein, possess intrinsic tyrosine phosphatase activity when expressed in bacteria and assayed in vitro using various substrates. The gene encoding hPTP1E has been partially characterized and was found to span approximately 250 kb and to consist of more than 40 exons. A number of mRNA transcripts of various lengths have been detected in the breast carcinoma cell line ZR75.1. These variant sequences are the product of alternative splicing of the primary transcript. Interestingly, in certain variant transcripts, termination codons are present which result in the premature termination of the reading frame upstream of the catalytic domain. The translated protein product would therefore have no phosphatase activity. Whether these mutated transcripts have contributed to the development of this tumor cell line remains an interesting question to study.

ACKNOWLEDGEMENTS

We thank Dr S. Ahmad and Mr R. Stocco and D. L'Abbo for technical assistance. The artwork was done by Mr R. Stocco and P. Bouchard.

REFERENCES

Ahmad, S., Banville, D., Zhao, Z., Fischer, E.H. & Shen., S.-H.(1993). A widely expressed human protein-tyrosine phosphatase containing src homology 2 domains. Proc. Natl. Acad. Sci. USA. 90, 2197-2201.

Brown-Shimer, S., Johnson, K.A., Hill, D.E., & Bruskin, A.M.(1992). Effect of protein tyrosine phosphatase 1B expression on transformation by the human neu oncogene. Cancer Res. 52, 478-482.

Cho, K.-O., Hunt, C.A. and Kennedy, M.B.(1992). The rat brain postsynaptic density fraction contains a homolog of the drosophila discs-large tumor supressor protein. Neuron **9**, 929-942.

Chretien, P., & Shen, S.H.(1992). Multiplex random priming of internal restriction fragments for DNA sequencing. DNA and Cell Biology, 11, 337-343.

Dice, J.F.(1987). Molecular determinants of protein half-lives in eukaryotic cells. FASEB J. 1, 349-357.

Feng, G.-S., Hui, C.-C., & Pawson, T.(1993). SH2-containing phosphotyrosine phosphatase as a target of protein tyrosine phosphatase. Science 259, 1607-1611.

Fischer, E.H., Charbonneau, H., & Tonks, N.K.(1991). Protein tyrosine phosphatase: A diverse family of intracellular and transmembrane enzyme. Science 253, 401-406.

Freeman, R.M., Plutzky, J. & Neel, B.G.(1992). Identification of a human src homology 2- containing protein-tyrosine-phosphatase: A putative homolog of drosophila corkscrew. Proc.Natl. Acad. Sci. USA. 89, 11239-11244.

Gu,M.J., York, D., Warshawsky,I. & Majerus, P.W.(1991). Identification, cloning, and expression of a cytosolic megakaryocte protein-tyrosine phosphatase with sequence homology to cytoskeletal protein 4.1. Proc. Natl. Acad. Sci. USA. 88, 5867-5871.

Kuhne, M.R., Pawson, T., Lienhard, G.E., & Feng, G.-S.(1993). The insulin receptor substrate 1 associates with the SH2-containing phosphotyrosine phosphatase Syp. J. Biol. Chem. 268, 11479-11481.

Lankes, W.T.& Furthmayr, H.(1991). Moesin: A member of the protein 4.1-talin-ezrin family of protein. Natl. Acad. Sci. USA. 88, 8297-8301.

Lechlcider, R.J., Freeman, R.M., Jr., & Neel, B.G.(1993). Tyrosyl phosphorylation and growth factor receptor association of the human corkscrew homologue, SH-PTP2. J. Biol. Chem. 268, 13434-13438.

Lechleider, R.J., Sigimoto, S., Bennett, A.M., Kashishan, A.S., Cooper, J.A., Shoelson, S.E.,Walsh, C.T.& Neel, B.G.(1993). Activation of the SH2-containing phosphotyrosine phosphatase SH-PTP2 by its binding site, phosphotyrosine 1009, on the human platelet-derived growth factor receptor. J. Biol. Chem. 268, 21478-21481.

Mustelin, T., Coggeshall, K.M.,& Altman, A.(1989). Rapid activation of the T-cell tyrosine protein kinase $pp56^{lck}$ by CD45 phosphotyrosine phosphatase. Proc. Natl. Acad. Sci. 86, 6302-6306.

Murry, A.W.(1992). Creative blocks: cell cycle checkpoints and feedback controls. Nature, 359, 599-562.

Pawson, T., & Gish, G.D.(1992). SH2 and SH3 domains: from structure to function. Cell 71, 359-362.

Perkins, L.A., Larsen, I. & Perrimon, N.(1992). Corkscrew encodes a putative protein tyrosine phosphatase that functions to transduce the terminal signal from the receptor tyrosine kinase torso. Cell 70, 225-236.

Plutzky, J., Neel, B.G. & Rosenberg, R.D.(1992). Isolation of a src homology 2-containing tyrosine phosphatase. Proc. Natl. Acad. Sci. USA. 89, 1123-1127.

Ruff, P., Speicher, D.W. & Husain-Chishti, A.(1991). Identification of a major palmitoylated erythrocyte membrane protein containing the src homology 3 motif. Proc. Natl. Acad. Sci. USA. 88, 6595-6599.

Shen, S.H., Bastien, L., Posner, B.I., & Chretien, P.(1991). A protein tyrosine phosphatase with sequence similarity to the SH2 domain of the protein-tyrosine kinases. Nature 352:736-739.

Shultz, L.D., Schweitzer, P.A., Rajan, T.V., Yi, T., Ihle, J.N., Matthews, R.J., Thomas, M.L., & Beier, D.R.(1993). Mutations at the murine motheaten locus are within the hematopoietic cell protein-tyrosine phosphatase (Hcph) gene. Cell 73, 1445-1454.

Studier, F. W., Rosenberg, A.H., Dunn, J.J., & Dubendorff, J.W.(1990). Use of T7 RNA polymerase to direct expression of cloned genes. Methods Enzymol. 185, 60-89.

Tonks, N.K., Diltz, C.D. & Fischer, E.H.(1988). Purification of the major phosphotyrosine phosphatases of human placenta. J. Biol. Chem. 263, 6722-6730.

Tsukita, S., Itoh, M., Nagafuchi, A., Yonemura, S. & Tsukita, S.(1993). Submembranous junctional plaque proteins include potential tumor suppressor molecules. J. Cell Biol. 123, 1049-1053.

Vogel, W., Lammers, R., Huang, J., & Ullrich, A.(1993). Activation of a phosphotyrosine phosphatase by tyrosine phosphorylation. Science 259, 1611-1614.

Walton, K.M., & Dixon, J.E.(1993). Protein tyrosine phosphatases. Annu. Rev. Biochem. 62, 101-120.

Willott, E., Balda, M.S., Fanning, A.S., Jameson, B., Van Itallie, C. & Anderson, J.M. (1993). The tight junction protein ZO-1 is homologous to the Drosophila discs-large tumor suppressor protein of the septate junction. Proc. Natl. Acad. Sci. USA. 90, 7834-7838.

Woods, D.F.& Bryant, P.J.(1991). The discs-large tumor suppressor gene of drosophila encodes a guanylate kinase homolog localized at septate junctions. Cell 66, 451-464.

Yang, Q. & Tonks, N.K.(1991). Isolation of a cDNA clone encoding a human protein-tyrosine phosphatase with homology to the cytoskeletal-associated proteins band 4.1, ezrin, and talin. Proc. Natl. Acad. Sci. USA. 88, 5949-5953.

Yi, T. & Ihle, J.N.(1993). Association of hematopoietic cell phosphatase with c-kit after stimulation with c-kit ligand. Mol. Cell. Biol. 13, 3350-3358.

Yi, T., Mui, A.L., Krysstal, G. & Ihle, J.N.(1993). Hematopoietic cell phosphatase associates with the interleukin-3 (IL-3) receptor beta chain and down-regulates IL--3-induced tyrosine phosphorylation and mitogenesis. Mol. Cell. Biol. 13, 7577-7586.

Yi, T., Cleveland, J.L., & Ihle, J.N.(1992). Protein tyrosine phosphatase containing SH2 domains: characterisation, preferential expression in hematopoietic cells, and localization to human chromosome 12p12-p13. Mol. Cell. Biol. 12, 836-846.

Zhao, Z., Bouchard, P., Diltz, C.D., Shen, S.-H., & Fischer, E.H.(1992). Purification and characterization of a protein tyrosine phoshatase containing SH2 domains. J. Biol. Chem. 268, 2816-2820.

Zheng, X.M., Wang,Y., & Pallen, C.J.(1992). Cell transformation and activation of pp60$^{c\text{-}src}$ by overexpression of a protein tyrosine phosphatase. Nature, 359, 336-338.

C-MYC IS ESSENTIAL FOR INITIATION OF DNA SYNTHESIS IN EGF-STIMULATED HEPATOCYTE CULTURES

George G. Skouteris
CellGene Ltd. Athens, Greece and
*Deutsches Krebsforschungszentrum
Department of Applied Tumor Virology
Laboratory of Virus-Host Cell interactions
Im Neuenheimer Feld 242,
D-69120 Heidelberg, Germany

INTRODUCTION

The *c-myc* gene is expressed at constitutively elevated levels in a variety of tumors, suggestive of an important role in multistage carcinogenesis (Alitalo et al.,1983). Introduction of *c-myc* into primary cells in culture under most circumstances does not cause transformation (Kaszmarec et al.,1985, Hann et al.,1985). This gene is expressed at a low constitutive level in growing cells and is down regulated in both quiescent and differentiating cells. *C-myc* is thus often grouped into the category of immediated early genes and it has been suggested that it participates in the cascade of the events that follows mitogenic stimulation of quiescent cells (Makino et al.,1984). This gene is regulated in a growth-dependent manner and when over-expressed abrogates the requirement for various growth stimuli (Shichiri et al.,1993). One intriguing aspect is that *c-myc* appears to negatively autoregulate its own expression. The effect was first observed in Burkitt's lymphoma and murine plasmacytoma

*present address

NATO ASI Series, Vol. H 88
Liver Carcinogenesis
Edited by G. G. Skouteris

cell lines, in which one gene copy is highly expressed by virtue of a chromosomal translocation or retroviral insertion, while the unaffected gene copy is significantly repressed (Marcu et al., 1992).

Much of the investigation on the mechanisms of hepatic growth has been done in partially hepatectomized animals in vivo and in hepatocytes in primary culture. Almost immediately after partial hepatectomy there are major changes in the binding capacity of transcription activators and in the expression of a relatively large number of genes. Many of the immediate early response gene products are themselves transcription activators and thus can multiply and propagate the initial gene activation processes (Nishida et al., 1994).

Liver regeneration after two thirds partial hepatectomy, induced a marked increase in *c-myc* expression (Makino et al.,1984).

It is thus far believed that the tumorigenic potential of *c-myc* is not correlated with the synthesis of <u>mutated</u> proteins, but the normally produced oncoprotein(s) regulate both the normal cellular proliferation and at the same time can induce changes leading to tumor formation (Moore et al., 1987, Persson et al.,1985).

The *c-myc* protein contains a basic-helix-loop-helix (bHLH) DNA binding and dimerization motif, hence it belongs to a large family of bHLH-transcription factors. The *c-myc* protein binds another bHLH protein, called MAX and this association is critical for the biological function of Myc.

Recent evidence indicates that *c-myc* may also regulate programmed cell death (Evan et al.,1992).

MATERIALS AND METHODS

Hepatocyte isolation and culture

Hepatocytes were isolated from adult male Wistar rats using the two-step collagenase perfusion method (Seglen P.O.1976). The final cell suspension containing more than 83% viable hepatocytes, was resuspended in Eagle's Minimum Essential Medium, containing non-essential aminoacids, insulin (10^{-7} M) and gentamycin (50 μg/ml). Hepatocytes were plated at 1.5 x 10^5 cells/cm^2/ml onto collagen coated Petri dishes (Skouteris et al., 1988). After two hours attachment, the medium was replaced and the cells were incubated at 37° C in 5% CO_2 and 95% humidified atmosphere.

Autoradiography.

[^{3}H methyl]-thymidine (5μCi/dish) was added to the cultures 24 hours before harvesting. Autoradiography proceeded as previously described (Chan et al.,1989) and the nuclear labeling index was evaluated by counting the number of intensively labeled nuclei in 250 cells/plate.

Hepatocyte labeling and immunoprecipitation conditions

Cultured hepatocytes were labeled with 100 μCi/ml of [^{35}S] methionine (sp.activ.>800 Ci/mmol) in methionine-free Dulbecco's modified Minimum Essential Medium (Gibco) or with 150 μCi/ml of 32Pi in ordinary culture medium, for 2 hours. For pulse-chase experiments, hepatocytes were starved for methionine for 30 minutes before the addition of the label (200 μCi/ml). The cells were pulsed for 5 or 20 minutes and the final cell pellet was homogenized in lysis buffer (LB) pH:7.4 containing 20 mM HEPES, 5 mM KCl, 5 mM $MgCl_2$, 0.5% Triton X-100, 0.1% sodium deoxycholate and 0.1 mM phenyl-methyl-sulfonyl fluoride (PMSF). Control experiments have shown that the immunoprecipitation reactions were performed in antibody excess. Culture supernatants were spun down at 10000 x g for 5 minutes at 4°C before the antibodies were added. Cell lysates and culture mediums were incubated at room temperature for three hours under gentle rocking (60 rpm) and at the end of the incubation, 40 μl of protein A-sepharose (1:1 slurry in LB). The beads were then resuspended in sample buffer containing 2-mercaptoethanol (5%,v/v), heated at 100°C for 8 minutes and loaded onto a discontinuous gel system (10% polyacrylamide separating gel,4.5% stacking gel) (Laemmli 1970).

Additions were made as described in the legends of the figures. All chemicals were purchased from Sigma Chemical Co. St. Louis Missouri and the radioactive compounds were from Amersham International.

Preparation of c-myc antibodies

The rabbit polyclonal antibody against human recombinant human c-myc protein was prepared as previously described (Watt et al., 1985). This polyclonal antibody was tested and shown to react with recombinant c-myc protein and also with c-myc protein(s) from Colo 320 HSR, Daudi, K-562 and other cell lines. The two monoclonal antibodies against c-myc peptides used in this study were the 9E10.2 and the CT14-GT.3 (Evan et al.,1985). The hybridomas secreting the above antibodies were from the American Tissue Type Collection and purified immunoglobulins were prepared as previously described (Evan et al. op. cit.).

Microinjection of hepatocytes

Hepatocytes were grown on coverslips under conditions previously described in the methods section. Microinjection was performed as previously described (Graessman and Graessman 1986).

RESULTS

We have examined the effects of transient expression of a transfected cellular oncogene, the *c-myc*, upon the DNA synthetic activity of primary rat hepatocytes stimulated with EGF.

The expression of the transfected *c-myc* oncogene in primary hepatocytes was unable to stimulate DNA synthesis unless EGF was present in culture. EGF exerted its stimulatory effect at doses similar to those used in normal hepatocyte cultures. The initiation of DNA synthesis in transfected hepatocytes followed the same time course as that observed in non-treated normal cultures (Fig.1a, b). In our experiments we have used two similar monoclonal antibodies raised against a C-terminus peptide corresponding to residues 408 through 432 (Fig.2).

Hepatocyte cultures transfected with the *c-myc* construct and labeled with ^{35}S-methionine and then immunoprecipitated with myc antibodies and analyzed on SDS-PAGE (Skouteris and Kaser, 1992). Four major MYC polypeptides ranging from 55-67 KDa were resolved on SDS-PAGE and the synthesis of myc polypeptides are significant throughout the culture time. The synthesis of hepatocyte *c-myc* polypeptides following the same pattern was further stimulated after the addition of EGF, despite the inclusion of proteolytic inhibitors in all the extraction buffers.

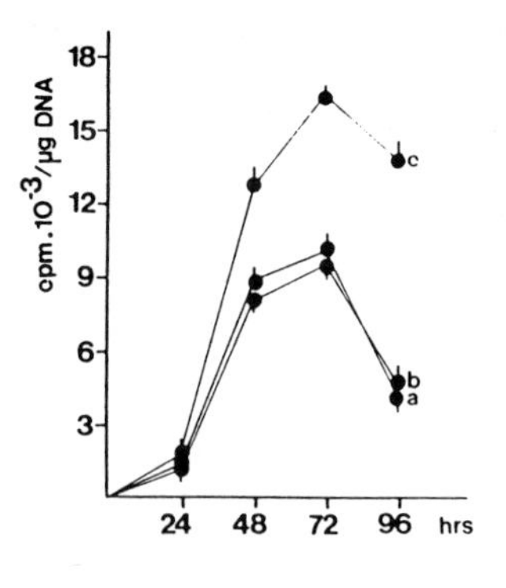

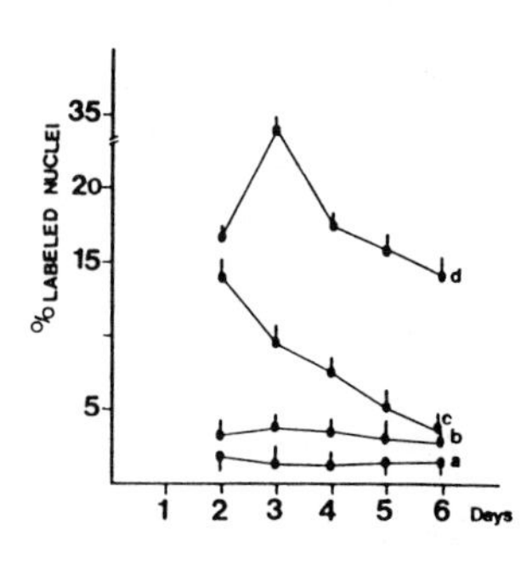

Fig.1 [A] DNA synthesis in primary hepatocyte cultures transfected with the pSV$_2$neo-c-myc oncogene: Primary hepatocytes were plated in 60 mm dishes and incubated with 50 ng/dish of EGF immediately after the transfection for various times. (a): hepatocytes transfected with 12 µg/dish of thc pSV$_2$-c-myc plasmid. (b) normal non-transfected hepatocytes and (c) hepatocytes transfected with 12 µg/dish of pSV$_2$c-myc plasmid.In all transfection mixtures 8 µg/dish of carrier DNA (liver DNA) was included. The points represent values +/- S.E.M. from quadruplicated dishes from at leat three independent experiments.

[B] Time course for EGF effect on DNA synthesis in hepatocytes transfected with pSV$_2$neo and pSV$_2$neo-c-myc plasmids.Hepatocyte cultures were labeled with ^{3}H-TdR as described and intensively labeled nuclei were counted (250 cells/plate).(a): Hepatocytes transfected with pSV$_2$ neo and treated without EGF or (b):with pSV$_2$ neo-c-myc as in (a). (c):hepatocyte cultures transfected with pSV$_2$neo and treated continuously with EGF (50 ng/dish) and (d):hepatocyte cultures transfected with pSV$_2$neo-c-myc construct as in (c). Points represent means+/- S.E.M. from at least two independent experiments.

Peptide G (aminoacid residue 408-439)

human c-myc (immunogen) Fig.2

A E E Q K L I S E E D L L R K R R E Q L K H K L E Q L R N S C A

CT14-G4 IgG1.k
Myc1-9E10 IgG1.k

In other studies by Miyamoto (1985), it has been reported that *c-myc* antigens migrated outside the 60-68 KDa range, in particular pp55 and p90 in Colo 320 and pp56 and pp120 in Xenopus oocytes.

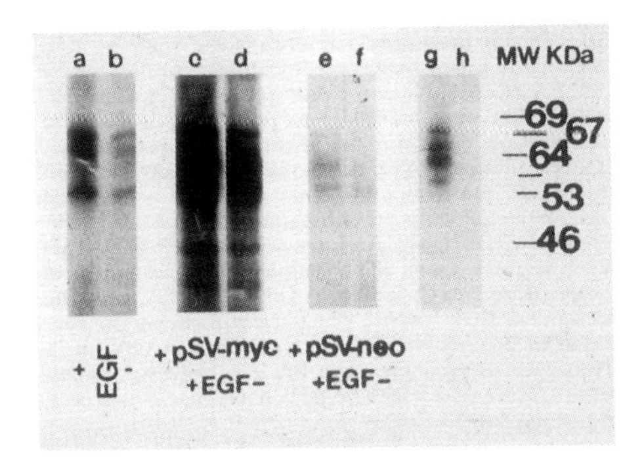

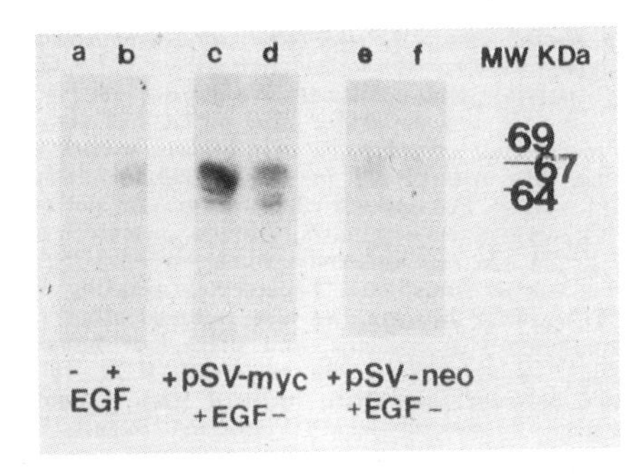

A B

Fig.3 [A]:Immunoprecipitated polypeptides are analysed onto 10% SDS-PAGE.
(a,b):hepatocyte polypeptides reacted with MYC antisera from normal hepatocyte cultures treated with or without EGF.
(c,d):immunoprecipitates from hepatocyte cultures transfected with the pSV_2-neo-c-myc and treated with or without EGF.
(e,f):lysates from pSV_2neo transfected cultures and treated as in (c,d).(a-f):hepatocyte lysates reacted with the CT14-GT.3 monoclonal antibody and (g):hepatocyte lysate as in (a) reacted with the polyclonal antiserum (from 60 hour cultures).The (h) lane is identical with (g) but the hepatocyte lysate reacted with pre-immune rabbit serum.

[B] :Immunoprecipitated polypeptides from ^{32}P-orthophosphate labeled hepatocyte lysates (48 hrs) reacted with MYC monoclonal antisera.(a,b): hepatocyte cultures labeled and treated with or without EGF (c,d):polypeptides from hepatocyte cultures transfected with the pSV_2neo-c-myc plasmid in the presence or not of EGF (e,f):lysates from hepatocytes transfected with the pSV_2neo construct and reacted with MYC antisera.
(a,b,c and e):hepatocyte lysates reacted with the CT14-GT.3 monoclonal antibody and (d,f):lysates reacted with the polyclonal antibody.

This suggests that post-translational modifications can alter the apparent molecular mass of MYC proteins depending upon the cellular background in which they are synthesized (Skouteris and Kaser op.cit.).
Immunocytochemical studies on normal and transfected hepatocytes using the CT14-GT.3 monoclonal antibody have shown that the great percentage of the hepatocyte population expressed MYC protein throughout the culture time. Other studies by Muakkasah-Kelly (1988) have shown similar results and have reported that the hepatocyte *c-myc* polypeptides migrated below 60KDa in SDS-PAGE and this was assigned to proteolytic degradation.
The only known modification of MYC proteins is phosphorylation. Hepatocyte cultures transfected with the plasmid constructs, were labeled with 100μCi of ^{32}P-orthophosphate and their lysates were analyzed onto 10% SDS-PAGE. Only two of four major synthesized polypeptides reacting with either the CT14-GT.3 or the polyclonal antibody were shown to undergo phosphorylation (Fig.3). These polypeptides migrated between 64-67 KDa on SDS-PAGE.

The differences in the degree of phosphorylation of the 64-67 KDa doublet in lysates originating from cultures transfected with the *c-myc* construct and treated with EGF may be assigned to the increased synthesis of the respective polypeptides as observed in the synthesis experiments.
CDM-supplemented normal hepatocyte cultures were grown on coverslips in the presence of 50 ng/dish of EGF. EGF was present immediately after cell attachment and hepatocytes were microinjected with the 9E10.2 anti-*c-myc* monoclonal antibody, at 0.5, 1,3,6 and 10 hours after the addition of EGF. The results are shown in the next figure, where we may see the effects of the microinjected antibody on the EGF-induced hepatocyte DNA synthesis.

Control experiments were performed with microinjection of serum albumin (Fig.4).

The microinjected *c-myc* monoclonal antibody at 2 μg/ml 0.5 hr and for up to six hours after the addition of EGF, caused a marked decrease in hepatocyte labeling index observed at 48 and 72 hours respectively. However, the monoclonal antibody had no significant effect on hepatocyte labeling index if microinjected 10 hours after the addition of EGF.

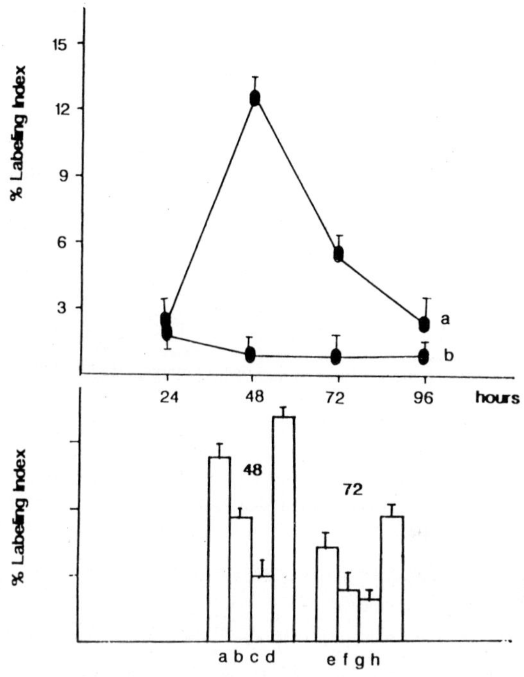

Fig.4 Labeling index of normal hepatocytes cultured in the presence of 50 ng/dish of EGF.[A].Hepatocytes cultured in the presence (a) or not (b) of EGF for 4 days.[B].Effects of microinjection of 9E10 anti-c-myc monoclonal antibody on hepatocyte labeling index at 48 and 72 hours in culture.
(a,e) : 0.5 hours after attachment.(b,f) : 1.0 h after attachment, (c,g) : 3.0 h after attachment
(d,h) : 6.0 " "

It has thus far been shown that the addition of growth factors such as EGF or TGFa in hepatocyte culture is accompanied by a transient increase in *c-myc* expression. The present results suggest that hepatocyte myc protein plays an essential role for the entry of the hepatocytes in DNA synthesis.

We then examined the effects of recombinant MYC protein microinjection at a concentration of 1 mg/ml on hepatocyte DNA synthesis. Microinjection of EGF-stimulated hepatocytes with recombinant MYC protein, superinduced the hepatocyte DNA synthesis (Fig.5). Similar results were obtained when the hepatocytes were transfected with a myc construct and the magnitude of the DNA synthesis was markedly increased.

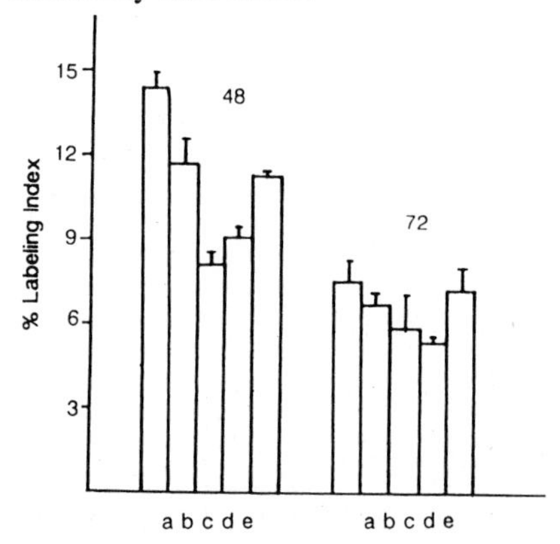

Fig.5 Effects of microinjected recombinant MYC protein on the labeling index of EGF-stimulated hepatocytes. Primary hepatocytes in EGF-stimulated culture were microinjected with the 9E10 anti-c-myc monoclonal antibody together with recombinant MYC protein (9E10 + rMYC).The cultures were then pulse-labeled with ^{3}H-TdR and the labeling index was measured at 48 and 72 hours post-attachment. a : at 0.5 h (9E10 + rMYC) b : at 1.0 h ,c : at 3.0 h ,d : at 6.0 h and e : at 10.0 h

Co-microinjection of the recombinant MYC protein at 1mg/ml together with the 9E10 anti-*c-myc* monoclonal antibody, restored the effect of EGF on hepatocyte DNA synthesis. This effect was observed during all time points of microinjection. This data supports further the idea that myc protein is essential during the transition of the hepatocyte from the G1 to the S phase.

Some hepatocyte cultures with plating density of 0.5 x 10^6 cells/ml/dish supplemented with EGF at 50 ng/dish were microinjected with MYC protein and the DNA synthesis was monitored for up to 5 days in culture. We noticed that hepatocytes undergo significant cell death in culture around the fifth day (results not shown). This resulted in a marked decrease in hepatocyte viability which dropped below 40% at the fifth day in culture. It has been suggested that constitutive high-level over-expression of *c-myc* may be cytotoxic. However, results from other laboratories studying a different cellular model, the Rat-1/myc cells which were infected with viruses encoding Myc-ER chimeras and have shown that *c-myc* may comprise a general component of the apoptotic pathway (Freytag et al., 1990, Penn et al., 1990). In addition to regulating genes mediating proliferation, myc also activates genes mediating apoptosis, then all cells expressing myc would necessarily be primed for programmed cell death. Successful proliferation would then presumably occur only if apoptosis was actively inhibited, perhaps by activation of other signal transduction pathways. Given the rapidity with which myc induces programmed cell death in Rat-1/myc-ER cells, it is believed that myc is directly involved in initiating apoptosis, presumably by regulating specific apoptotic genes. We have been at present unable to explore the roles of myc as an inducer of programmed cell death in hepatocytes.

We have performed a line of experiments to explore the roles of myc protein on the S phase entry of the hepatocytes isolated from regenerating rat liver.

At 3, 6, 10 and 20 hours after a 2/3 partial hepatectomy hepatocytes were isolated and plated in culture in a Chemically defined medium not supplemented with EGF (Fig.6).

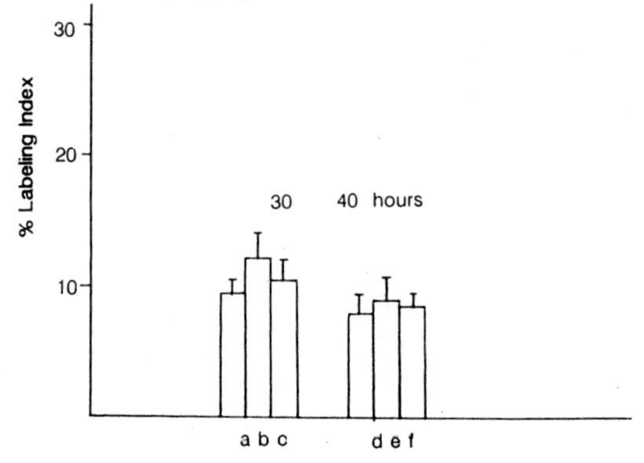

Fig.6 Labeling index of hepatocytes isolated at 3, 6 and 10 hours post-hepatectomy, cultured in the presence of EGF and microinjected with the 9E10 monoclonal antibody. Hepatocytes were microinjected with the 9E10 antibody immediately after attachment, labeled with ^{3}H-TdR and the labeling index was measured.
a ,d : 3 hours
b ,e : 6 hours
c ,f : 10 hours

In hepatocytes isolated 20 hours after partial hepatectomy and cultured for up to additional 40 hours, the labeling index was as high as 24% estimated at 3 hours post-plating. At seven hours post-plating, the labeling of the hepatocytes was maintained at almost the same levels and dropped to a final value around 8% when measured at 20 hours. At 35 hours in culture less than 1% of the hepatocytes were labeled (Fig.7).
Hepatocytes were microinjected with the 9E10 anti-*c-myc* antibody immediately after the attachment and the labeling index was measured.
As we may see, no significant differences in the labeling index between injected and non-injected hepatocytes was observed.

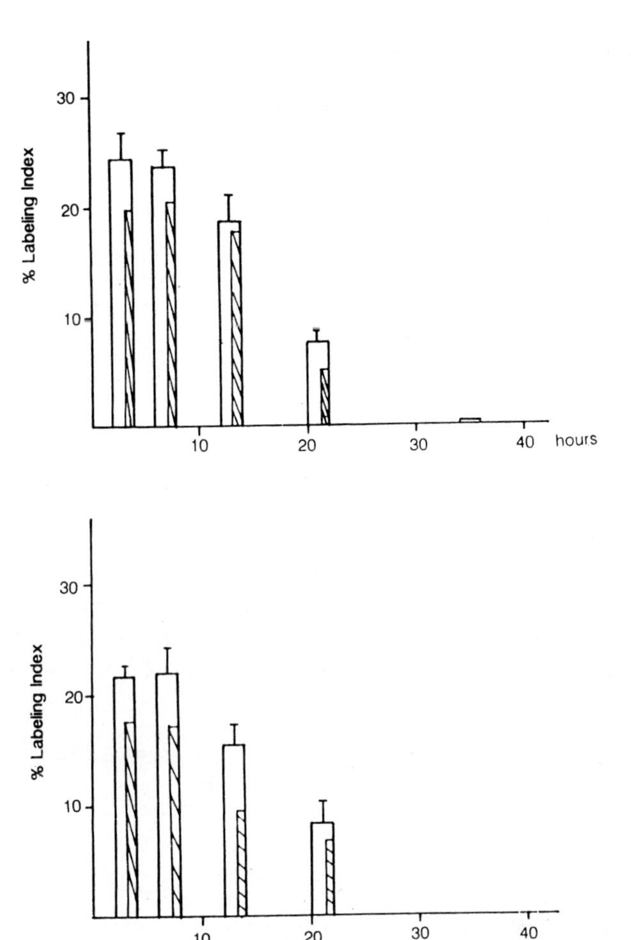

Fig.7 Labeling index of hepatocytes isolated at 20 hours after a 2/3 partial hepatectomy. [A]. Hepatocytes were isolated 20 hours post-hepatectomy and cultured in DMEM not supplemented with EGF. [B]. Hepatocytes were isolated and cultured as in [A] and were also microinjected with the 9E10 anti-c-myc antibody, 2 hours after attachment.

Microinjection of hepatocytes originating from sham operated animals with the anti-*c-myc* antibody had no effect in modifying the labeling index, which, however, was below 3% in all time points. This data suggests that the *c-myc* protein seems not to be essential for hepatocyte progress into S phase when the isolated hepatocytes have in vivo traversed the state of quiescence and are in the G1 towards the pre-synthetic phase.

The shaded columns represent the labeling index of the hepatocytes plated at 2.5×10^5 /dish, whereas the ordinary plating in all other experiments was 6×10^5 /dish.

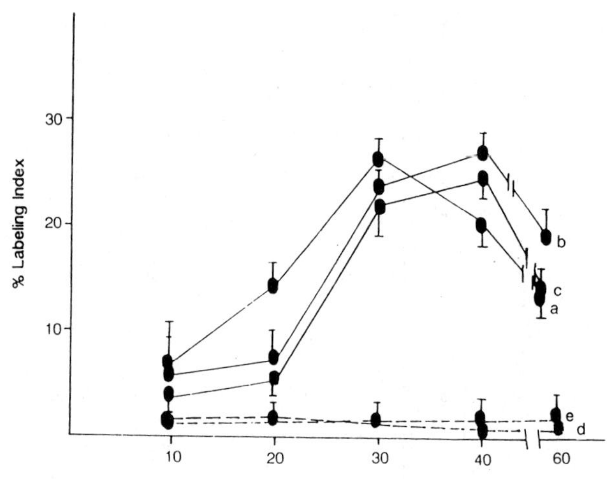

Fig.8 Labeling index of hepatocytes cultured in the presence of EGF and isolated 3, 6 and 10 hours after partial hepatectomy.
a : hepatocytes isolated 3 hours after P.H.
b : " " 6 " "
c : " " 10 " "
d ,e : " " 3 and 6 " and cultured with no EGF present.

Microinjection of hepatocytes with albumin did not affect either the cell viability or the labeling of the cells throughout the culture time.

In another line of experiments, hepatocytes were isolated at 3, 6 and 10 hours after a 2/3 partial hepatectomy and plated in chemically defined medium supplemented with or without EGF. Following a pulse labeling with tritiated thymidine, the labeling index was measured and the results are shown in the figure 8.

DISCUSSION

Those cultures which were not supplemented with EGF have shown a low labeling index observed in all time points. As we see, the labeling index was below 2% throughout the culture time. However, in the presence of 50 ng of EGF per dish, the hepatocytes isolated at all three time times after partial hepatectomy have shown, as expected, significant incorporation of thymidine, in principle between 24 and 42 hours in culture. Hepatocytes isolated at 3, 6 and 10 hours after partial hepatectomy were plated and microinjected immediately after attachment with either the CT14 or the 9E10 anti-*c-myc* antibodies. The labeling index measured at 30 hours in culture was dramatically decreased down to 9, 11.2 and 10.6 respectively. At 40 hours in culture the microinjected hepatocytes have shown a further decrease in their labeling index, which was below 9% for cells isolated at all three time points. Microinjection of the hepatocytes with albumin did not again affect both their viability nor their labeling index.

These data suggests that in hepatocytes primed to proliferate and isolated after partial hepatectomy, the *c-myc* protein seemed to be crucial for their traverse from G1 to the pre-S and the synthetic phase.

On the other hand, from experiments we have carried out with hepatocytes isolated at 3,6, 10 and 20 hours after partial hepatectomy not shown here and labeled in suspension with ^{35}S methionine or cysteine, it seemed that *c-myc* protein synthesis was significantly increased only between 3 to 7 hours post-hepatectomy. During all other time points the *c-myc* protein was continuously present at almost similar amounts. This data is in agreement with expression studies carried out with growth factor-stimulated hepatocytes which correlate the traverse of the hepatocyte from quiescence to the G1 phase with an increase in *c-myc* expression and the subsequent synthesis of the protein. In EGF, TGF-a and HGF belong to the mitogenic substances capable to increase DNA synthesis in quiescent hepatocytes. During liver regeneration, thc activation of transcription factors and protooncogenes precede the molecular phenomena of growth factor-receptor signalling. Therefore, over-expression of *c-myc* in cultured hepatocytes after growth factor stimulation or after transfection of an exogenous oncogene cannot be regarded as one of the events capable to initiate hepatocyte DNA synthesis.

In proliferating cells, *c-myc* expression is usually elevated, but this increase has also been observed in non-proliferating cultures. Recent studies by Etienne (1988) with primary hepatocytes have shown that *c-myc* is constantly expressed throughout the cell cycle. In hepatocytes either primed in vivo after a regenerating stimulus, stimulated in vitro with growth factors, or even in those resting in quiescence, the constitutive expression of *c-myc* which is confirmed by many laboratories. Therefore, the increase in *c-myc* expression and protein synthesis observed within the first 8 hours after an in vivo or in vitro growth stimulus, may regulate the traverse of the "primed" hepatocytes to the S phase of the cell cycle rather than being involved in the traverse of hepatocytes from quiescence to the G1 phase.

Acknowledgements: We would like to thank Drs. R. Watt, A. Schatzman and M. Rosenberg, Smith Kline and French for their kind pOTs-myc gift.

REFERENCES

Alitalo K.,Schwab M.,Lin CC.,Varmus H.E. and Bishop J.M. (1983) Homogeneously staining chromosomal regions contain amplified copies of an abundantly expressed cellular oncogene (c-myc)in malignant neuroendocrine cells from human colon carcinoma. Proc.Natl.Acad.Sci.U.S.A. 80:1711-1717.

Chan K.,Kost DP. and Michalopoulos GK.(1989) Multiple sequential periods of DNA synthesis and quiescence in primary hepatocyte cultures maintained on the DMSO-EGF on/off protocol. J.Cell Physiol. 141:584-590.

Etienne PL., Baffet G.,Desvergene B.,Boisnard-Rissel M.,Glaise D., and Guguen-Guillouzo C. (1988) Transient expression of c-fos and constant expression of c-myc in freshly isolated and cultured normal adult rat hepatocytes. Oncogene Res. 3:255-262.

Evan G.,Lewis G.K.,Ramsey G. and Bishop J.M.(1985)Isolation of monoclonal antibodies specific for human c-myc proto-oncogene product. Mol.Cell Biol. 5:3610-3616.

Evan GI.,Wyllie AH.,Gilbert CS.,Littlewood TD.,Land H.,Brooks M.,Waters CM.,Penn L. and Hancock DC. (1992) Cell 69:119-128.

Freytag SO.,Dang CV.,and Lee WMF. (1990) Definition of the activities and properties of c-myc required to inhibit cell differentiation. Cell Growth Diff. 1:339-343.

Graessman A. and Graessman M. (1983) Microinjection of tissue cultured cells. Meth.Enzymol.30:482-492.

Hann S.R. and Eisenman R.(1984)Proteins encoded by the human c-myc oncogene: Differential expression in neoplastic cells. Mol.Cell.Biol.4:2486-2497.

Hann S.R.,Thompson C.B. and Eisenman R.(1985) C-myc oncogene protein synthesis is independent of the cell cycle in human and avian cells. Nature 314:366-369.

Kaszmarec L.,Hyland J.K.,Watt R.,Rosenberg M.and Baserga R.(1985) Micro-injected c-myc protein as a competence factor. Science 228:1313-1315.

Laemmli U.K.(1970)Cleavage of structural proteins during the assembly of the head of the bacteriophage T4. Nature 227:680-685.

Marcu KB.,Bossone SA., and Patel AJ. (1992) Myc function and regulation. Ann.Rev.Biochem. 61:809-60.

Makino R.,Hayashi K. and Sugimura T.(1984)C-myc transcript is induced in rat liver at a very early stage of regeneration or after cycloheximide treatment. Nature 310:697-698.

Michalopoulos G.K. (1990)Liver regeneration:Molecular mechanisms of growth control. FASEB J.4:176-187.

Miyamoto C.,Smith GE., Farell-Towt J.,Chizzonite R., Summers MD., and Ju G. (1985) Production of human c-myc protein in insect cells infected with a baculovirus expression vector. Mol.Cell Biol.5:2860-2865.

Moore JP.,Hancock DC., Littlewood TD.,and Evan GI.(1992) A sensitive and quantitative enzyme-linked immunosorbence assay for the c-myc and N-myc oncoproteins. Oncogene Res. 2:65-80.

Muakkassah-Kelly SF.,Jans DA.,Lydon N.,Bieri F.,Wchter F.,Bentley P.,and Staubli W. (1988) Electroporation of cultured hepatocytes with the c-myc gene potentiates DNA synthesis in response to EGF. Exp.Cell Res. 178:296-306.

Nakamura T.,Nishizawa T.,Hagiya M.,Seki T.,Shiminishi M.,Sugimura A.,Tashiro K. and Shimizu S.(1989) Molecular cloning and expression of human hepatocyte growth factor. Nature 342:440-443.

Penn LJZ., Brooks MW., Laufer EM.,and Land H. (1990) Negative autoregulation of c-myc transcription. EMBO J. 9:1113-1121.

Persson H.,Gray H.E., and Godeau F. (1985)Growth dependent synthesis of c-myc encoded proteins:Early stimulation by serum factors in synchronized mouse 3T3 cells. Mol.Cell.Biol. 5:2903-2912.

Seglen P.O.(1976)Preparation of isolated rat liver cells. Meth.Cell.Biol.13:29-83.

Shichiri M.,Hanson KD. and Sedivy JM. (1993) Effects of c-myc expression on proliferation,quiescence and the G0 to G1 transition in nontransformed cells. Cell Growth diff. 4:93-104.

Skouteris G.G.,Ord M.G. and Stocken L.A.(1988). Regulation of the proliferation of primary rat hepatocytes by eicosanoids. J.Cell.Physiol. 135:516-520.

Skouteris GG. and Kaser MR. (1992) Expression of exogenous c-myc oncogene does not initiate DNA synthesis in primary rat hepatocyte cultures. J.Cell Physiol. 150:353-359.

Watt R.A.,Schatzman A.R. and Rosenberg M.(1985)Expression and characterization of the human c-myc DNA-binding protein.Mol.Cell.Biol. 5:448-456.

CELLULAR BIOLOGY OF THE RAT HEPATIC STEM CELL COMPARTMENT

Snorri S. Thorgeirsson,
Ritva P. Evarts,
Kozo Fujio, and
Zongyi Hu
Laboratory of Experimental Carcinogenesis
Division of Cancer Etiology
National Cancer Institute
Bethesda, MD 20892 USA

INTRODUCTION

The existence of hepatic stem cells has been, and no doubt will continue to be, a matter of considerable controversy. This controversy is partly fueled by the fact that cell turnover in the liver is very slow and the two major types of hepatic epithelial cells, hepatocytes and biliary epithelia, are capable of proliferation and can, at least in a healthy liver, meet replacement demands of cellular loss from these two differentiated populations. The best example of the capacity of adult hepatocytes and bile epithelial cells to proliferate is seen after partial hepatectomy in rats and mice, in which the compensatory hyperplasia of these cells in the remaining lobes restore the liver mass. The increased use and success of liver transplantation in clinical medicine have shown that these animal models correctly reflect the capacity of the human liver to regenerate (Van Thiel *et al.*, 1989). What then is the evidence that there exists a stem cell compartment in the liver? The existence of hepatic stem cells was first postulated by Wilson and Leduc in 1958 based on experiments involving liver regeneration in the mouse after chronic injury induced with a methionine-rich basal diet mixed with an equal amount of bentonite (Wilson and Leduc, 1958).

NATO ASI Series, Vol. H 88
Liver Carcinogenesis
Edited by G. G. Skouteris

The authors concluded that "prolonged and severe injury to the liver may make direct restoration by division of pre-existing parenchymal cells impossible, and that, when this occurs, the new parenchyma is derived from the indifferent cholangiole cells."

There is now increasingly robust experimental evidence in support of the presence of a pluripotent cell compartment in the liver (Germain *et al.*, 1988; Shiojiri *et al.*, 1991; Vandersteenhoven *et al.*, 1990; Vos and Desmet, 1992; Hsia *et al.*, 1992; Grisham and Porta, 1974; Sell, 1990). This compartment can under certain conditions function as a stem cell compartment and provide the needed progeny for regeneration of the hepatic parenchyma (Evarts *et al.*, 1987a; Sigal *et al.*, 1992). In the adult rat specific conditions can be utilized to induce proliferation of a distinct population of small epithelial cells in the portal area of the liver (Farber, 1956; Lemire *et al.*, 1991). These cells, conventionally described as oval cells, are characterized by ovoid nuclei and basophilic cytoplasma (Farber, 1956), and display features of both bile duct cells and fetal hepatoblasts/hepatocytes (Lemire *et al.*, 1991; Evarts *et al.*, 1987b; Evarts *et al.*, 1990). There are three experimental systems, two in the rat and one in the mouse, in which it has been conclusively demonstrated that oval cells are capable of differentiation into hepatocytes (Evarts *et al.*, 1987a; Lemire *et al.*, 1991; Factor and Radaeva, 1993). The developmental potential of oval cells is however not restricted to hepatic lineages. Oval cells can differentiate into intestinal-type epithelia, and have been implicated in the development of pancreatic tissues in the liver (Evarts *et al.*, 1987a; Lemire *et al.*, 1991; Tatematsu *et al.*, 1985; Kimbrough *et al.*, 1972; Rao *et al.*, 1986; Fig.1). The observation that subpopulations of proliferating oval cells phenotypically similar to early hepatoblasts originate in or around the ductular structures in the portal area, strongly support the notion that the hepatic stem cell compartment resides in these structures (Shiojiri *et al.*, 1991; Sell, 1990; Sigal *et al.*, 1992). Furthermore, present evidence clearly indicates that the hepatic stem cell compartment functions as a facultative stem cell compartment being activated when the parenchymal cells are functionally compromised and unable to proliferate in response to growth stimuli (Shiojiri *et al.*, 1991; Evarts *et al.*, 1987a; Grisham, 1980; Thorgeirsson, 1993).

In this paper we will review our recent results on the localization and growth factor involvement in the activation of hepatic stem cells as well as the lineage commitment of these cells in the rat liver.

RESULTS AND DISCUSSION

Localization of Hepatic Stem Cells

The experimental system used to initiate proliferation and differentiation of oval cells in rat liver involves the administration of acetylaminofluorene (AAF) to male Fischer 344 rats (approximately 150 g) by gavage five times over a week period, at the end of which a two-third partial hepatectomy (PH) is performed (Evarts *et al.*, 1987a). After one day recovery, AAF administration is continued for four days resulting in a total dose of 13.5 mg AAF per rat. Animals are then sacrificed at specified times after the operation. In this experimental system, the AAF/PH model, a rapid and extensive proliferation of oval cells takes place after the PH; first in the portal area and later these cells expand into the liver acinus and differentiate into small basophilic hepatocytes (Fig. 2; Evarts *et al.*, 1987a). The powerful activation of the stem cell compartment seen in the AAF/PH model is a consequence of a close to complete mitoinhibitory effect of AAF upon the adult hepatocytes that prevents the regeneration from the remaining liver tissue (Evarts *et al.*, 1987a; Tatematsu *et al.*, 1984). Similarly, following liver injury induced by D-galactosamine, another experimental system used to activate the hepatic stem cell, liver parenchyma are replaced by oval cells that differentiate into hepatocytes (Lemire *et al.,* 1991).

In the adult liver, the lining cells of the canals of Hering are thought to represent a pluripotent cell compartment (Germain *et al.*, 1988; Shiojiri *et al.*, 1991; Vandersteenhoven *et al.*, 1990; Vos and Desmet, 1992; Hsia *et al.*, 1992; Grisham and Porta, 1974; Sell, 1990). A stem cell nature of undefined periductal cells has also been proposed (Sell, 1990). Furthermore, the origin of oval cells from any component of the biliary tree has been suggested (Lenzi *et al.*, 1992). We have recently observed that proliferation of desmin-positive Ito cells is closely associated with the early stages of oval cell proliferation in the AAF/PH model (Evarts *et al.*, 1992). Since we can identify the early population of oval cells by use of the monoclonal antibody OV-6 (Evarts *et al.*, 1992) and thereby discriminate between replicating oval cells and desmin-positive Ito cells, we have attempted to identify and localize the cell population that first responds to the growth stimulus provided by PH in the AAF/PH model (Evarts *et al.*, 1993). Results from a combination of immunohisto-chemistry with OV-6 and desmin antibodies and autoradiography following [^{3}H]thymidine administration are shown in Fig. 3. Both OV-6 and desmin-positive cells were labelled with [^{3}H]thymidine already at 4 hr after PH. The thymidine labelled cells were present either as individual cells embedded in the periportal matrix or as a part of ductules in close proximity to the portal vein. The large ducts in the periportal space remained unlabelled 12 hr after the PH but increased number of labelled OV-6 positive ductular cells as well as desmin-positive periportal cells were observed (Fig. 3; Evarts *et al.*, 1993). By 72 hr, the majority of the cells in the periportal area were labelled, including approximately one-half of the cells in the large duct but the hepatocytes remained unlabelled (Fig. 3). In addition to the thymidine labelled cells in the periportal area, we also observed both OV-6 and desmin-positive thymidine labelled cells in the Glisson capsule at this early time point (Evarts *et al.*, 1993).

The present results are in agreement with data obtained in other models showing proliferation of ductular and periductular cells at early stages of stem cell activation (Sell, 1990; Lenzi *et al.*, 1992; Lesch *et al.*, 1970). The observation that both OV-6 and desmin-positive thymidine labelled cells are also seen in the Glisson capsule shortly after PH in the AAF/PH model may suggest that these cells could be part of the pluripotent cell population activated in this experimental system. However, no infiltration of OV-6 positive cells into the liver acinus is observed in the vicinity of the Glisson capsule in the AAF/PH model (Evarts *et al.*, 1993). It seems therefore unlikely the cells of the Glisson capsule can contribute progenitors for the expanding oval cell population in this model.

Although the data from the detailed time course of hepatic stem cell activation in the AAF/PH model has given us new insights into the close association between a mesenchymal cell population (i.e. Ito cells) and the emerging oval cells (vide infra), we still have not resolved whether the hepatic stem cell compartment comprises only the bile ductular cells or includes a nondescript periductular cell population. Nevertheless, our data clearly show that the majority of thymidine labelled OV-6 positive cells first observed after

PH in the AAF/PH model reside in the bile ductules (Evarts *et al.*, 1993). Moreover, at the time when few of the OV-6 positive cells in the large bile ducts become labelled with thymidine the ductular derived OV-6 positive and thymidine labelled "oval" cells have already started to infiltrate into the liver acinus (Evarts *et al.*, 1993). We conclude that the major source of oval cells, at least in the AAF/PH model, is derived from the lining cells of the biliary ductules, and that these cells constitute the dormant/facultative hepatic stem cell compartment.

Growth Factors Involved in Hepatic Stem Cell Activation

During normal hepatic regeneration as well as during renewal from the stem cell compartment, several growth factors appear to affect the proliferation and differentiation of hepatic cells (Evarts *et al.*, 1992; Hu *et al.*, 1993a; Marsden *et al.*, 1992). We have, therefore, addressed the question whether the same growth factors known to be involved in normal hepatic regeneration are also involved in the regeneration from the stem cell compartment.

There are three "primary" growth factors associated with normal liver regeneration, namely transforming growth factor-alpha (TGF-α), hepatocyte growth factor (HGF), and acidic fibroblast growth factor (aFGF) (Michalopoulos, 1990). Each one of these growth factors is also capable of inducing replication of primary hepatocytes in vitro (Michalopoulos, 1990). In addition, transforming growth factor-beta 1 (TGF-β1) is also expressed during hepatic regeneration, and it has been proposed that TGF-β1 may provide at least part of the negative growth signals controlling the liver size following the compensatory hyperplasia that occurs after loss of liver mass (Mead and Fausto, 1989).

The first cells entering DNA synthesis following PH in the AAF/PH model are the OV-6 and desmin-positive cells in the periportal area (Fig. 4). Coincident with the appearance of these cells an increase in the expression of TGF-α, HGF, and TGF-β1 is observed whereas increased expression of aFGF is first seen 24 hr later (Fig. 4). All the growth factors are then expressed at high levels throughout the period of expansion and differentiation of the oval cells and return to levels seen in normal liver at the end of the regeneration process. The cellular distribution of the growth factor transcripts differs; TGF-α and aFGF transcripts are found both in Ito cells and oval cells (Evarts *et al.*, 1992; Marsden *et al.*, 1992) whereas the HGF transcripts are only found in Ito cells (Hu *et al.*, 1993a). The TGF-β1 transcripts are located mainly in Ito cells, but the early population of oval cells also contain the TGF-β1 transcripts (Evarts *et al.*, 1990). The data on cellular distribution of all the receptors corresponding to the growth factors has revealed that all are located on oval cells (Lenzi *et al.*, 1992; Lesch *et al.*, 1970; Hu *et al.*, 1993b).

The most straightforward interpretation of these data is that the same primary growth factors that are involved in liver regeneration from existing differentiated parenchyma also are involved in regeneration from the stem cell compartment. In fact, a slight and transient increase in the expression of the 2.1 kb transcript of alpha-fetoprotein (AFP), an indicator of liver stem cell activation, is observed following PH of normal rat liver (Nakatsukasa H, Evarts RP, Thorgeirsson SS, unpublished results). These observations support the role and possible importance of growth factors in stem cell activation. The data further suggest that the stem cell compartment may be transiently activated during regeneration following PH of a normal healthy liver.

We have recently discovered a novel ligand/receptor system, the stem cell factor (SCF)/c-kit system, that may be uniquely involved in the earliest stages of hepatic stem cell activation (Fujio *et al.*, 1994). In the AAF/PH model the expression of both SCF and c-kit is seen prior to the expression of AFP (Fig. 5), and the levels of both the SCF and the c-kit transcripts decline prior to those of TGF-α, aFGF, HGF, and TGF-β1 (Fig. 4). We have also shown that in contrast to TGF-α, HGF, aFGF, and TGF-β1, the SCF/c-kit system is only slightly and transiently activated in regeneration following PH in normal liver (Fujio *et al.*, 1994). The SCF/c-kit signal transduction system is believed to play a fundamental role in the survival, proliferation, and migration of stem cells in hematopoiesis, melanogenesis, and gametogenesis (Morrison-Graham and Takahashi, 1993). It appears that in all cases SCF and c-kit are involved in the early stages of stem cell activation. In the hemopoietic stem cell system it has also been demonstrated that SCF in combination with selective multipotential colony-stimulating factors can influence the relative frequency of progenitor cells committed to various lineages (Metcalf, 1991). Whether the SCF/c-kit system in the early hepatic stem cell population interacts with other hepatic growth factors in such a way as to influence the frequency of lineage commitment of progenitor cells is at present not known. However, this exciting possibility can now be experimentally tested.

Further studies are clearly needed to define both the cellular and molecular biology of the hepatic stem cell compartment. However, using lineage markers such as cytokeratins and AFP in combination with activation of the growth factor/receptor systems discussed above may provide a fruitful approach to study the mechanisms involved in the activation and differentiation of the hepatic stem cell and oval cell compartments.

REFERENCES

Evarts RP, Hu Z, Fujio K, Marsden ER, Thorgeirsson SS (1993) Activation of hepatic stem cell compartment in the rat: role of transforming growth factor α, hepatocyte growth factor, and acidic fibroblast growth factor in early proliferation. Cell Growth Differen 4:555-561

Evarts RP, Nagy P, Marsden E, Thorgeirsson SS (1987a) A precursor-product relationship exists between oval cells and hepatocytes in rat liver. Carcinogenesis 8:1737-1740

Evarts RP, Nagy P, Marsden E, Thorgeirsson SS (1987b) In situ hybridization studies on expression of albumin and α-fetoprotein during the early stage of neoplastic transformation in rat liver. Cancer Res 47:5469-5475

Evarts RP, Nakatsukasa H, Marsden ER, Hsia C-C, Dunsford HA, Thorgeirsson SS (1990) Cellular and molecular changes in the early stages of chemical hepatocarcinogenesis. Cancer Res 50:3439-3444

Evarts RP, Nakatsukasa H, Marsden ER, Hu Z, Thorgeirsson SS (1992) Expression of transforming growth factor-α in regenerating liver and during hepatic differentiation. Mol Carcinog 5:25-31

Factor VM, Radaeva SA (1993) Oval cells - hepatocytes relationships in Dipin-induced hepatocarcinogenesis in mice. Exp Toxicol Pathol 45: 239-244

Farber E (1956) Similarities in the sequence of early histological changes induced in the liver of the rat by ethionine, 2-acetylaminofluorene, and 3'-methyl-4-dimethylaminoazobenzene. Cancer Res 16:142-148

Fujio K, Evarts RP, Hu Z, Marsden ER, Thorgeirsson SS (1994) Expression of stem cell factor and its receptor, c-kit, during liver regeneration from putative stem cells in adult rat liver. Lab Investig (In press).

Germain L, Flouin M-J, Marceau N (1988) Biliary epithelial and hepatocytic cell lineage relationships in embryonic rat liver as determined by the differential expression of cytokeratins, α-fetoprotein, albumin, and cell surface-exposed components. Cancer Res 48:4909-4918

Grisham JW (1980) Cell types in long-term propagable cultures of rat liver. Annals NY Acad Sci 349:128-137

Grisham JW, Porta EA (1974) Origin and fate of proliferating hepatic ductal cells in the rat: electron microscopic and autoradiographic studies. Exp Mol Pathol 2:242-261

Hsia CC, Evarts RP, Nakatsukasa H, Marsden ER, Thorgeirsson SS (1992) Occurrence of oval cells in hepatitis B virus associated human hepatocarcinogenesis. Hepatology 67:427-433

Hu Z, Evarts RP, Fujio K, Marsden ER, Thorgeirsson SS (1993a) Expression of hepatocyte growth factor and c-met gene during hepatic differentiation and liver development in the rat. Am J Pathol 142:1823-1930

Hu Z, Evarts RP, Fujio K, Marsden ER, Thorgeirsson SS (1993b) Expression of transforming growth factor-alpha/epidermal growth factor receptor, hepatocyte growth factor/c-met, acidic fibroblast growth factor/fibroblast growth factor receptors during hepatocarcinogenesis. Proc Am Assoc Cancer Res 34:149 (abstract 887)

Kimbrough RD, Linder RE, Gaines TB (1972) Morphological changes in liver of rats fed polychlorinated biphenyls. Arch Environ Health 25:354

Lemire JM, Shiojiri N, Fausto N (1991) Oval cell proliferation and the origin of small hepatocytes in liver injury induced by D-galactosamine. Am J Pathol 139:535-552

Lenzi R, Liu MH, Tarsetti F, Slott PA, Alpini G, Xhai WR, Paronetto F, Lenzen R, Tavolini N (1992) Histogenesis of bile duct-like cell proliferating during ethionine hepatocarcinogenesis. Lab Invest 66:390-402

Lesch R, Reutter W, Keppler D, Decker K (1970) Liver restitution after acute galactosamine hepatitis: autoradiographic and biochemical studies in rats. Exp Mol Pathol 12:58-69

Marsden ER, Hu Z, Fujio K, Nakatsukasa H, Thorgeirsson SS, Evarts RP (1992) Expression of acidic fibroblast growth factor in regenerating liver during hepatic differentiation. Lab Invest 67:427-433

Mead JE, Fausto N (1989) Transforming growth factor α may be a physiological regulator of liver regeneration by means of autocrine mechanisms. Proc Natl Acad Sci USA 86:1558-1562

Metcalf D (1991) Lineage commitment of hemopoietic progenitor cells in developing blast cell colonies: influence of colony-stimulating factors. Proc Natl Acad Sci USA 88:11310-11314

Michalopoulos G (1990) Liver regeneration: molecular mechanisms of growth control. FASEB J 4:176-187

Morrison-Graham K, Takahashi Y (1993) Steel factor and c-kit receptor: from mutants to a growth factor system. BioEssays 15:77-84

Rao MS, Bendayan RD, Kimbrough RD, Reddy JK (1986) Characterization of pancreatic-type tissue in the liver of rats induced by polychlorinated biphenyls. J Histochem Cytochem 34:197-201

Sell S (1990) Is there a liver stem cell? Cancer Res 50:3811-3815
Shiojiri N, Lemire JM, Fausto N (1991) Cell lineage and oval cell progenitors in rat liver development. Cancer Res 51:2611-2620
Sigal HS, Brill S, Fiorino AS, Reid LM (1992) The liver as a stem cell and lineage system. Am J Physiol 26:G139-G148
Tatematsu M, Ho RH, Kaku T, Ekem JK, Farber E (1984) Studies on the proliferation and fate of oval cells in the liver of rats treated with 2-acetylaminofluorene and partial hepatectomy. Am J Pathol 114:418-430
Tatematsu M, Thohru K, Medline A, Farber E (1985) Intestinal metaplasia as a common option of oval cells in relation to cholangiofibrosis in livers of rats exposed to 2-acetylaminofluorene. Lab Invest 52:354
Thorgeirsson SS (1993) Hepatic stem cells. Am J Pathol 142:1331-1333
Van Thiel DH, Gavaler JS, Kam I, Francavilla A, Polimeno L,Schade PR, Smith J, Diven W, Penkrot RJ, Starzl TE (1989) Rapid grwoth of an intact human liver transplanted into a recipient larger than the donor. Gastroenterology 93:1414-1419
Vandersteenhoven AM, Burchette J, Michalopoulos G (1990) Characterization of ductular hepatocytes in end-stage cirrhosis. Arch Pathol Lab Med 114:403-406
Vos R, Desmet V (1992) Ultrastructural characteristics of novel epithelial cell types identified in human pathologic liver specimens with chronic ductular reaction. Am J Pathol 140:1441-1450
Wilson JW, Leduc EH (1958) Role of cholangioles in restoration of the liver of the mouse after dietary injury. J Pathol Bacteriol 76:441-449

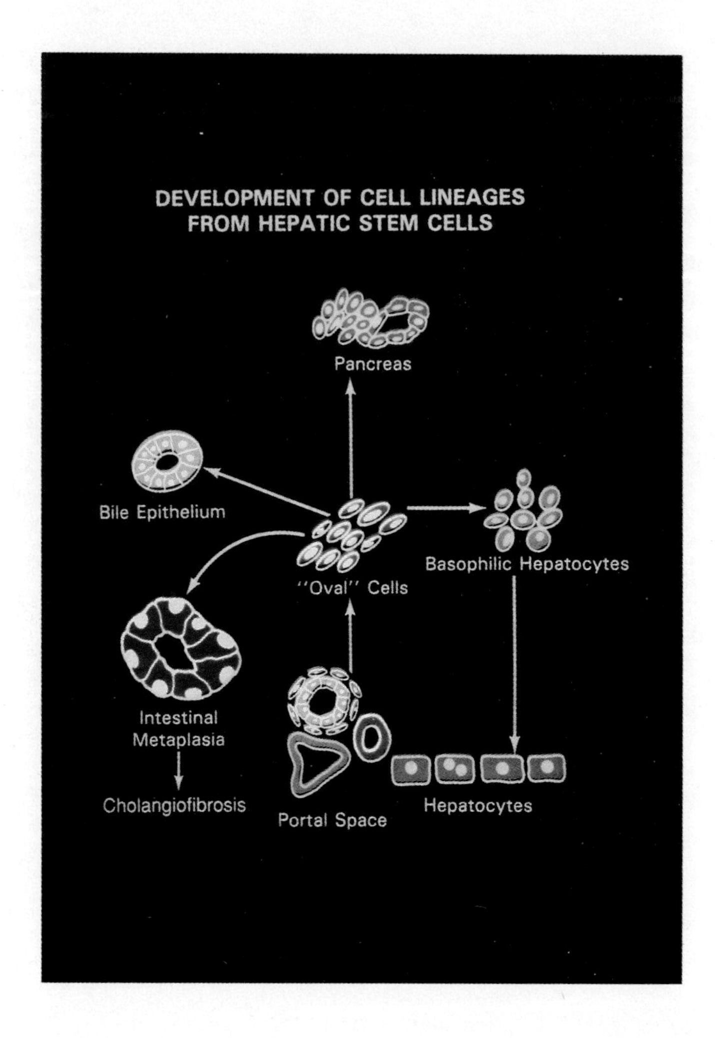

FIG. 1. Schematic representation of the development of cell lineages from hepatic stem cells.

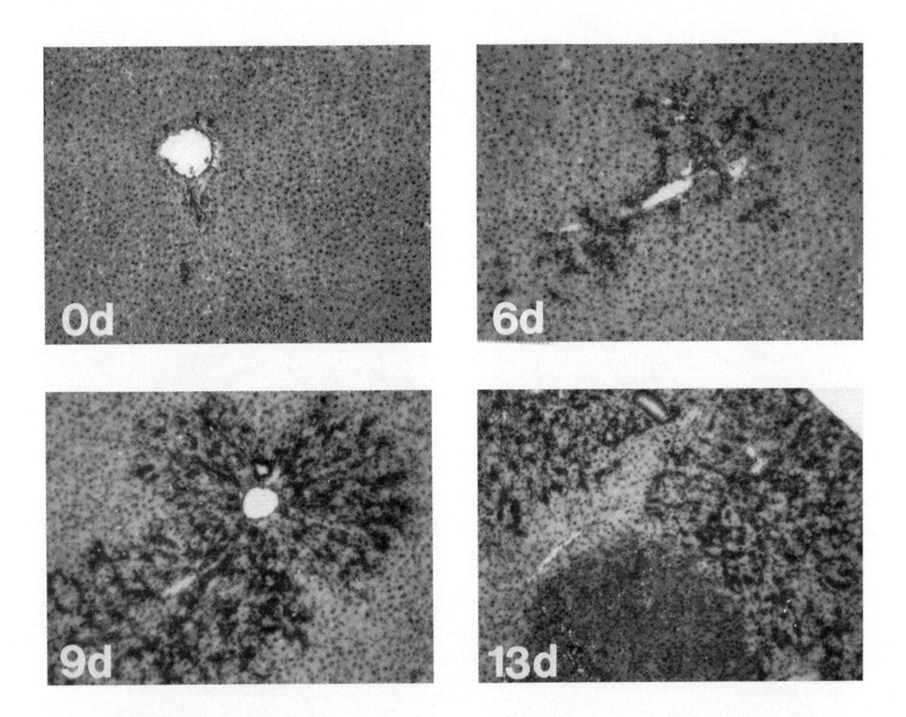

FIG. 2. Development and expansion of the oval cell population in the AAF/PH model. Oval cells are identified by γ-GT staining. The time of partial hepatectomy is day 0 (0d), and 6d, 9d, and 13d are days following the operation.

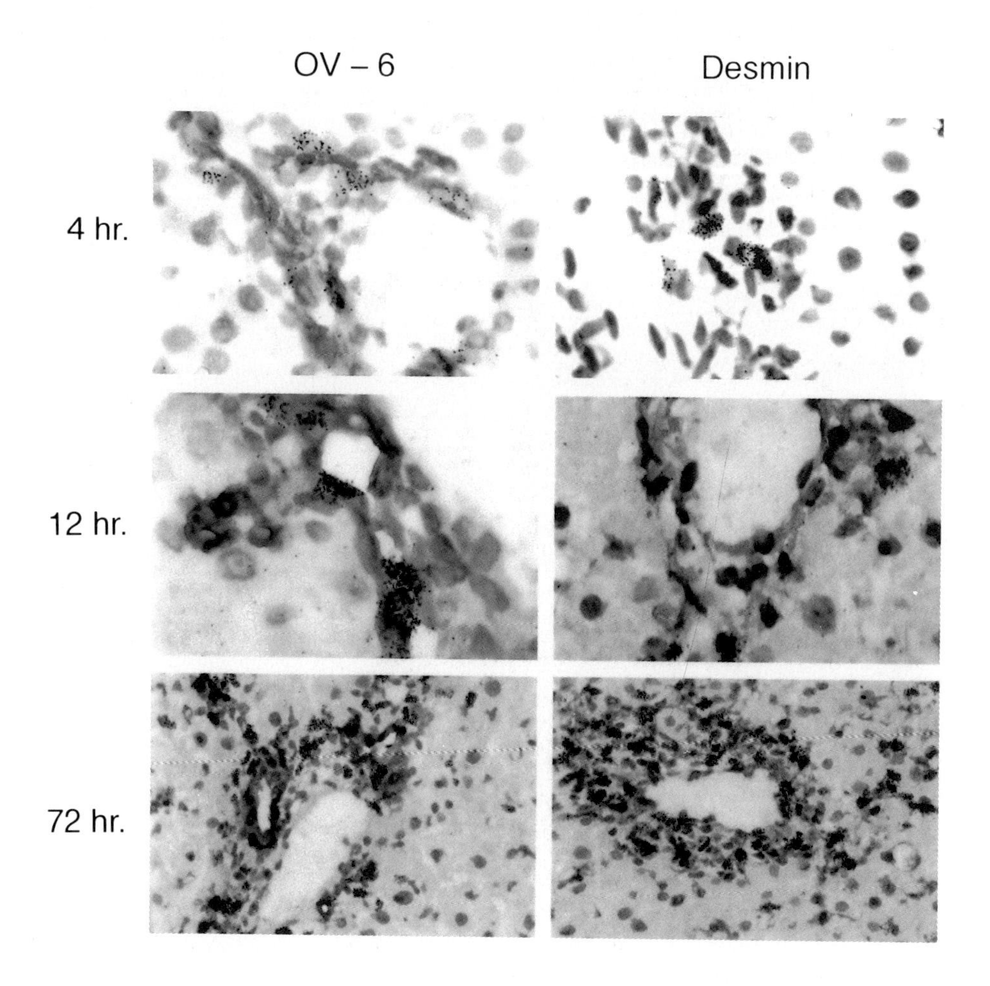

FIG. 3. [^{3}H]Thymidine labelled OV-6 and desmin-positive cells at early time points in the AAF/PH model. [^{3}H]Thymidine was administered i.p. (1 μCi/b body weight) to the animals 2 hr before sacrifice.

A. Times in hours after the operation

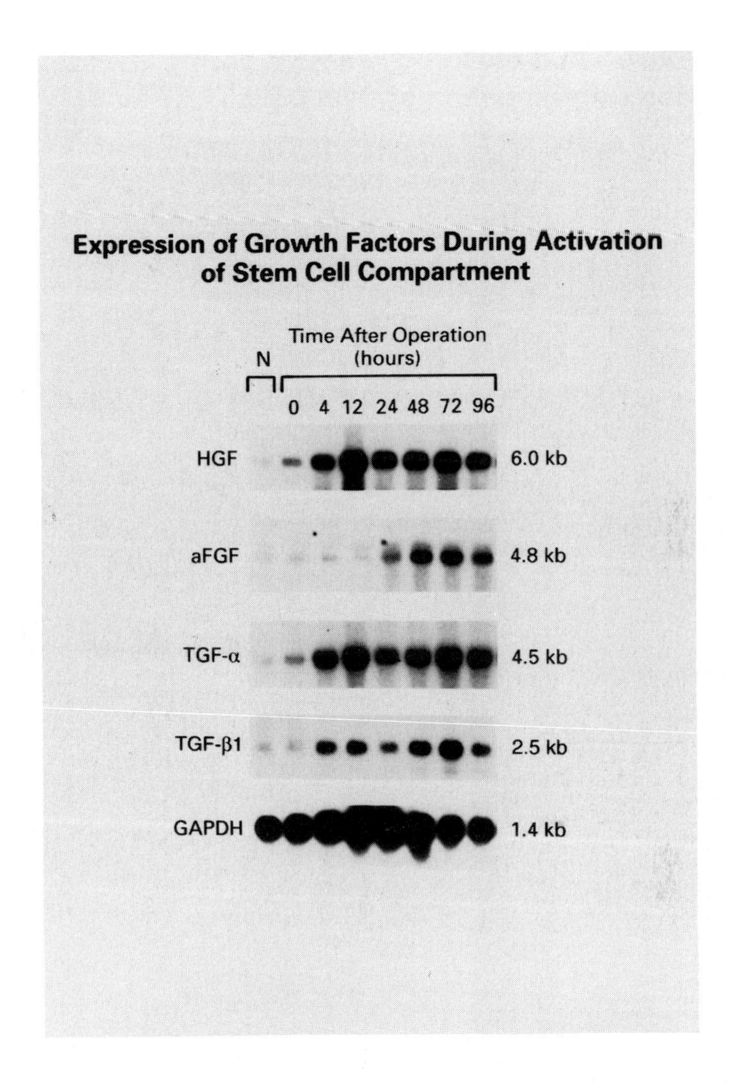

FIG. 4. Northern blot analysis of HGF, aFGF, TGF-α, and TGF-β1 expression during activation of the hepatic stem cell compartment. Poly(A)+ RNA was isolated at indicated time points and analyzed by Northen blot analysis (5 μg/mRNA) and hybridized with [^{32}]P-labelled riboprobes. Glyceraldehyde phosphatedehydrogenase (GAPDH) was used as control. N = normal liver.

B. Times in days after the operation

Expression of Growth Factors during Growth and Differentiation of Oval Cells (AAF Model)

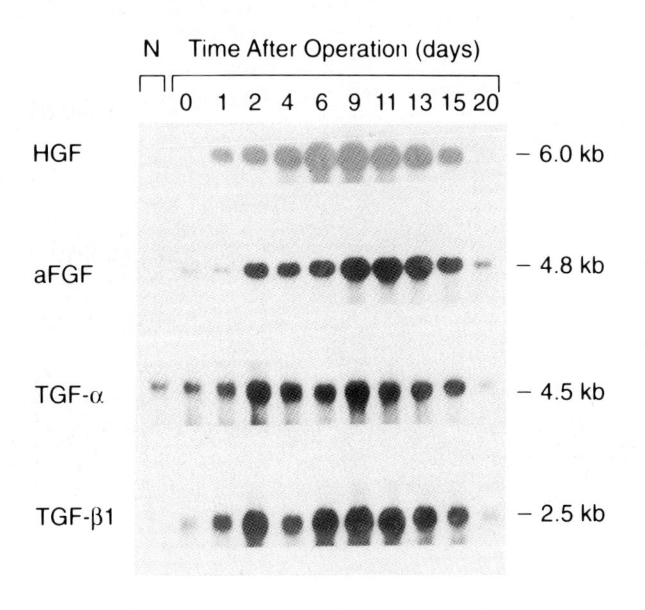

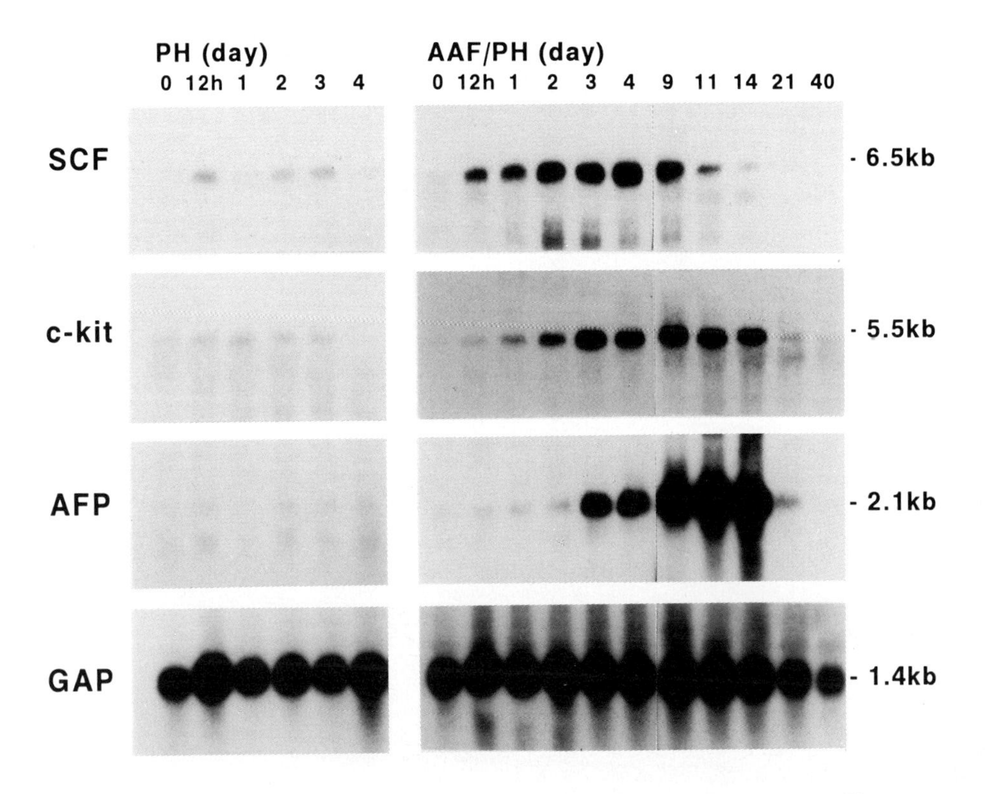

FIG. 5. Northern blot analysis of SCF, c-kit, and AFP expression during liver regeneration. Rats were subjected to simple partial hepatectomy (PH) and AAF/PH treatment and sacrificed as indicated after PH. Poly(A)+ RNA (5 μg mRNA) was analyzed by Northern blot analysis.

CHANGES IN PROTEIN EXPRESSION DURING OVAL CELL PROLIFERATION IN THE LIVER

T.William Jordan, Irene E. Nickson and Hua Feng
School of Biological Sciences
Victoria University of Wellington
PO Box 600
Wellington
New Zealand

INTRODUCTION

Two-dimensional polyacrylamide gel electrophoresis (2DE) is a powerful technique for separation of proteins. This technology can now be used for analysis of protein expression and variation, and for identification of proteins. In this paper we briefly review some applications of 2DE in liver biology and present data on changes in protein expression in rat liver during oval cell proliferation.

Protein separation by 2DE

2DE, in the form which is most widely used, relies on a first dimension separation in a polyacrylamide tube gel, followed by transfer of the developed tube gel containing the separated proteins to a polyacrylamide slab gel for a second dimension separation. The most widely used form of 2DE was introduced by O'Farrell (1975) who combined an isoelectric focusing separation in the presence of protein solubilising agents in the first dimension tube gel, with a separation based on molecular weight differences in the second

NATO ASI Series, Vol. H 88
Liver Carcinogenesis
Edited by G. G. Skouteris

dimension, sodium dodecyl sulfate (SDS) containing, slab gel. Because the proteins were separated on the basis of two independent parameters, isoelectric point and molecular weight, complex mixtures of proteins separated over the whole region of the second dimension slab gel and protein spot sizes were generally small with little smearing, particularly in the acidic and neutral regions of the gel. One drawback of this method was that proteins with more alkaline isoelectric points did not separate well, because of collapse of the pH gradient at the basic end of the tube gel. O'Farrell overcame this problem by introducing the technique of non-equilibrium pH gradient electrophoresis which gave good separation of alkaline proteins (O'Farrell *et al.* 1977).

O'Farrell (1975) calculated that his technique was capable of resolving more than a 1000 proteins on a single gel, and that protein species representing as little as 10^{-4} to 10^{-5}% of the total protein could be detected and quantified. Others have since increased the sensitivity and resolution. Notably, Donald Young and others (Young, 1984; Levensen *et al.* 1992) have developed giant 2DE gel systems which have about 4 times the area of the traditional 16-20 cm gels. Several thousand proteins can be detected using such systems and it has been estimated that about 50% of all the proteins present in a cell can be detected in this way.

The 2DE pattern of tissue proteins gives a global view of protein production and processing at each point in time. Changes in the pattern may result from altered gene expression or protein processing, organelle expansion or loss, and from changes in cell populations due to cell proliferation or death. 2DE has been widely used in studies of liver cell biology and pathology including the effects of mutation (Giometti *et al.* 1987), adaption of cells to culture (Colbert *et al.* 1984), gene expression (Baumann *et al.* 1983) and toxic challenge (Anderson *et al.* 1986, 1992; Jordan and Pedersen 1986). 2DE has also been used to show that only a few proteins are newly expressed during neoplastic transformation of liver (Wirth *et al.* 1986; Huber *et al.* 1988).

Detection and Measurement of Separated Proteins

A large range of detection methods can be used to visualise proteins which have been separated by 2DE. Staining with Coomassie blue detects the most abundant proteins. The intensity of staining usually has a semi-quantitative or quantitative relationship to the amount of each protein (Anderson *et al.* 1985; Burgess-Cassier *et al.* 1989). Silver staining gives more sensitive detection but is often non-quantitative (Guevara *et al.* 1982).

Autoradiographic or fluorographic detection of radioactively labelled proteins is commonly used in combination with 2DE. Labelling techniques which minimise changes in the charge or size of proteins are desirable so that the labelled proteins separate like the native proteins on 2DE. Biosynthetic labelling of proteins in cultured cells, by incubation with $^{32}PO_4$ or isotopically (commonly ^{3}H, ^{14}C, ^{35}S) labelled amino acids (Lecocq *et al.* 1990), or in vitro translation of isolated mRNA in the presence of radioactive amino acids, are often used to label proteins prior to their separation by 2DE. Alternatively, proteins can be labelled by reductive methylation with [^{3}H] or [^{14}C]formaldehyde in the presence of sodium cyanoborohydride (Wheeler *et al.* 1986, 1988). This reaction results in incorporation of radioactive methyl residues into the amino acid side chains of proteins and peptides. Migration of the labelled proteins during 2DE is usually unaltered, because the amino nitrogen can still ionise and the addition of methyl groups does not substantially change the molecular weight of the derivatised proteins.

Differences between two protein samples separated by 2DE can often be detected by visual inspection of the gels. However the introduction of computerised densitometry and software packages which carry out analysis of quantitative and qualitative differences between protein samples separated on two dimensional gels has greatly increased the power of 2DE (for examples see Anderson *et al.* 1985, Doz and Gorg, 1991; Garrels, 1989). Sample handling and electrophoresis techniques which minimise differences in protein degradation, and maximise uniform separation and detection of proteins, are necessary prerequisites for quantitative 2DE.

2DE Maps and Databases

Individual proteins migrate to a position on O'Farrell-type 2DE gels which are defined by isoelectric point (determined by amino acid composition and post-translational modification) and molecular weight. Maps of proteins separated by 2DE have been produced for human plasma (Anderson and Anderson, 1991), for organs including liver (Anderson *et al.* 1991; Giometti *et al.* 1992; Hochstrasser *et al.* 1992), and for cultured and transformed cells including liver cells (Wirth *et al.* 1992). The intention of these mapping programmes is to produce data bases which will allow ready identification of proteins separated by 2DE. A large number of proteins have been identified from 2DE gels, either by reaction with antibodies or by the acquisition of peptide sequence data for proteins separated by 2DE (Aebersold and Leavitt 1990; Celis *et al.* 1991). The long term goal is to relate the positional information for identity, and the quantitative information about abundance, of individual proteins with nucleotide data bases. It may then be possible to readily interpret data between the genome maps and the protein expression maps (Celis *et al.* 1991) for organisms, and their component cells and organelles.

Changes in protein expression during oval cell proliferation

We have been using 2DE to examine protein variation related to liver damage and repair. This has included studies of protein change in the liver of rats fed a choline-deficient, ethionine-supplemented, (CDE) diet. In these animals oval cells appear to proliferate around the portal tracts, and subsequently migrate to populate the hepatic lobules (Shinozuka *et al.* 1978). Oval cells are epithelial cells which are intermediate in size between the hepatocytes and intrahepatic biliary epithelial cells (BEC) which normally populate the liver. Oval cells may give rise to both hepatocytes and BEC during some forms of liver renegeration (Alison *et al.* 1993, Evarts *et al.* 1989) and may also be precursors of hepatocellular carcinomas and cholangiocarcinomas (Sell and Dunsford 1989; Evarts *et al.* 1990).

There are now many descriptions of oval cell proliferation using immunocytochemistry or in situ hybridisation to detect changes in protein or mRNA, as described elsewhere in this volume. Our studies have as their goal characterisation of the proteins of isolated oval cells. We have been using 2DE to carry out this analysis because of the potential to measure presence or absence, and quantitative change, for many proteins. We would like to identify abundant oval cell proteins, and then analyse the way in which expression of these proteins varies during oval cell proliferation. Further information about oval cell function might be gained by identifying the major proteins of these cells. In addition, if oval cells are derived from a BEC-like cell in the smallest bile ducts (Yang *et al.* 1993) the earliest proliferating oval cells may share many proteins in common with normal BEC. Similarly, the more mature oval cells which have migrated into the lobules may begin to preferentially express the proteins of the terminal cells in their differentiation pathways.

MATERIALS AND METHODS

Oval cell proliferation was induced in male Sprague-Dawley rats maintained on a CDE pellet diet (obtained from ICN Biochemicals, Cleveland, Ohio, USA). Animals were killed after feeding the CDE diet for 2, 4, or 6 weeks and samples of liver, and isolated hepatocytes were prepared by collagenase perfusion (Cordiner and Jordan 1983), and analysed by 2DE as previously described (Frazer *et al.* 1985). Our 2DE methods are based on those of O'Farrell (1975). Briefly, isoelectric focusing of liver or hepatocyte homogenates was carried out in 100 by 2 mm tube gels, containing 2% Triton X-100, 9 M urea, 4% Pharmalytes, and 5% glycerol. Focusing was at 400 V for 20 h, followed by 800 V for 1 h. Focused gels were equilibrated in 62.5 mM Tris-HCl buffer, pH 6.8, containing 10% (w/v) glycerol, 5% (w/v) SDS and 50 mM dithiothreitol prior to transfer to the top of 140 by 100 mm polyacrylamide slab gels. The slab gels were cast as a 3.5% stacking gel above a 10 or 12% resolving gel. Electrophoresis was for approximately 5 h at 20 mA constant current. The reservoir buffer (pH 8.3) contained 0.1% SDS, 25 mM Tris and 0.192 M glycine. The separated proteins (200-400 µg protein per gel) were detected by staining the gels in 0.12% Coomassie blue R-250 in methanol-acetic acid-water (5:2:5 by vol). Under these circumstances approximately 200 proteins could be readily examined on each gel. Some proteins were identified by comparison with the 2DE map for rat liver proteins (Anderson *et al.* 1991).

Individual gels of liver or hepatocyte proteins were prepared from livers collected at intervals after CDE feeding. The gels were examined visually to detect changes in protein patterns. A number of proteins were chosen for quantitative analysis by densitometry of the dye stained gels using a Molecular Dynamics (Sunnyvale, California, USA) computing densitometer with ImageQuant™ software, version 3.0. The amount of each protein was measured using the ImageQuant™ software and for each individual protein the quantitative data was calculated after subtraction of a background density value for the region of that protein spot.

Results

Histological analysis of haematoxylin and eosin stained sections of livers from animals which had been maintained on the CDE diet showed progressive accumulation of oval cells. At 2 weeks, oval cells were present in the portal tracts and immediately surrounding parenchyma. By 4 weeks oval cells was widely dispersed throughout the lobules and by 6 weeks oval cells appeared to be more numerous than hepatocytes.

The purity of isolated cell populations was examined by phase contrast light microscopy, and by flow cytometry using forward and right angle light scatter. Hepatocyte populations isolated from the livers of control, or 2 week CDE, rats contained less than 5% contamination with nuclei or non-parenchymal cells. However the hepatocyte preparations obtained at 4 or 6 weeks after feeding CDE were more heterogeneous. Although hepatocyte populations were purified by several cycles of 50g centrifugal sedimentation there was substantial contamination with oval cells in samples prepared from the livers of animals which had been maintained on the CDE diet for 4-6 weeks. When the isolated cell populations were examined by flow cytometry the hepatocyte fractions isolated from livers at 4-6 weeks after feeding showed great variability in size and granularity (data not shown).

2DE of liver proteins

Figure 1 shows the 2DE separations of liver homogenate proteins from one control rat and from one rat which had been maintained on the CDE diet for 4 weeks. The proteins on these gels had been separated by isoelectric focusing in the horizontal dimension and by electrophoresis in SDS gels in the vertical dimension. This resulted in separation of proteins with isoelectric points between 3.5 and 8 (acid on the left, base on the right) and molecular weights between approximately 100,000 (top of the gel) and 20,000 (bottom of the gel).

Changes in the abundance of several proteins occurred after CDE feeding. Analysis of these differences was carried out by assigning each protein a number and visually inspecting replicate gels and gels from several animals to detect consistent differences. Some proteins were also selected for quantitative analysis by computerised densitometry. This included the numbered proteins in Figures 1 and 2.

Visual inspection of the gels showed that a number of proteins varied consistently in the livers from CDE fed animals. Changes in abundance were detected for approximately 10% of the proteins (26 losses and 7 gains). Some of the differences are illustrated in Figures 1 and 2, and data for the quantitative variation of four proteins is presented in Figure 3. Protein spot number 9 (albumin) was substantially depressed from control levels at 2-6 weeks and protein spots 6 and 15 showed progressive increases during the period of CDE feeding. Some other varying proteins are arrowed in Figure 1. Many proteins, like spot 13, showed no detectable change in abundance.

An analysis of protein variation was also carried out using hepatocytes isolated from the livers of control and CDE fed rats. This was done to examine whether any of the varying proteins in the liver were non-hepatocytic, and thus might represent expression of unique proteins in the proliferating oval cells. All of the protein variation which had been detected in the liver however was also represented in the hepatocyte fractions (Figure 4).

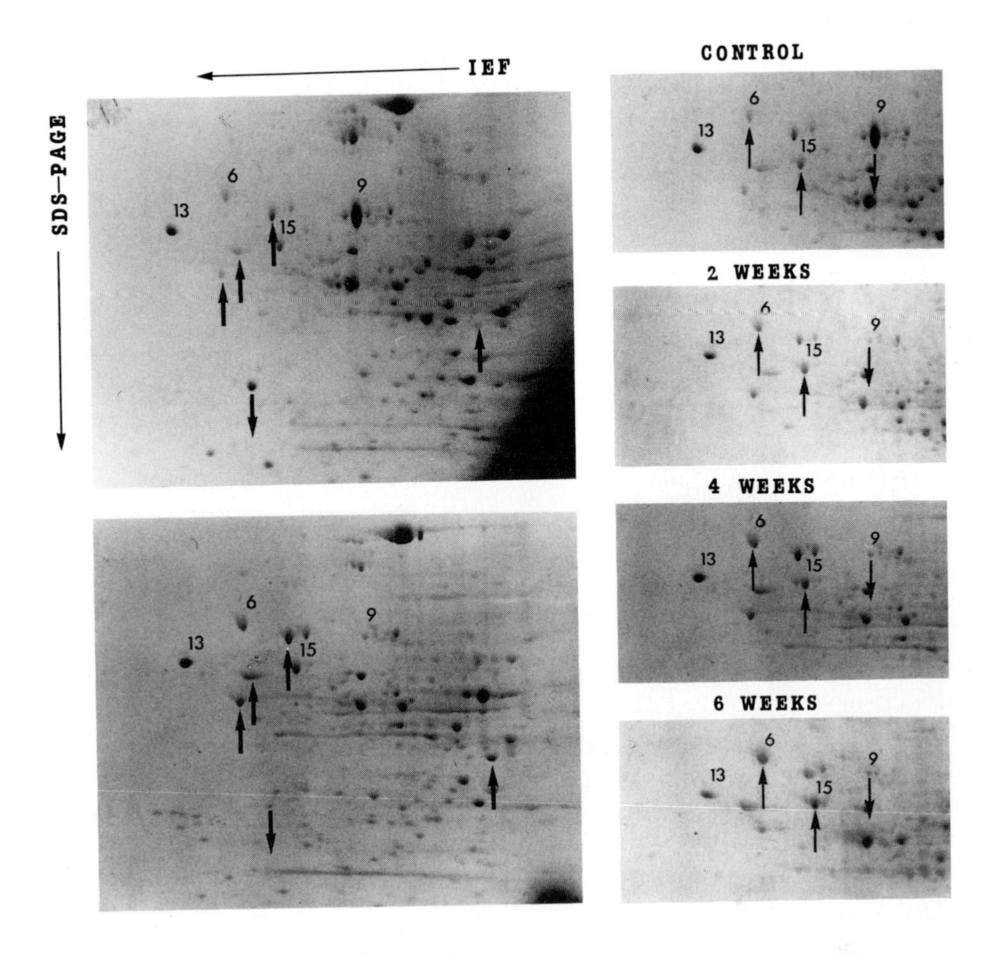

Figure 1. 2DE of liver proteins from a control rat (top) and from a rat maintained on the CDE diet for 4 weeks (bottom). Numbers to the upper right of protein spots show proteins whose amounts are quantitated in Figure 3. Arrows indicate some other proteins which increased (↑) or decreased (↓) in amount during CDE feeding.

Figure 2. Regions of 2DE gels showing changes in liver protein patterns in rats maintained on the CDE diet for up to 6 weeks. Gains (↑) or losses (↓) in protein amount are indicated by arrows for the proteins which are numbered in Figure 1.

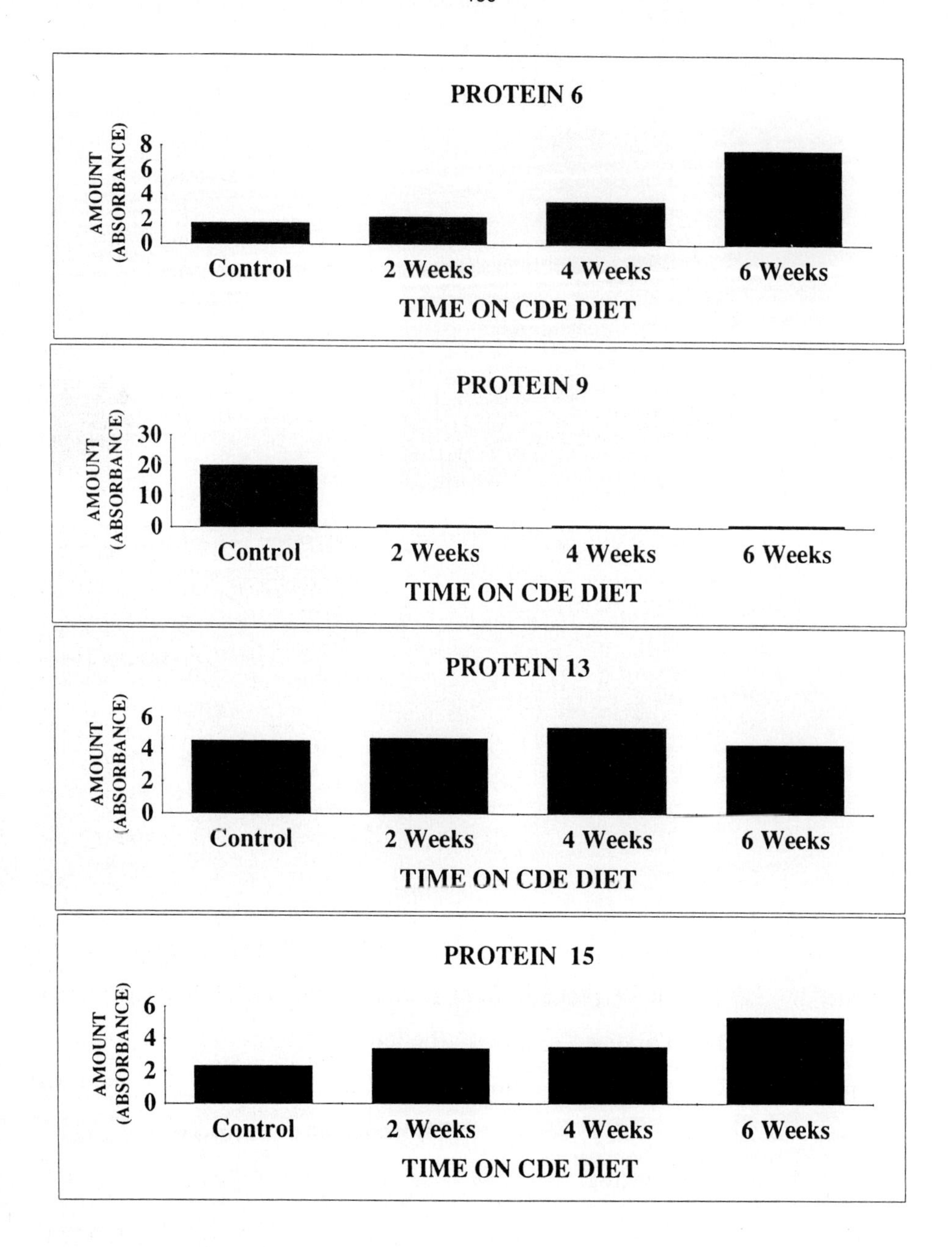

Figure 3. Quantitative 2DE analysis of 4 liver proteins at intervals during CDE feeding.

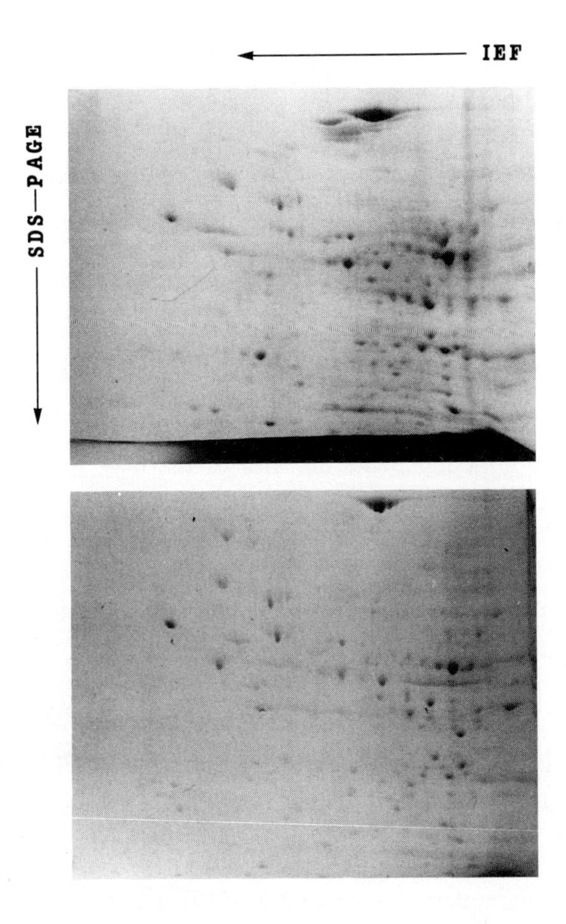

Figure 4. 2DE of proteins in isolated hepatocyte preparations from a control rat (top), and from a rat maintained on the CDE diet for 4 weeks (bottom). The protein patterns are similar to those of liver (Figure 1).

DISCUSSION

We have detected changes in the abundance of several proteins in the livers of rats fed a CDE diet. Some of these changes may represent modulated expression of hepatocellular genes, and others may reflect changing cell populations as oval cells proliferate and give rise to their differentiation products. Our studies indicated that all of the major changes which were detected in the liver were also present in the isolated hepatocyte populations. In general, the protein profile of liver is dominated by the contribution of hepatocytes which normally make up 90% of the liver mass. However our findings can not be used to exclude the possibility that some of the liver protein changes were due to oval cell proliferation. This qualification is necessary because the hepatocyte populations which were isolated 4-6 weeks after CDE feeding could not be separated completely from oval cells by the differential centrifugation techniques which we used. It is therefore possible that some of the change which we detected represents the accumulation of oval cells. Resolution of this problem will require isolation of purer cell populations, possibly by centrifugal elution or flow cytometry. We believe that more definite conclusions about the relative contributions of hepatocytes and oval cells to the changing patterns of liver protein expression could be obtained by using hepatocyte or oval cell populations with less than 10-20% contamination by the other cell type.

Altered hepatocellular gene expression is a common phenomena in many forms of liver injury, including the early stages of hepatocarcinogenesis (Alison *et al.* 1993; Huber *et al.* 1988; Wirth *et al.* 1986). Albumin for example is commonly lost (Evarts *et al.* 1990). We are more interested however in proteins which show de novo expression during oval cell proliferation as a subset of these proteins may be uniquely derived from oval cells. We did not detect such proteins in the experiments which are reported here but we plan to use more sensitive detection methods to survey a much larger number of liver and isolated cell proteins. If oval cell specific proteins can be detected it may be possible to gain further information about the functions of oval cells by sequencing these proteins. In addition it should be possible to use the sequence data to design probes for further analysis of oval cell production and differentiation, both by in situ hybridisation or immunocytochemistry, and by analysis of protein and mRNA changes in the isolated cells.

Acknowledgements

This work was supported by grants from the Wellington Cancer Society and the Medical Research Council of New Zealand.

REFERENCES

Aebersold R, Leavitt J (1990) Sequence analysis of proteins separated by polyacrylamide gel electrophoresis: Towards an integrated data base. Electrophoresis 11: 517-527

Alison MR, Poulsom R, Jeffery R, Anilkumar TV, Jagoe R, Sarraf CE (1993) Expression of hepatocyte growth factor mRNA during oval cell proliferation in the rat liver. J Pathol 171: 291-299

Anderson NL, Anderson NG (1991) A two-dimensional gel database of human plasma proteins. Electrophoresis 12: 883-906

Anderson NL, Copple DC, Bendele RA, Probst GG, Richardson FC (1992) Covalent protein modifications and gene expression changes in rodent liver following administration of methapyrilene: A study using two-dimensional electrophoresis. Fundam Appl Toxicol 18: 570-580

Anderson NL, Esquer-Blasco R, Hofmann J-P, Anderson NG (1991) A two-dimensional gel database of rat liver proteins useful in gene regulation and drug effects. Electrophoresis 12: 907-930

Anderson NL, Nance SL, Tollaksen SL, Giere FA, Anderson NG (1985) Quantitative reproducibility of measurements from Coomassie Blue-stained two-dimensional gels: Analysis of mouse liver protein patterns and a comparison of BALB/c and C57 strains. Electrophoresis 6: 592-599

Anderson NL, Swanson M, Giere FA, Tollaksen S, Gemmell A, Nance S, Anderson NG (1986) Effects of Aroclor 1254 on proteins of mouse liver: Application of two-dimensional electrophoretic protein mapping. Electrophoresis 7: 44-48

Baumann H, Firestone GC, Burgess TC, Gross KW, Yamamoto KR, Held WA (1983) Dexamethasone regulation of alpha 1-acid glycoprotein and other acute phase reactants in rat liver and hepatoma cells. J Biol Chem 258: 563-570

Burgess-Cassier A, Johansen JJ, Santek DA, Ide JR, Kendrick NC (1989) Computerised quantitative analysis of Coomassie-blue stained serum proteins separated by two-dimensional electrophoresis. Clin Chem 35: 297-2304

Celis JE *et al.* (1991) The master two-dimensional gel database of human AMA cell proteins: Towards linking protein and genome sequence and mapping information (Update 1991). Electrophoresis 12: 765-801

Colbert, RA, Amatruda JM, Young DA (1984) Changes in the expression of hepatocyte protein gene-products associated with adaption of cells to primary culture. Clin Chem 30: 2053-2058

Cordiner SJ, Jordan TW (1983) Inhibition by sporidesmin of hepatocyte bile acid transport. Biochem J 212: 197-204

Evarts RP, Nagy P, Nakatsukasa H, Marsden E and Thorgeirsson SS (1989) In vivo differentiation of rat liver oval cells into hepatocytes. Cancer Res 49: 1541-1547

Evarts RP, Nakatsukasa H, Marsden ER, Hsia C-C, Dunsford HA and Thorgeirsson SS (1990) Cellular and molecular changes in the early stages of chemical hepatocarcinogenesis in the rat. Cancer Res 50: 3439-3444

Doz P, Gorg A (1991) Ready-made gels, immobilized pH gradients and automated procedures for high resolution two-dimensional electrophoresis. Science Tools 35: 1-5

Frazer IH, Mackay IR, Jordan TW, Whittingham S, Marzuki S (1985) Reactivity of anti-mitochondrial autoantibodies in primary biliary cirrhosis: Definition of two novel mitochondrial polypeptide autoantigens. J Immunol 135: 1739-1745

Garrels JI (1989) The QUEST system for quantitative analysis of two-dimensional gels. J Biol Chem 264: 5269-5282

Giometti CS, Gemmell MA, Nance SL, Tollaksen SL, Taylor J (1987) Detections of heritable mutations as quantitative changes in protein expression. J Cell Biol. 262: 12764-12707

Giometti CS, Taylor J, Tollaksen SL (1992) Mouse liver protein database: A catalog of proteins detected by two-dimensional electrophoresis. Electrophoresis 13: 970-991

Guevara J, Johnston DA, Ramagali LS, Martin BA, Capetillo S, Rodriguez, LV (1982) Quantitative aspects of silver deposition in proteins resolved on complex polyacrylamide gels. Electrophoresis 3: 197-205

Hochstrasser *et al.* (1992) Human liver protein map: A reference database established by microsequencing and gel comparison. Electrophoresis 13: 992-1001

Huber BE, Wirth PJ, Miller MJ, Glowinski IB (1988) Comparison of gene expression in preneoplastic and neoplastic rat liver to adult, fetal, regenerating and tumour-promoted liver. Cancer Res 48: 2382-2387

Jordan TW, Pedersen JS (1986) Sporidesmin and gliotoxin induce cell detachment and perturb microfilament structure in cultured liver cells. J Cell Sci 85: 33-46

Lecocq R, Lamy F, Dumont JE (1990) Use of two-dimensional electrophoresis and autoradiography as a tool in cell biology: The example of the thyroid and the liver. Electrophoresis 11: 200-212

Levenson RM, Anderson GM, Cohn JA, Blackshear PJ (1992) Giant two-dimensional gel electrophoresis. Methodological update and comparison with intermediate-format gel systems. Electrophoresis 11: 269-279

Miller MJ, Schwartz DM, Thorgeirsson SS (1988) Inter- and intraclonal variability of polypeptides synthesized in a rat hepatoma cell line. J Biol Chem 263: 11227-1123

O'Farrell PH (1975) High resolution two-dimensional electrophoresis of proteins. J Biol Chem 250: 4007-4021

O'Farrell PZ, Goodman HM, O'Farrell PH (1977) High resolution two-dimensional electrophoresis of basic as well as acidic proteins. Cell 12: 1133-1142

Sell S, Dunsford HA (1989) Evidence for the stem cell origin of hepatocellular carcinoma and cholangiocarcinoma. Am J Pathol: 134 1347-1363

Shinozuka H, Lombardi B, Sell S, Immarino RM (1978) Early histological and functional alterations of ethionine liver carcinogenesis in rats fed a choline-deficient diet. Cancer Res 38: 1092-1098

Wheeler TT, Loong PC, Jordan TW, Ford HC. (1986) A double-label two-dimensional gel electrophoresis procedure specifically designed for serum or plasma analysis. Analyt Biochem 159: 1-7

Wheeler TT, Jordan TW, Ford HC (1988) A double-label two-dimensional electrophoresis procedure for the analysis of membrane proteins. Electrophoresis 9: 279-287

Wirth PJ, Benjamin T, Schwartz DM, Thorgeirsson SS. (1986) Sequential analysis of chemically induced hepatoma development in rats by two dimensional electrophoresis. Cancer Res 46: 400-413

Wirth PJ, Luo L-d, Fujimoto Y, Bisgaard HC (1992) Two-dimensional electrophoretic analysis of transformation-sensitive polypeptides during chemical, spontaneously, and oncogene-induced transformation of rat liver epithelial cells. Electrophoresis 13: 305-320

Yang L, Faris RA, Hixson DC (1993) Characterisation of a mature bile duct antigen expressed on a subpopulation of biliary ductule cells but absent from oval cells. Hepatology 18: 357-366

Young, DA (1984) Advantages of separations on "giant" two-dimensional gels for detection of physiologically relevant changes in the expression of protein gene-products. Clin Chem 30: 2104-2108

STEM CELL ACTIVATION IN THE ACETYLAMINOFLUORENE-TREATED REGENERATING RAT LIVER: A BILE DUCTULAR REACTION ?

T.V. Anilkumar, Matthew Golding, Catherine Sarraf, El-Nasir Lalani, Richard Poulsom and Malcolm Alison.
Department of Histopathology
Royal Postgraduate Medical School
Du Cane Road, London W12 ONN
UK.

INTRODUCTION

The ability of the liver to regenerate in response to the loss of hepatocytes is widely recognised, and this is usually accomplished by the triggering of normally proliferatively quiescent hepatocytes into the cell cycle (Wright and Alison 1984; Alison 1986). Liver regeneration is commonly studied in the rat after a two-thirds partial hepatectomy. However, when hepatocyte regeneration is impaired, then relatively undifferentiated cells emerge from the portal space and take over the burden of regenerative growth. These potential stem cells are called **oval cells** (Sell 1990) and they are seen in chronically damaged human liver (De Vos and Desmet 1992), in galactosamine poisoned rats (Lemire et al 1991; Dabeva and Shafritz 1993) and when hepatocyte regeneration is prevented by the presence of cytotoxic carcinogens (Alison and Hully 1991). The antiproliferative effects of chemical carcinogens were recognised by Haddow in 1935 and this property has been exploited in the development of the so-called **'Resistant Hepatocyte'** model of carcinogenesis (Solt et al 1977) whereby rats injected with the genotoxic chemical diethylnitrosamine developed basophilic liver cell foci shortly after being partially hepatectomised while being fed 2-acetylaminofluorene (2-AAF). 2-AAF prevents hepatocyte regeneration, but initiated oval cells can escape (*resist*) this inhibition and divide and eventually form dysplastic foci (Alison and Hully 1991). By simply omitting the initial exposure to diethylnitrosamine, the model has been adapted to study the differentiation of oval cell progeny into both hepatocytes and biliary epithelia (Tatematsu et al 1985; Evarts et al 1987, 1992).

NATO ASI Series, Vol. H 88
Liver Carcinogenesis
Edited by G. G. Skouteris

As in other tissues and organs, growth control in the liver seems to be largely exerted through the availability of growth factors and the level of expression of the appropriate receptor molecules (Alison and Wright 1993). A wide variety of molecules influence liver regeneration (Michalopoulos 1990), and there is good evidence that TGF-α and TGF-β1 are strong players in the stimulation and curtailment respectively of hepatocyte replication (Fausto 1991; Webber et al 1993).

The search for a humoral regulator of liver growth has increasingly concentrated on the role of **hepatocyte growth factor**, also called scatter factor (HGF-SF). HGF is synthesised from a single precursor molecule of 728 amino acids, which is proteolytically processed to form the mature HGF, a heterodimeric molecule composed of a 69KDa α chain and a 34KDa β chain. The α chain contains four kringle domains (double loop structures held together by disulphide bonds) reminiscent of various proteases involved in coagulation and fibrinolysis. The β chain has a 37% homology with the β chain of plasmin. HGF is currently considered to be a pleiotropic factor (Strain 1993), produced by mesenchymal cells, influencing epithelial cell growth, motility and morphogenesis. HGF was originally termed *'scatter factor'* (Stoker and Perryman 1985) because the protein from the conditioned medium of human embryonic fibroblasts was able to enhance the local motility (scatter) of a variety of epithelial cells *in vitro*. HGF and scatter factor are identical. The usual target cells for the bioassay of HGF scattering activity are Madin-Darby canine kidney (MDCK) cells, and when these cells are grown in collagen gels, HGF can also act as a morphogen promoting nephrogenesis (Montesanto et al 1991). In the liver, HGF is not produced by hepatocytes, but by non-parenchymal sinusoid-lining cells, principally Ito cells (Hu et al 1993), while systemically administered HGF is not only primarily taken up by the liver but is also predominantly located in periportal hepatocytes (Liu et al 1994), the cells which respond most extensively to inflicted damage (Alison 1986).

Interest in liver stem cells has reached almost fever pitch (Aterman 1992; Travis 1993; Thorgeirsson 1993; Marceau 1994), but the exact origin of these cells is not resolved. This paper describes a prominent ductular reaction emanating from preexisting bile ducts occurring in the AAF-treated rat after partial hepatectomy, with morphological evidence that these cells can differentiate into hepatocytes. Moreover these cellular changes were accompanied by increased expression of HGF mRNA in the liver, and we show that sinusoid-lining cells expressing HGF mRNA become most abundant in the periportal regions immediately adjacent to the emerging ductular cells.

MATERIALS AND METHODS

Treatment of animals

Male Fischer rats weighing 200g were used; they were maintained on standard pelleted chow and had access to water *ad libitum*. To inhibit hepatocyte cell proliferation all rats received daily oral gavage of 0.2ml of 1% 2-AAF for a period of up to 14 days. After 7 days on this regime all rats underwent a two-thirds partial hepatectomy under diethyl ether anaesthesia; animals were killed under diethyl ether anaesthesia by exsanguination *via* the hepatic vein, in groups of four at daily intervals up to 8 days after partial hepatectomy. All rats received a single intraperitoneal injection of bromodeoxyuridine (BrdUrd) at a dosage of 50mg per Kg body weight one hour before death. For routine light microscopy (haematoxylin and eosin staining) and *in situ* hybridisation studies, thin (2-3mm) liver slices were immersion fixed in neutral buffered formalin for 24hr before processing and embedding in paraffin wax. For the immunocytochemical detection of BrdUrd, other liver slices were fixed in Carnoy's solution for 4hr before processing and embedding, while liver slices used for the immunocytochemical detection of intermediate filament proteins and alpha foetoprotein were fixed in Methacarn for 24hr before processing.

Electron microscopy.

Tissue samples, not exceeding $1mm^3$ in volume were fixed in 2% glutaraldehyde for 2 hours. After washing in phosphate buffer (pH 7.2), tissues were osmicated and dehydrated in acidified 2, 2- dimethoxypropane (DMP) before routine embedding in TAAB resin. 1μm sections were cut and stained with toluidine blue for observation at light microscope level and selection of relevant blocks, followed by ultrathin sections of approximately 100nm, collected on nickel grids and stained with uranyl acetate and lead citrate, for observation on a Philips CM-10 electron microscope.

In situ hybridisation for detection of HGF mRNA

Blocks of tissue were stored at room temperature without special precautions.

Probes - Plasmids containing cDNA for regions of rat hepatocyte growth factor (*Eco* RI subclone from RBC1[25] in pBluescript SK vector; kindly provided by Dr T Nakamura, Kyushu University, Fukuoka 812, Japan) and rat β-actin (produced by Dr. R Chinery, ICRF/RCS Histopathology Unit) were used to produce antisense riboprobes labelled internally with ^{35}S (~800 Ci/mmol; Amersham). The orientation of the clones was verified by sequencing. Rat HGF probe of approximately 1.4 kb was generated from *Eco* RV linearised plasmid using T3 RNA polymerase and was used without alkali hydrolysis. Rat β-actin probes of approximately 240 bases were generated from *Eco* RI linearised plasmid also with T3 RNA polymerase.

Hybridisation - The method used was essentially that of Senior et al (1988) with minor modifications. Sections were hybridised to 1x 10^6 cpm of either probe in 20μl of hybridisation mixture and incubated overnight at 55°C. Subsequently, un-hybridised probe was destroyed by digestion with RNAse A, and stringency washes carried out to reduce non-specific binding. Slides were dipped in photographic emulsion and exposed at 4°C for approximately 40 days before development. Sections were counterstained weakly by Giemsa's method. Patterns of hybridisation were studied under dark-field, reflected light, conditions.

Immunocytochemistry

For the detection of intermediate filament proteins, tissues were fixed for 24 hr in Methacarn, processed routinely and embedded in paraffin wax. After blocking for endogenous peroxidase, 3 μm sections were incubated for 10 mins with normal serum from the donor species of the secondary antibody followed by a 1 hr incubation with the primary antibody (see Table 1). After rinsing with PBS, biotinylated rabbit anti-mouse immunoglobulins (Dako) diluted 1:200 in PBS was applied for 45 mins, sections were then rinsed with PBS and incubated with horseradish peroxidase conjugated Streptavidin/Biotin Complex (Strep.ABC/ HRP; [Dako]) for 30 mins. The sections were rinsed with PBS and peroxidase developed for 2-7 mins with 0.05% diaminobenzidine (DAB) and 0.03% H_2O_2 in PBS, counterstained with Harris' haematoxylin, dehydrated, cleared and mounted in Pertex.

Table 1

Antibody	Specificity	Source	Dilution in PBS
LE 41	Cytokeratin 8	ICRF	1:10
LP2K	Cytokeratin 19	ICRF	1:5
Clone V9 (V-6630)	Vimentin	Sigma	1:2000

To visualize BrdUrd incorporation immunostaining was performed on sections of 5μm thickness which were mounted on poly L-lysine-coated glass slides. Immunostaining was carried out as previously described (Sarraf et al 1991). Briefly, sections were deparaffinized with CNP, hydrated in 100% ethanol, and endogenous peroxidase inactivated by immersing the slides in 98.4% methanol/1.6% hydrogen peroxide for 45min. DNA was denatured by immersing the slides in 2 N HCl for 1 hr, followed by rinsing in PBS. Immunostaining was performed using a mouse monoclonal antibody to BrdUrd (Dako, U.K. Ltd), applied in a 1:20 dilution to the sections and left overnight at 4°C. Negative controls used PBS in place of the the primary antibody. All sections were then stained by the indirect immunoperoxidase method. The sections were covered with a 1:50 dilution of peroxidase conjugated rabbit, anti-mouse antibody (Dako) and were incubated for 1 hr at room temperature. The sections were washed three times in PBS for 5 min each. The peroxidase reaction was developed using the diaminobenzidine method, and sections were counterstained with Cole's haematoxylin for 1 min.

In Situ End Labelling (ISEL) for the detection of apoptotic cells.

This is a modification of the protocol from Wijsman et al (1993). Tissues were prepared as above and 3 μm sections cut and mounted onto silane coated slides. After blocking for endogenous peroxidase, sections were incubated for 1 hr at 37°C with a reaction buffer (50 mM Tris, 5 mM $MgCl_2$, 10 mM β-Mercaptoethanol, 0.005% BSA [Fraction V]) containing a final concentration of 0.01 mM each of dATP, dGTP, dCTP (Pharmacia); 0.3 μM Biotin-16-dUTP (Boehringer Mannheim) and 2.5 Units/ml of Klenow fragment of DNA polymerase 1 (Pharmacia). Sections were rinsed well with distilled water followed by PBS and then incubated with Strep.ABC/ HRP (Dako) for 30 min at room temperature. Sections were developed with DAB, counterstained and mounted as above.

Results

At one day after partial hepatectomy (PH) there was evidence of the emergence of small basophilic cells with a high nuclear:cytoplasmic ratio in and around the portal tracts. These so-called oval cells became more numerous with time and by 8 days long strings of these cells could be seen fanning out into the periportal and midzonal parenchyma (Figure 1).

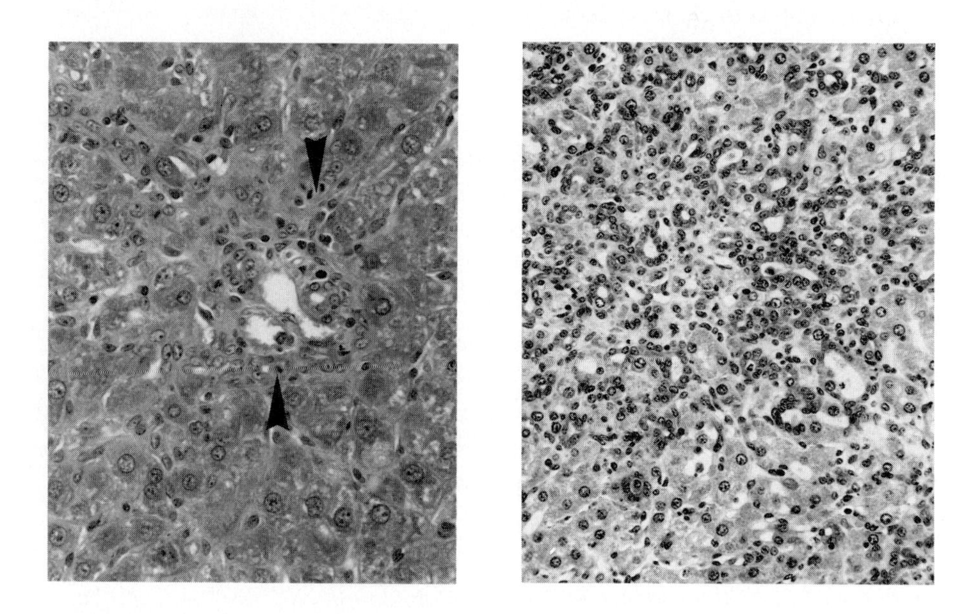

Figure 1. At 5 days after PH (LH side) the portal areas are noticeably hypercellular with mitotic activity apparent (arrows) and by 8 days (RH side) long cords of cells are seen with numerous ductular profiles.

BrdUrd incorporation was observed in occasional bile ductular cells as early as 6hr after PH and by 24hr many ductular cells were in DNA synthesis; at later times BrdUrd incorporation mainly occurred in the advancing cell cords (Figure 2).

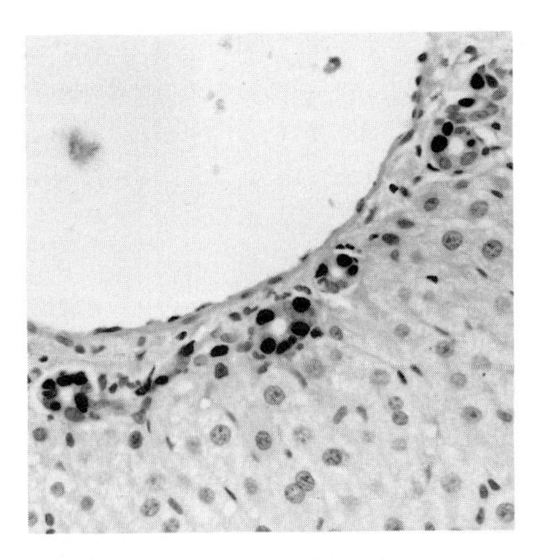
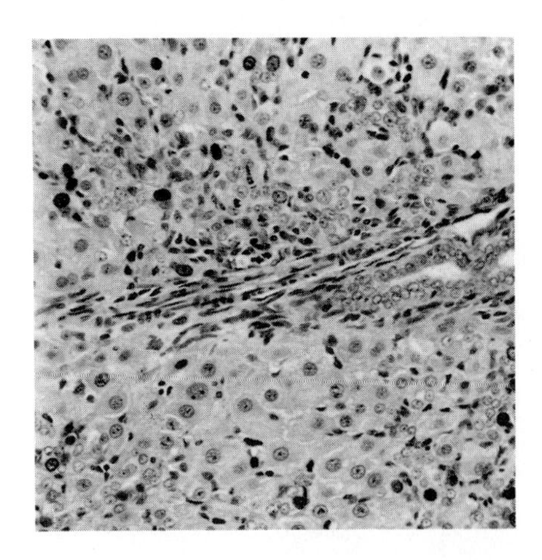

Figure 2. At 1 day after PH (LH side) most BrdUrd incorporation occurred in preexisting bile ducts but at 7 days (RH side) labelling was seen in the cells spreading out between the hepatic plates.

In the normal control liver, cytokeratin 19 immunoreactivity was restricted to bile ductular epithelial cells, but all oval cells stained intensely and the pattern of staining at later times after PH clearly demonstrated that most oval cells were in fact organised into long ducts (Figure 3). A bile duct injection of such an animal with a blue pigmented gelatin medium resulted in the filling of the lumens of these new structures strongly suggesting they were sprouted from preexisting bile ducts.

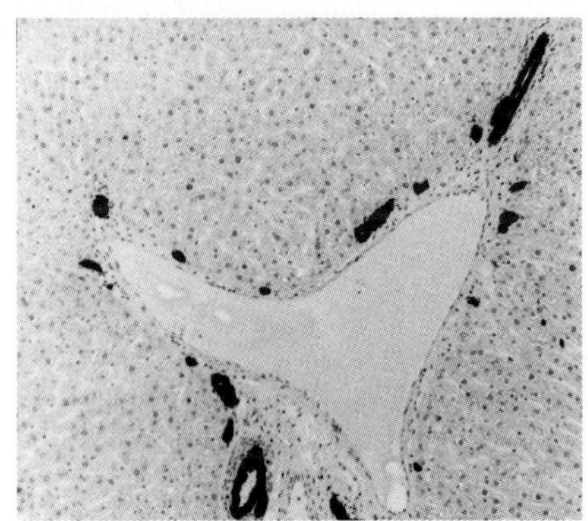
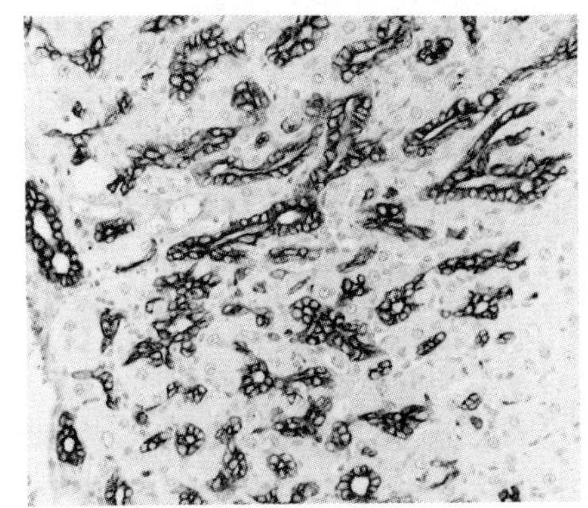

Figure 3. Cyokeratin 19 immunoreactivity. In the normal liver (LH side) bile duct epithelium is intensely positive, while at 8 days after PH the branching ductular nature of the oval cells can be readily appreciated by the strong staining (RH side).

In the normal liver staining with the antivimentin antibody was restricted to some of the sinusoid-lining cells, fibroblasts in the portal space and the smooth muscle of blood vessel walls. However, oval cells were very strongly stained, though when these cells appeared to differentiate they lost vimentin immunoreactivity (Figure 4).

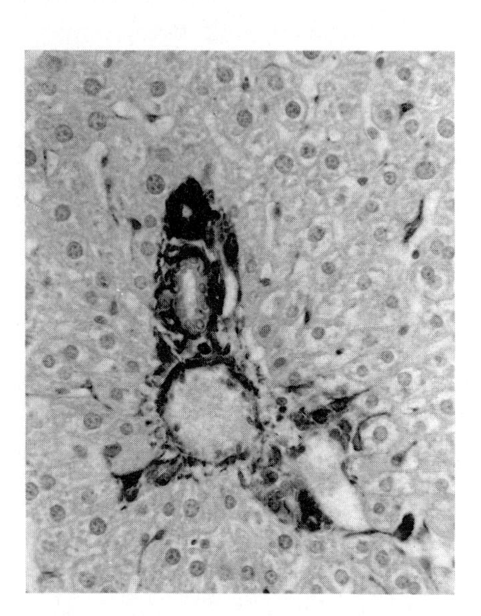

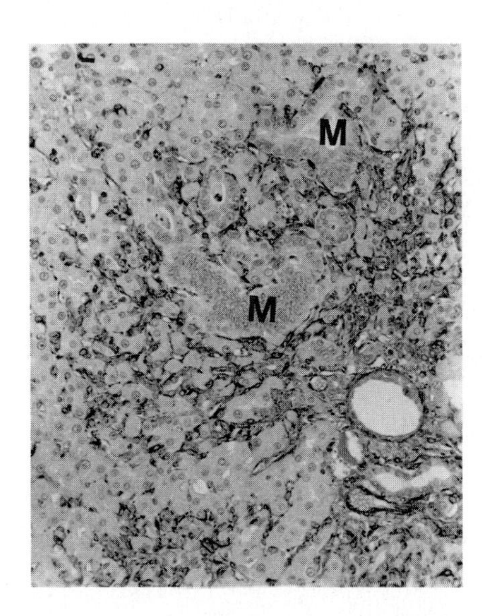

Figure 4. Vimentin immunoreactivity in normal liver (LH side) contrasting with strong staining of the strings of cells at 8 days after PH (RH side). Note the loss of staining in the apparently metaplastic areas (M) where differentiation into columnar epithelium reminiscent of gastric foveolar epithelium has taken place.

Cytokeratin 8 immunoreactivity in the normal liver was present not only in biliary epithelium but also in periportal hepatocytes, albeit at a low level. After PH all oval cells showed strong staining (Figure 5), and, as with vimentin immunostaining, cytokeratin 8 immunoreactivity appeared to diminish in cells undergoing differentiation (Figure 6). This differentiation did not occur simultaneously in all ductular cells, but rather occurred focally. These events at 7-8 days after PH were accompanied by substantial cell death in the ductular population with dead and dying cells being shed into the duct lumen, seemingly undergoing apoptosis from the evidence of well preserved organelles in the lumenal fragments (Figure 7).

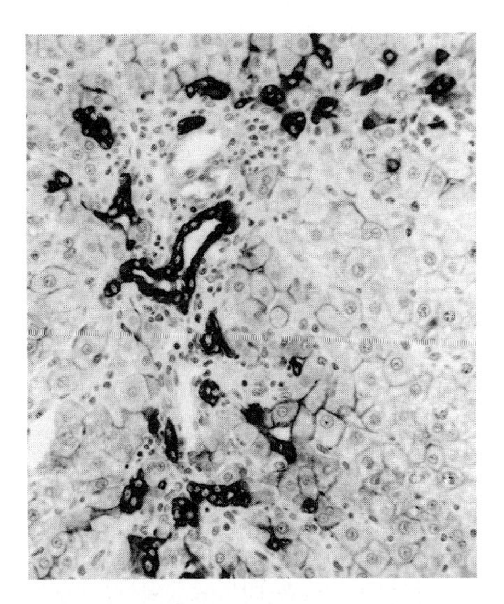

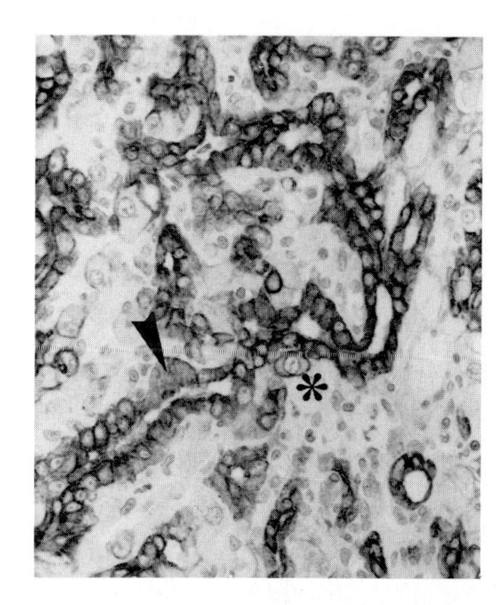

Figure 5. Cytokeratin 8 immunoreactivity. At 5 days after PH (left) the emerging oval cells are strongly stained, and at 8 days long tortuous ducts invade the parenchyma with the lining cells showing focal differentiation (arrow) and apparent budding of immunopositive cells (*) to the outside.

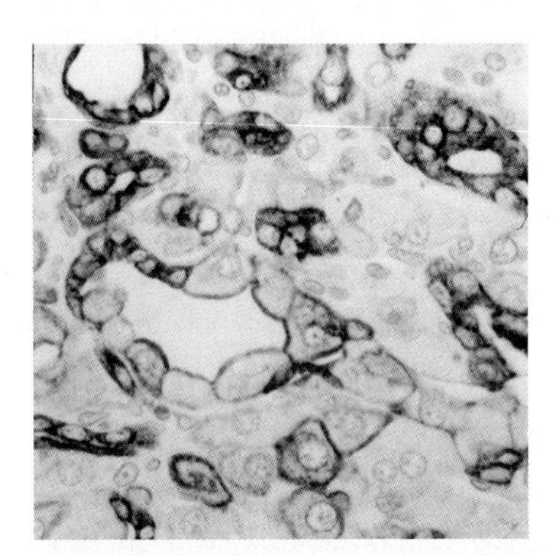

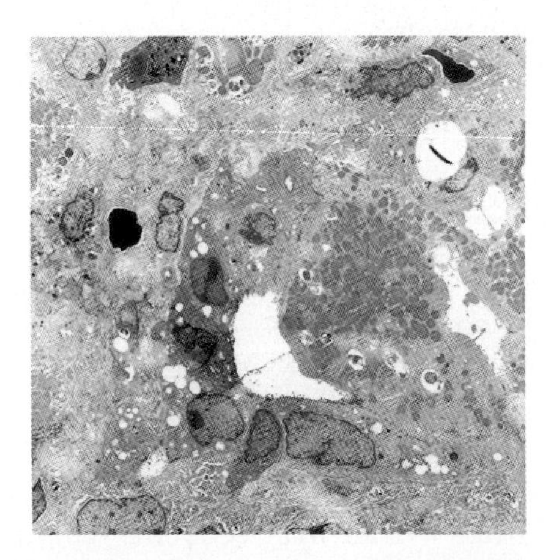

Figure 6. Transverse sections of oval cells organised into ducts amongst the hepatic parenchyma at 8 days after PH. Viewed by light microscopy (LH side) some of the lining cells appear to be differentiating into hepatocytes and simultaneously losing the intense cytokeratin 8 immunoreactivity of ductular cells. At electron microscope level (RH side) the hepatic nature of these hypertrophied cells can be readily seen by their abundant mitochondria.

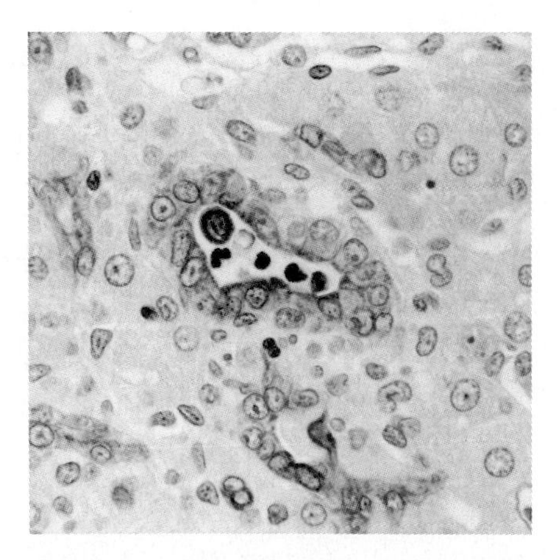

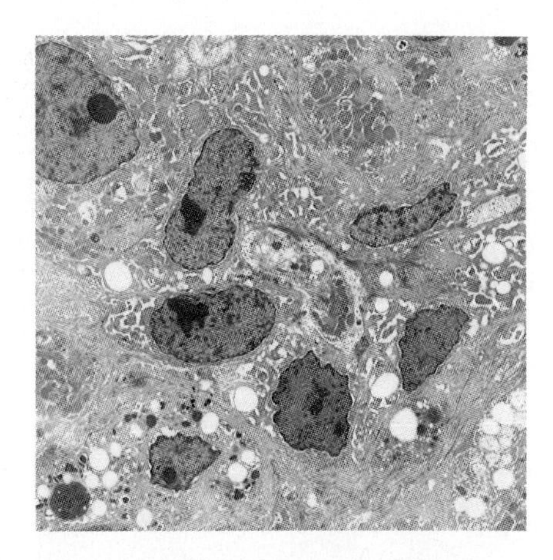

Figure 7. At 8 days after PH many dead cells are seen in the ducts. These are easily seen by ISEL (LH side) with the dark brown DAB final reaction product localised over the nucleus, and electron microscopy (RH side) confirms the presence of membrane-bound fragments containing well preserved organelles.

HGF mRNA expression was only seen in sinusoid-lining cells. Sinusoid-lining cells expressing HGF transcripts became more numerous with time after PH, and by 7 days were far more numerous in the periportal regions (Figure 8) immediately adjacent to the cords of oval cells. Neither oval cells nor hepatocytes expressed HGF mRNA, however there was abundant intact mRNA in oval cells as judged by β-actin mRNA expression in these cells.

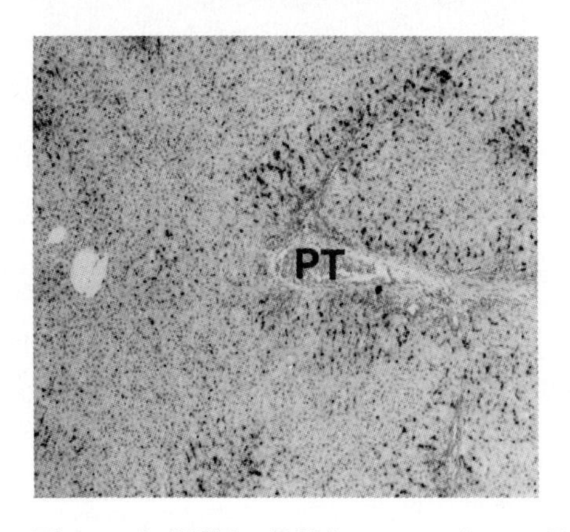

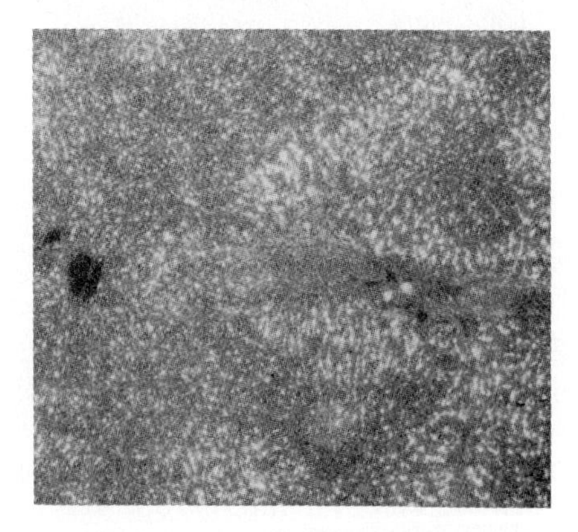

Figure 8. HGF mRNA expression at 7 days after PH as viewed by bright field (LH side) and dark field reflected light (RH side) illumination. Note preferential accumulation of labelled cells around the portal tracts (PT).

DISCUSSION

This paper describes the response of the rat liver to PH when the proliferative response of differentiated hepatocytes is prevented by the chronic oral administration of 2-AAF. An absence of proliferative activity in normal hepatocytes was confirmed. However, small cells with a high nuclear:cytoplasmic ratio spread out from the portal tracts, and by 8 days after PH there was morphological evidence of differentiation of these cells into either hepatocytes or columnar epithelium. These small cells are commonly referred to as oval cells (Aterman 1992), *potential* liver stem cells which appear not to participate in normal cell renewal. The origin of these cells has been keenly debated, with periductular cells, transitional duct cells, terminal bile ductules and bile ducts themselves being variously proposed as the normal sites for these progenitor cells. Current thought is tending to move away from a single location for these cells, adopting the more flexible view that any component of the biliary tree can give rise to oval cells (Lenzi et al 1992).

Liver parenchymal cells have a simple cytokeratin composition expressing only one cytokeratin pair, cytokeratins 8 and 18 (Van Eyken et al 1987). Intrahepatic bile ducts on the other hand, also express cytokeratins 7 and 19 in addition to 8 and 18. In the present study the expanding population of oval cells stained strongly with antibodies raised against each of these four cytokeratins. These observations imply an ancestry of oval cells from biliary epithelium, though phenotypic traits are not necessarily markers of histogenesis since, for example, biliary cytokeratins can be expressed by hepatocytes undergoing ductular metaplasia (Van Eyken et al 1989) or neoplastic transformation (Van Eyken et al 1988b).

Routine haematoxylin and eosin staining revealed relatively little about the spatial organisation of the emerging oval cells, though many oval cells formed ductular profiles by 8 days after PH (Figure 1), and such structures have commonly been referred to as either "disorganised bile ductular cells" or "pseudoducts". In liver disease at least three types of bile ductular reaction can be distinguished (Popper 1990); 1) a proliferation of ductules at the parenchymal border due to mechanical obstruction of large bile ducts, 2) a ductular metaplasia of hepatocytes due to chronic damage and inflammation and 3) a sprouting of bile ductules which invade the hepatic parenchyma. Using a combination of anticytokeratin antibodies and the bile duct injection of pigmented medium it was quite clear that the oval cells observed in the present study fitted the third category, ie. largely a collection of long, branching ductular structures sprouted from preexisting bile ducts and continuous with them. However the emergence of some oval cells from the portal areas either singly or in small groups is not discounted. The connection of ductular oval cells to portal bile ducts is in itself not a novel finding and has been observed under a variety of conditions. For example Dunsford et al (1985) recorded the phenomenon in rats fed AAF in a choline-deficient diet, and likewise Lenzi et al (1992) feeding rats ethionine in a choline-devoid diet (CDE diet) observed long strings of cytokeratin 19 positive cells apparently connected to portal bile ducts, and physiological measurements confirmed a massive increase in biliary tree volume coincident with the oval cell proliferation. Rats treated chronically with carbon tetrachloride to induce cirrhosis develop a similar ductular reaction in the parenchyma, and casts of the biliary tree also confirmed a connection to preexisting bile ducts (Masuko et al 1964).

However, the present experiments extend these observations to suggest that the epithelial cells forming these sprouted ducts can differentiate *in situ*. Transformation of small ductular cells to a tall columnar epithelium with occasional interspersed goblet cells was a common occurrence and has been noted by others (Tatematsu et al 1985), but we also observed that some lining cells progressively enlarged and accumulated increased numbers of large mitochondria typical of hepatocytes (Figure 6). This focal differentiation was accompanied by a high incidence of apoptotic cell death (see review by Alison and Sarraf 1992), with the affected cells budding off into the duct lumen. Of course, we should add the caveat that morphology alone is not evidence for functional hepatocytic differentiation (Thorgeirsson 1993), nevertheless these ductular cells synthesised alpha foetoprotein (data not shown), an event which signals a hepatic commitment in endodermal cells of the ventral foregut during development (Shiojiri et al 1991). High levels of serum alpha foetoprotein in severe forms of acute hepatitis are a favourable prognostic sign, perhaps indicative of regeneration from the stem cell compartment? Interestingly, we observed that this apparent hepatic differentiation of ductular cells occurred focally, and inspection of a micrograph published by Onoe et al (1975) illustrates a similar phenomenon in a model of hepatocarcinogenesis. Feeding rats with the azo dye 3-methyldimethylaminobenzene (3-MeDAB) these authors reckoned few oval cells were organised into ducts, but they clearly show a duct partly lined by small undifferentiated cells and partly lined by cells resembling hepatocytes with large mitochondria and well developed endoplasmic reticulum. Labelling proliferating oval cells with tritiated thymidine in the model employed in the present studies, Evarts et al (1987) showed that the label shortly appears in small basophilic hepatocytes suggesting a direct lineage relationship. Likewise *in vitro* studies have suggested that hepatocytes are derived from biliary ductular cells (Grisham 1980), while intrahepatic transplantation of a diploid oval cell-like cell line into Fischer rats caused their differentiation into hepatocytes, at least when judged by morphological criteria (Coleman et al 1993).

It has not been possible to identify with any degree of certainty the exact location of the first cells which respond to PH. BrdUrd labelled bile ductular and periductular cells can be found as early as 6hr after PH, though at 24hr small bile ducts appeared to be the most proliferatively active cells (Figure 2). At later times after PH the advancing sprouts are the sites of cell proliferation and the parent ducts are relatively quiescent. Largely similar findings have been reported by Evarts et al (1993) suggesting small bile ducts are the prime movers in the initiation of the oval cell response in this model. Cells with the features of oval cells are seen proliferating in human liver after submassive liver necrosis (Koukoulis et al 1992), and Seki et al (1993) have noted proliferating portal bile ducts in acute viral hepatitis in man, while in chronic hepatitis they also noted cell proliferation in cytokeratin 19-positive strings of ductular cells in areas of confluent necrosis - findings mirrored in the present experimental study.

Normal bile ducts express cytokeratins 7,8, 18 and 19, while hepatocytes have only cytokeratins 8 and 18. Cytokeratin 19 immunoreactivity proved a useful way of tracing the tortuous and branched ductular structures within the hepatic parenchyma, and a similar approach can be used to locate oval cells in human liver (Hsia et al 1992). Cytokeratin 8 immunoreactivity was common to both ductular cells and hepatocytes, though the intracellular patterns of staining were distinctly different. In ductular cells there was intense diffuse cytoplasmic staining, whereas in periportal hepatocytes the immunostaining was distinctly membranous confirming previous studies (Van Eyken et al 1988a). Cultured rat hepatocytes under a differentiating environment also have dense cytokeratin filament networks at the cell periphery (Marceau et al 1985), but mitogenic stimuli induce a filament rearrangement from the cell periphery to the cytoplasm (Baribault et al 1989). Similarly pure cultures of rat hepatocytes have a uniformally even distribution of cytokeratin immunoreactivity in the cytoplasm, whereas hepatocytes co-cultured with rat liver epithelial cells (presumed to originate from primitive biliary epithelia) promotes expression of liver-specific genes and the organisation of dense cytokeratin fibrils at the cell periphery (Baffet et al 1991). In the present study it was quite clear that ductular cells showing morphological evidence of hepatocyte differentiation showed a distinct change in staining from the ductular pattern (cytoplasmic) to the hepatocyte pattern (membranous), lending further support to the idea that hepatic differentiation is occurring in these cells.

Of further interest was the emergence of vimentin immunoreactivity in the proliferating ductular cells, and its subsequent loss as cells differentiated. Vimentin is the intermediate filament characteristic of mesenchymal cells, but is known to be present in some carcinomas. Coexpression of intermediate filament classes is rare, though cultured mesothelial cells switch from cytokeratin to vimentin synthesis during the rapid growth phase (Connell and Rheinwald 1983). Vimentin expression has been noted in hyperplastic bile ducts resulting from biliary obstruction in the rat (Milani et al 1989), in cultures of hepatic stem cells (Bisgaard and Thorgeirsson 1991) and in areas of some human hepatoblastomas (Van Eyken et al 1990). On the other hand, Alpini et al (1992) failed to immunocytochemically detect vimentin in strings of cytokeratin 19 positive cells induced in the rat by feeding a CDE diet. The cause of this discrepancy with the present observations is not clear, but it could be related to tissue fixation since we used Methacarn for this purpose as opposed to formalin or snap-frozen material. The reason for vimentin expression is unknown though conceivably it could be related to the motility of the cells.

Our data support the observations of many others that HGF mRNA is more abundant in the liver after PH, but extend those observations substantially by showing that the increased expression is variable within the liver. HGF mRNA-positive cells were almost twice as abundant in the periportal *versus* the centrilobular regions after 7 days, despite there being a widespread distribution of HGF-positive sinusoidal lining cells across the liver during the previous days (Alison et al 1993b). This focal accumulation of HGF-producing sinusoid lining cells in the immediate vicinity of oval cell activation strongly supports the involvement of local HGF production in the processes of proliferation, migration and probably also duct formation since HGF can induce tubule formation in cultures of nonparenchymal epithelial liver cells (Johnson et al 1993). The synthesis of TGF-α also increases shortly after PH (Evarts et al 1993) and has been localised by immunocytochemistry principally to the new ductular structures (Alison et al 1993a).

In conclusion, when liver damage occurs and the liver mass can not be renewed from existing hepatocytes, stem cell-derived progeny, commonly referred to as oval cells, proliferate and ultimately differentiate, apparently into several distinct lineages. In this particular model these oval cells are ducts sprouted from and continuous with intrahepatic bile ducts in the portal areas.

Acknowledgements
This work was supported by a research grant from the *A*SSOCIATION FOR *I*NTERNATIONAL *C*ANCER *R*ESEARCH.

REFERENCES

Alison MR (1986) Regulation of liver growth. Physiol Rev 66: 499-541
Alison MR, Hully JR (1991) Stimulus dependent phenotypic diversity in the resistant hepatocyte model. In: Columbano A, Feo F, Pascale R, Pani P (eds). Chemical carcinogenesis 2. Modulating factors. New York: Plenum, pp563-577
Alison MR, Nasim MM, Anilkumar TV, Sarraf CE (1993a) Transforming growth factor-α immunoreactivity in a variety of epithelial tissues. Cell Prolif 26: 449-460
Alison MR, Sarraf CE (1992) Apoptosis: - a gene-directed programme of cell death. J R Coll Phys 26: 25-35
Alison MR, Wright NA (1993) Growth factors and growth factor receptors. Brit J Hosp Med 49: 774-788
Alison, MR, Poulsom R, Jeffery R, Anilkumar TV, Jagoe R, Sarraf CE (1993b) Expression of hepatocyte growth factor mRNA during oval cell activation in the rat liver. J Pathol 171: 291-299
Alpini G, Aragona E, Dabeva M, Salvi R, Shafritz DA,Tavoloni N (1992) Distribution of albumin and alpha-fetoprotein mRNAs in normal, hyperplastic, and preneoplastic rat liver. Am J Pathol 141: 623-632
Aterman K (1992) The stem cells of the liver - a selective review. J Cancer Res Clin Oncol 118: 87-115
Baffet G, Loyer P, Glaise D, Corlu A, Etienne P-L, Guguen-Giullouzo C (1991) Distinct effects of cell-cell communication and corticosteroids on the synthesis and distribution of cytokeratins in cultured rat hepatocytes. J Cell Sci 99: 609-615
Baribault H, Blouin R, Bourgon L, Marceau N (1989) Epidermal growth factor-induced selective phosphorylation of cultured rat hepatocyte 55-kD cytokeratin before filament reorganization and DNA synthesis. J Cell Biol 109: 1665-1676
Bisgaard HC, Thorgeirsson SS (1991) Evidence for a common cell of origin for primitive epithelial cells isolated from rat liver and pancreas. J Cell Physiol 147: 333-343

Coleman WB Wennerberg AE, Smith GJ, Grisham JW (1993) Regulation of the differentiation of diploid and some aneuploid rat liver epithelial (stemlike) cells by the hepatic microenvironment. Am J Pathol 142: 1373-1381

Connell ND, Rheinwald JG (1983) Regulation of the cytoskeleton in mesothelial cells: reversible loss of keratin and increase in vimentin during rapid growth in culture. Cell 34: 245-253

Dabeva MD, Shafritz DA (1993) Activation, proliferation, and differentiation of progenitor cells into hepatocytes in the D-galactosamine model of liver regeneration Am J Pathol 143: 1606-1620

De Vos R, Desmet V (1992) Ultrastructural characteristics of novel epithelial cell types identified in human pathologic liver specimens with chronic ductular reaction. Am J Pathol 140: 1441-1450

Dunsford HA, Maset R, Salman J, Sell S (1985) Connection of ductlike structures induced by a chemical hepatocarcinogen to portal bile ducts in the rat liver detected by injection of bile ducts with a pigmented barium gelatin medium. Am J Pathol 118: 218-224

Evarts RP, Hu Z, Fujio K, Marsden ER, Thorgeirsson SS (1993) Activation of hepatic stem cell compartment in the rat: role of transforming growth factor α, hepatocyte growth factor, and acidic fibroblast growth factor in early proliferation. Cell Growth Diff 4: 555-561

Evarts RP, Nagy P, Marsden E, Thorgeirsson SS (1987) A precursor-product relationship exists between oval cells and hepatocytes in rat liver. Carcinogenesis 8: 1737-1740

Evarts RP, Nakatsukasa H, Marsden ER, Hu Z, Thorgeirsson SS (1992) Expression of transforming growth factor-alpha in regenerating liver and during hepatic differentiation. Molecular Carcinogenesis 5: 25-31

Fausto N (1991) Protooncogenes and growth factors associated with normal and abnormal liver growth. Dig Dis Sci 36: 653-658

Grisham JW (1980) Cell types in long-term propagable cultures of rat liver. Ann NY Acad Sci 349: 128-137

Haddow A (1935) Influence of certain polycyclic hydrocarbons on the growth of the Jensen rat sarcoma. Nature 136: 868-869

Hsia CC, Evarts RP, Nakatsukasa H, Marsden ER, Thorgiersson SS (1992) Occurrence of oval cells in hepatitis B virus associated human hepatocarcinogenesis. Hepatology 16: 1327-1333

Hu Z, Evarts RP, Fujio K, Marsden ER, Thorgeirsson SS (1993) Expression of hepatocyte growth factor and c-*met* genes during hepatic differentiation and liver development in the rat. Am J Pathol 142: 1823-1830

Johnson M, Koukoulis G, Matsumoto K, Nakamura T, Iyer A (1993) Hepatocyte growth factor induces proliferation and morphogenesis in nonparenchymal epithelial liver cells. Hepatology 17: 1052-1061

Koukoulis G, Rayner A, Kai-Chan T, Williams R, Portmann B (1992) Immunolocalization of regenerating cells after submassive liver necrosis using PCNA staining. J Pathol 166: 359-368

Lemire JM, Shiojiri N, Fausto N (1991) Oval cell proliferation and the origin of small hepatocytes in liver injury induced by D-galactosamine. Am J Pathol 139: 535-552

Lenzi R, Liu MH, Tarsetti F, Slott PA, Alpini G, Zhai W-R, Paronetto F, Lenzen R, Tavoloni N (1992) Histogenesis of bile duct-like cells proliferating during ethionine hepatocarcinogenesis. Evidence for a biliary epithelial nature of oval cells. Lab Invest 66: 390-402

Liu M-L, Mars WM, Zarnegar R, Michalopoulos GK (1994) Uptake and distribution of hepatocyte growth factor in normal and regenerating adult rat liver. Am J Pathol 144: 129-139

Marceau N (1993) Epithelial cell lineages in developing, restoring, and transforming liver: evidence for the existence of a 'differentiation window'. Gut 35: 294-296

Marceau N, Baribault H, Leroux-Nicollet I (1984) Dexamethasone can modulate the synthesis and organisation of cytokeratins in cultured differentiating rat hepatocytes. Can J Biochem Cell Biol 63: 448-457

Masuko K, Rubin E, Popper H (1964) Proliferation of bile ducts in cirrhosis. Arch Path 78: 421-431

Michalopoulos GK (1990) Liver regeneration: molecular mechanisms of growth control. Faseb J 4: 176 - 187

Milani S, Herbst H, Schuppan D, Niedobitek G, Kim KY, Stein H (1989) Vimentin expression of newly formed rat bile duct epithelial cells in secondary biliary fibrosis. Virchows Archiv A Pathol Anat 415: 237-242

Montesanto R, Matsumoto K, Nakamura T, Orci L (1991) Identification of a fibroblast-derived epithelial morphogen as hepatocyte growth factor. Cell 67: 901-908

Onoe T, Kaneko A, Dempo K, Ogawa K, Minase T (1975) Alpha-foetoprotein and early histological changes of hepatic tissue in DAB-hepatocarcinogenesis. Ann NY Acad Sci 259: 168-180

Popper H (1990) The relation of mesenchymal cell products to hepatic epithelial systems. In 'Progress in Liver Diseases', H Popper and F Schaffner (eds) Vol IX. WB Saunders Co. Philadelphia, pp27-38

Sarraf CE, McCormick CSF, Brown G, Price YE, Hall PA, Lane DP, Alison MR (1991) Proliferating cell nuclear antigen immunolocalization in gastrointestinal epithelia. Digestion 50: 85-91

Seki S, Sakaguchi H, Kawakita N, Yanai A, Kuroki T, Kobayashi K (1993) Analysis of proliferating biliary epithelial cells in human liver disease using a monoclonal antibody against DNA polymerase α. Virchow Arch A Pathol Anat 422: 133-143

Sell S (1990) Is there a liver stem cell? Cancer Res 50: 3811-3815

Senior PV, Critchley DR, Beck F, Walker RA, and Varley JM (1988) The localization of laminin mRNA and protein in the postimplantation embryo and placenta of the mouse: an *in situ* hybridisation and immunocytochemical study. Development 104: 431-446.

Solt DB, Medline A, Farber E (1977) Rapid emergence of carcinogen induced hyperplastic lesions in a new model for the sequential analysis of liver carcinogenesis. Am J Pathol 88: 595-618
Shiojiri N, Lemire JM, Fausto N (1991) Cell lineages and oval cell progenitors in rat liver development. Cancer Res 51: 2611-2620
Stoker M, Perryman M (1985) An epithelial scatter factor released by embryo fibroblasts. J Cell Sci 77: 209 - 203
Strain AJ (1993) Hepatocyte growth factor: another ubiquitous cytokine. J Endocrinol 137: 1-5
Tatematsu M, Kaku T, Medline A, Farber E (1985) Intestinal metaplasia as a common option of oval cells in relation to cholangiofibrosis in liver of rats exposed to 2-acetylaminofluorene. Lab Invest 52: 354-430
Thorgeirsson SS (1993) Hepatic stem cells. Am J Pathol 142: 1331-1333
Travis J (1993) The search for liver stem cells picks up. Science 259: 1829
Van Eyken P, Sciot R, Callea F, Ramaekers F, Schaart G, Desmet VJ (1990) A cytokeratin-immunohistochemical study of hepatoblastoma. Hum Pathol 21: 302-308
Van Eyken P, Sciot R, Desmet VJ (1988a) Intrahepatic bile duct development in the rat: a cytokeratin-immunohistochemical study. Lab Invest 59: 52-59
Van Eyken P, Sciot R, Desmet VJ (1989) A cytokeratin-immunohistochemical study of cholestatic liver disease: evidence that hepatocytes can express "bile duct-type" cytokeratins. Histopathology 15: 125-135
Van Eyken P, Sciot R, Paterson A, Callea F, Kew MC, Desmet VJ (1988b) Cytokeratin expression in hepatocellular carcinoma: an immunohistochemical study. Hum Pathol 19: 562-568
Van Eyken P, Sciot R, Van Damme B, de Wolf-Peeters C, Desmet VJ (1987) Keratin immunohistochemistry in normal human liver. Cytokeratin pattern of hepatocytes, bile ducts and acinar gradient. Virchows Arch A Pathol Anat 412: 63-72
Webber EM, Fitzgerald MJ, Brown PI, Bartlett MH, Fausto N (1993) Transforming growth factor-α expression during liver regeneration after partial hepatectomy and toxic injury, and potential interactions between transforming growth factor-α and hepatocyte growth factor. Hepatology 18: 1422-1431
Wijsman JH, Jonker RR, Keijzer R, Van de Velde CJH, Cornlisse CJ, Van Dierendonck JH (1993) A new method to detect apoptosis in paraffin sections: in situ end-labelling of fragmented DNA. J Histochem Cytochem 41: 7-12
Wright NA, Alison MR (1984) The biology of epithelial cell populations. Clarendon Press, Oxford

APOPTOSIS AND ITS ROLE IN HEPATIC CARCINOGENESIS BY NON-GENOTOXIC AGENTS

R.Schulte-Hermann, W.Bursch, B.Grasl-Kraupp, W.Huber, B.Ruttkay-Nedecky and A.Wagner
Institut fur Tumorbiologie-Krebsforschung
Borschkegasse 8a
A-1090 Vienna
Austria

INTRODUCTION

Until recently cell death occurring during carcinogenesis and in manifest tumors has been considered as passive, degenerative phenomenon due to toxicity and to insufficient supply of nutrients, oxygen etc. However, during the last decades increasing evidence has accumulated to show that apoptosis, a type of active cell death, occurs in neoplasia in addition to necrosis (Kerr et al.,1972, Bursch et al.,1984, Sarraf and Bowen 1988, Farber et al.,1972). During regression of hormone dependent tumors evidence of active cell death was also obtained (Lanzerotti et al.,1972, Gullino 1980, Szende et al.,1989, Bursch et al.,1991).

Apoptosis is a type of active cell death defined on a morphological and functional basis. Characteristically, it begins with condensation of the cytoplasm as well as of nuclear chromatin which abuts to the nuclear membrane; cell and nucleus fragment into apoptotic bodies which are then phagocytosed by neighboring cells. Lysosomes have not been found to play a major role during early stages of apoptosis although they are involved in the breakdown of

NATO ASI Series, Vol. H 88
Liver Carcinogenesis
Edited by G. G. Skouteris

apoptotic bodies within the phagocytosing cells, and no inflammatory responses are usually seen (Kerr 1971, Kerr et al., 1972, Bursch et al.,1985).

While apoptosis in recent years has become widely known in biological sciences, other types of active cell death are known to exist with features clearly different from condensation events during apoptosis. In 1973 Schweichel and Merker in embryonal tissues observed cell death characterized by an early activation of lysosomes and formation of autophagic vacuoles. Similar findings have been made in insect tissue during metamorphosis (Lockshin and Williams 1965, Lockshin and Beaulaton 1974, Zakeri et al., 1993) and more recently in neuronal tissue (Clarke 1990). The lysosomal or type II cell death is probably involved in regression of mammary tumors after withdrawal of estrogens (Lanzerotti et al.,1972) and has been demonstrated recently in the human mammary cancer cell line MCF7 during anti-estrogen treatment (Bursch et al., submitted b).

Active cell death is a genetically encoded process of cell destruction complementary but opposite to mitosis. It serves to eliminate excessive or damaged cells. Active cell death is under control of the growth regulatory network of the organism. An important functional property is that it is inhibited by tissue specific mitogens and can be triggered when these mitogens are withdrawn (Bursch et al., 1984, Bursch et al., 1992).

We became interested in active cell death during studies on the mechanisms of action of non-genotoxic liver carcinogens. These agents include phenobarbital, chlorinated hydrocarbons such as DDT, hexachlorocyclohexane, TCDD, steroid hormones e.g. ethinylestradiol and some progestins, peroxisome inducers such as clofibrate or nafenopin etc. These agents stimulate liver growth and hepatic functions such as drug metabolism or peroxisomal fatty acid oxidation. During long term treatment they promote the development of liver tumors from pre-existing initiated cells (Schulte-Hermann 1974, Schulte-Hermann 1985, Pitot and Sirica 1980). We found that these liver mitogens and non-genotoxic carcinogens are able to inhibit apoptosis in normal liver and in putative pre-stages of liver cancer (Bursch et al., 1984, see below).

Occurrence and characteristics of active cell death in the liver

In the rodent liver active cell death morphologically appears as the condensation type or apoptosis. It's incidence is very low in normal resting liver (in adult rats or mice approx. 2-4/ 10.000 hepatocytes). Prolonged fasting or severe starvation, mitogen withdrawal after a preceding period of mitogen treatment and certain toxins, increase the rate of apoptosis (Bursch et al.,1984, Grasl-Kraupp et al., submitted, Bursch et al.,1992, Columbano et al., 1985). The inhibition by liver mitogens was used to estimate the duration of the histologically visible part of apoptosis. For this purpose the kinetics of disappearance of apoptotic bodies in a regressing hyperplastic liver after CPA treatment was closely followed, and an average duration of 3 hours was found (Bursch et al.,1990). The estimate allows computation of the rate of apoptosis from histological counts. We obtained rates of apoptosis in regressing liver of 0.5%/h and in liver foci of 5%/h (Bursch et al., 1990).

Studies on thymocytes and lymphocytes have suggested that the characteristic condensation of nuclear chromatin during apoptosis results from activation of an endonuclease and degradation of chromatin into mono- and oligo-nucleosomal fragments (Wyllie 1980). Gel electrophoresis of these fragments may show so called "DNA ladders" which have been advocated as general biochemical markers of apoptosis. However, during liver apoptosis after mitogen withdrawal or after treatment with TGFβ1 (see below) no DNA ladders could be demonstrated; also *in vitro* primary hepatocytes stimulated to undergo extensive apoptosis by TGFβ1 did not reveal detectable oligo-nucleosomal fragments (Oberhammer et al.,1993a, 1993b, 1993c). Recent evidence suggests that during chromatin condensation in apoptosis much larger DNA fragments 50 and 300 kbp long are formed, probably by detachment of chromatin loops from the nuclear scaffold (Filipski et al.,1990, Brown et al.,1993, Oberhammer et al.,1993b). Further degradation of chromatin during later stages of apoptosis may yield DNA fragments of (oligo)nucleosomal or irregular size (depending on the cell type). It is of interest that DNA fragmentations can be demonstrated on tissue sections by nick translation *in situ* (Oberhammer et al., 1993c) but it should be noted that these tests are not specific for apoptosis or any type of active cell death.

Genes and factors regulating hepatocyte apoptosis

Studies with non-hepatocytic systems provided some information on genes involved in the control of apoptosis. In general, there appear to be genes which favour apoptosis such as myc, bax and p53, others are inhibitory, e.g. bcl 2 or its homologue in the nematode caenorhabditis elegans, ced 9 (Buttyan et al.,1988, Fanidi et al.,1992, Hockenberry et al.,1990, Hengartner et al.,1992, McDonnell 1993). Genes such as myc or fos, which are also important during preparation for cell replication, may serve to bring cells into a state of increased functional performance. In thymocytes, p53 seems to be involved in apoptosis induction by genotoxic chemicals and radiation. However, glucocorticoid induced thymocyte apoptosis does not appear to depend on p53 (Clarke et al.,1993, Lowe et al., 1993).

The TRPM 2 gene was originally discovered during regression by apoptosis of prostate after castration (Leger et al., 1988). Later it was also found expressed in the liver during involution of mitogen-induced hyperplasia but not during massive cell elimination after intoxication with CCl_4. Further studies revealed that the enhanced TRPM 2 expression during liver involution was not specifically associated with the few individual cells undergoing apoptosis but rather occurred in many or all hepatocytes. It was therefore concluded that the role of this gene may be a more general one namely to protect cells in the involuting organ from complement induced lysis (Bursch et al., submitted a).

Some growth regulating factors controlling apoptosis in the liver have been identified. We were interested in transforming growth factor β1 (TGFβ1) because it is a member of a family of peptides with growth inhibitory functions (Roberts et al.,1988). It is known to inhibit hepatocyte DNA synthesis *in vitro* and *in vivo* after partial hepatectomy (Carr et al.,1986, Russel et al.,1988), and its physiological function in regenerating liver has been considered to serve as a stop signal for the enhanced proliferative activity. We have found that apoptotic hepatocytes in regressing rat liver showed positive immuno-staining for TGFβ1 precursor protein (antibodies were kindly provided by M.Sporn, Bethesda) (Bursch et al.,1993). TGFβ1 also induced apoptosis *in vitro* when injected a few hours before sacrifice (Oberhammer et al.,1992, Oberhammer et al.,1993a). The apoptosis-inducing effect of TGFβ1 was potentiated in rats whose livers were in the state of regression due to mitogen withdrawal after a preceding mitogen treatment period. It was concluded that TGFβ1 preferentially acts upon cells already committed for apoptosis by pre-existing hyperplasia. Therefore TGFβ1 may not be a primary signal triggering apoptosis but rather serve as a permissive factor (Oberhammer et al.,1992). It was also of interest that some seemingly intact hepatocytes showed positive immunostaining for pre-TGFβ1 and that the incidence correlated with the incidence of apoptosis. This suggests that hepatocytes preparing for apoptosis start to exhibit pre-TGFβ1 a few hours before begin of chromatin condensation (Bursch et al.,1993). Therefore staining for pre-TGFβ1 may provide a marker for pre-apoptotic cells. Hepatocytes undergoing the necrotic type of cell death as occurring after intoxication did not stain for pre-TGFβ1. This stain may therefore be useful to discriminate between apoptosis and necrosis (Bursch et al.,1993).

Another member of the TGFβ family of polypeptides is activin. This peptide, like TGFβ1, was found to induce apoptosis in intact rat liver as well as in isolated hepatocytes. Activin showed only 1/10 the activity of TGFβ1, and its presence for 24 h was required to induce apoptosis (Schwall et al.,1993). These studies support the view that TGFβ1 and related peptides are involved in the regulation of the balance between cell proliferation and cell death in the liver. TGFβ1 has also been found to induce apoptosis in some hepatoma cell lines (Lin and Chou 1992, Fukuda et al.,1993). A further factor able to trigger hepatocyte apoptosis is tumor necrosis factor (TNF) the effect of which is known to be mediated by the receptor coded for by the oncogene fas (Ogasawara et al.,1993).

Apoptosis in stages of hepatocarcinogenesis

Carcinogenesis in the liver as in other organs is known to occur in stages including initiation, promotion, progression and tumor growth and spreading (Pitot and Sirica 1980, Schulte-Hermann 1985, Farber and Cameron 1980). Initiation is believed to be induced through genotoxic effects of carcinogens and may result from mutation of (a) crucial gene(s). In most human cancers it occurs for unknown reasons; also in rodent liver "spontaneous" initiation has been demonstrated (Schulte-Hermann et al.,1983, Kraupp-Grasl et al.,1991). Promotion results in formation and growth of clones of preneoplastic cells. Numerous studies have shown that replication is essential in these stages of carcinogenesis; we have found that likewise apoptosis is important.

Multistage carcinogenesis can be studied particularly well in rodent liver. In this organ putative initiated cells can be identified as single cells or as foci (clones?) of cells by histochemical and immunocytochemical means. We and others have found that these foci show much higher rates of DNA synthesis and mitosis and grow more rapidly that normal liver. Non-genotoxic carcinogens, such as phenobarbital, hexachlorocyclohexane and many others enhance the growth rate of these foci (Schulte-Hermann et al.,1981, Schulte-Hermann 1985, Schulte-Hermann et al.,1990). Surprisingly, detailed studies on the relation between cell replication (indicated by DNA synthesis, LI-measurement) and growth rate of the foci showed much less growth than expected. Conversely treatment with phenobarbital enhanced the growth rate of the foci but had little effect on the extent of DNA replication. These paradoxons were resolved by the discovery of apoptosis in foci (Bursch et al.,1984). Without promoter treatment there was quite a high rate of apoptosis in foci counterbalancing much of the enhanced cell replication. Treatment with phenobarbital inhibited apoptosis in foci and thereby favoured their growth. Obviously phenobarbital and similar tumor promoters act like trophic hormones or mitogens do on their target tissue (see above). Withdrawal of the promoter resulted in a dramatic increase of apoptosis in foci, explaining how promotion can be reversible (Bursch et al.,1984, Schulte-Hermann 1985, Schulte-Hermann et al.,1990). The occurrence of apoptosis in hepatic foci was subsequently confirmed and extended by other groups (Columbano et al.,1984, Garcea et al.,1989).

The occurrence of apoptosis in foci has important implications. One is that initiated cells become extinct with high probability without giving rise to observable foci (Moolgavkar et al.,1990, Luebeck et al.,1991). We have studied the fate of single putative initiated cells by using glutathione-S-transferase P (GST-P) as a marker. In untreated young rat liver the number of GST-P positive cells was almost zero. After a single dose of the genotoxic carcinogen N-nitrosomorpholin numerous single GST-P positive cells appeared, followed by formation of small GST-P foci. After a plateau phase the number of these single cells declined again to approx. 20% of the maximum, and there was evidence of apoptosis in some of these cells (Grasl-Kraupp, Wagner, Schulte-Hermann, manuscript in preparation). This experimental finding fits the mathematical prediction that the majority of initiated cells are extinguished (Moolgavkar et al.,1989, Luebeck et al.,1991).

In order to test whether initiated cells or cell clones can actually be eliminated we used aged rats in which substantial numbers of "spontaneous" preneoplastic foci were present. No genotoxic initiator was administered. The animals were subjected to food restriction for 3 months. The fasting regimen induced in normal liver and in putative preneoplastic foci were present. No genotoxic initiator was administered. The animals were subjected to food restriction for 3 months. The fasting regimen induced in normal liver and in putative preneoplastic foci a decrease of DNA replication and an increase of apoptosis, both effects being much more pronounced in foci than in normal liver. As a result, at the end of the fasting period the number of normal liver cells (indicated by total liver DNA) had decreased by 15%, while foci had declined by 90%. Subsequent treatment with the liver tumor promoter nafenopin supported that many initiated clones had been initiated (Grasl-Kraupp et al., submitted). These findings suggest that fasting shifts the balance between cell proliferation and apoptosis in both normal and preneoplastic liver tissue in favour of apoptosis, but more strongly in foci. This observation may provide a new explanation for the old finding in both experimental animals and human populations that restricted feeding can lower cancer rates while overfeeding may enhance them.

Finally, even in manifest tumors the rate of apoptosis determines to a large part the rate of growth (Kerr et al.,1972, Sarraf and Bowen 1988, Bursch et al.,1991). More recently we have observed in rat liver tumors (adenomas and carcinomas) induced by nafenopin not only higher rates of DNA replication as in normal liver but also higher rates of apoptosis. Many of these tumors are still dependent on the promoting effect of nafenopin: cessation of treatment resulted in a rapid regression of adenomas and even of some carcinomas. This correlated with an increase of DNA synthesis (Kraupp-Grasl, Ruttkay-Nedecky, Schulte-Hermann, manuscript in preparation). Thus even in liver carcinoma, cells may still depend on the presence of a promoter as a survival factor.

Conclusions

1) In preneoplastic and neoplastic liver cells the rates of cell replication and of apoptosis are both higher than in normal liver cells. Shifts in the balance favoring cell proliferation may result in tumor promotion and accelerate tumor development. Shifts favoring active cell death may result in elimination of initiated cells and regression of preneoplastic clones and even of tumors.

2) Non-genotoxic carcinogens which are liver tumor promoters act as mitogens and as survival factors for (pre)neoplastic cells. Their carcinogenic potency depends on the presence and number of initiated cells. This is in clear contrast with mutagenic effects by which genotoxic carcinogens may generate initiated cells. This is in clear contrast with mutagenic effects by which genotoxic carcinogens may generate initiated cells and produce cancer.

REFERENCES

Brown DG., Sun X.-M. and Cohen GM. (1993) Dexamethasone-induced apoptosis involves cleavage of DNA to large fragments prior to internucleosomal fragmentation. J.Biol.Chem. 268:3037-3039.

Bursch W., Fesus L. and Schulte-Hermann R. (1992) Apoptosis ("programmed" cell death) and its relevance in liver injury and carcinogenesis, in Tissue specific toxicology :Biochemical mechanisms, Dekant W. and Neuman HG. eds., Academic Press Ltd., London, 1992, chap.5.

Bursch W., Lauer B., Timmermann-Troisiener T., Barthel G., Schuppler J. and

Schulte-Hermann R. (1984) Controlled cell death (apoptosis) of normal and putative neoplastic cells in rat liver following withdrawal of tumor promoters. Carcinogenesis 5:53-58.
Bursch W., Taper HS., Lauer B. and Schulte-Hermann R. (1985) Quantitative histological and histochemical studies on the occurrence and stages of controlled cell death (apoptosis) during regression of rat liver hyperplasia. Virch. Arch. Cell Pathol. 50:153-166.
Bursch W., Paffe S., Putz B., Barthel G. and Schulte-Hermann R. (1990) determination of the length of the histological stages of apoptosis in normal liver and in altered hepatic foci of rats. Carcinogenesis 11:847-853.
Bursch W., Oberhammer F., Jirtle RL., Askari M., Sedivy R., Grasl-Kraupp B. and Purchio AF. (1993) Transforming growth factor β1 as a signal for induction of cell death by apoptosis. Br. J. Cancer 67:531-536.
Bursch W., Liehr JG., Sirbasku D., Putz B., Taper H. and Schulte-Hermann R. (1991) Control of cell death (apoptosis) by diethylstilbestrol in an estrogen dependent kidney tumor. Carcinogenesis 12:855-860.
Bursch W., Oberhammer F. and Schulte-Hermann R. (1992) Cell death by apoptosis and its protective role against disease. Tr.Pharmacol.Sci. 13:245-251.
Bursch W., Gleeson T., Kleine L. and Tenniswood M. Expression of testosterone-repressed prostate message (TRPM-2) mRNA during growth and regression of rat liver. Submitted a.
Bursch W., Kienzl H., Ellinger A. and Schulte-Hermann R. Cell death in cultured human mammary carcinoma cells (MCF-7) after treatment with the antiestrogens tamoxifen and ICI 164 384.Submitted b.
Buttyan R., Zakeri Z., Lockshin RA. and Wolgemuth D. (1988) Cascade induction of c-fos, c-myc and heat shock 70 K transcripts during regression of the rat vetral prostate gland. Mol. Endocrinol. 2:650-657.
Carr BI., Hayashi I., Branum EL. and Moses HL.(1986) Inhibition of DNA synthesis in rat hepatocytes by platelet-derived TGF-β1.Cancer Res. 46:2330-2334.
Clarke AR.,Purdie CA., Harrison DJ., Morris RG., Bird CC., Hooper ML. and Wyllie AH. (1993) Thymocyte apoptosis induced by p53-dependent and independent pathways. Nature 362: 849-852.
Clarke PGH. (1990) Developmental cell death: morphological diversity and multiple mechanisms. Anat. Embryol. 181:195-213.
Columbano A., Ledda-Columbano GM., Rao PM., Rajalakshmi S. and Sarma DSR. (1984) Occurrence of cell death (apoptosis) in preneoplastic and

neoplastic liver cells: A sequential study. Am.J.Pathol. 116:441-446.
Columbano A., Ledda-Columbano GM., Coni PP., Faa G., Liguori C., Santa Cruz G. and Pani P. (1985) Occurrence of cell death (apoptosis) during the involution of liver hyperplasia. Lab.Invest. 52:670.
Fanidi A., Harrington EA. and Evan GI. (1992) Cooperative intaraction between c-myc and bcl-2 proto-oncogenes. Nature 359:554-556.
Farber E., Verbin RS. and Lieberman M. (1972) Cell suicide and cell death in a symposium on mechanisms of toxicology. ALdrige N. ed. McMillan, New York, pp.163-173.
Farber E. and Cameron R. (1980) The sequential analysis of cancer development. Adv.Cancer Res. 31:125-225.
Filipski J., Leblanc J., Youdale T., Sikorska M. and Walker PR.(1990) Periodicity of DNA folding in higher order chromatin structures. The EMBO J. 9:1319-1327.
Fukuda K., Kojiro M. and Chiu JF. (1993) Induction of apoptosis by transforming growth factor β1 in the rat hepatoma cell line McA-RH7777: A possible association with tissue transglutaminase expression. hepatology 18:945-953.
Garcea R., Daino L., Pascale R., Simile MM., Puddu M., Frassetto S., Cozzolino P., Seddaiu MA., Gaspa L. and Feo F. (1989) Inhibition of promotion and persistent nodule growth by S-adenosyl-L-methionine in rat liver carcinogenesis: role of remodeling and apoptosis. Cancer Res. 49:1850-1856.
Grasl-Kraupp B., Bursch W., Ruttkay-Nedecky B., Wagner A., Lauer B. and Schulte-Hermann R. Food reduction eliminates preneoplastic cells through apoptosis and antagonizes carcinogenesis in rat liver, submitted.
Gullino PM. (1980) The regression process in hormone-dependent mammary carcinomas, in:Hormones and Cancer iacobelli et al., Eds., Raven Press, New York, pp.494-499.
Hengartner MO., Ellis RE. and Horvitz HR. (1992) Caenorhabditis elegans gene ced-9 protects cells from programmed cell death. Nature 356:494-499.
Hockenberry D., Nunez G., Milliman C., Schreiber RD. and Korsmeyer SJ. (1990) Bcl-2 is an inner mitochondrial membrane protein that blocks programmed cell death. Nature 348:334-336.
Kerr JFR. (1971) Shrinkage necrosis: a distinct mode of cellular death. J. Pathol. 105:13-20.
Kerr JFR., Wyllie AH. and Currie AR. (1972) Apoptosis: a basic biological

phenomenon with wide-ranging implications in tissue kinetics. J. Cancer 26:139-157.

Kraupp-Grasl B., Huber W., Taper H. and Schulte-Hermann R.(1991) Increased susceptibility of aged rats to hepatocarcinogenesis by the peroxisome proliferator nafenopin and the possible involvement of altered liver foci occurring spontaneously. Cancer Res. 51:666-671.

Lanzerotti LH. and Gullino PM. (1972) Activity and quantity of lysosomal enzymes during mammary tumor regression. Cancer Res. 32:2679-2685.

Leger J., Le Guellec R. and Tenniswood PR. (1988) Treatment with antiandrogens induces an androgen repressed gene in the rat vetral prostate. The prostate 13:131-142.

Lin JK. and Cou CK. (1992) In vitro apoptosis in the human hepatoma cell line induced by transforming growth factor β1. Cancer Res. 52:385-388.

Lockshin RA. and Williams CM. (1965) Programmed cell death. I. Cytology of degeneration in the intersegmental muscles of the Pernyi silk moth. J. Insect. Physiol. 11:123-133.

Lockshin RA. and Beaulaton J.(1974) Programmed cell death.cytochemical evidence for lysosomes during the noemal breakdown of the intersegmental muscles. J.Ultrastruct. Res. 46:43-62.

Lowe SW., Schmitt EM., Smith SW., Osborne BA. and Jacks T. (1993) p53 is required for radiation-induced apoptosis in mouse thymocytes. Nature 362:847-849.

Luebeck EG., Moolgavkar SH., Buchmann A. and Schwarz M. (1991) Effects of polychlorinated biphenyls in rat liver: Quantitative analysis of enzyme-altered foci. Toxicol.Appl. Pharmacol. 111:469-484.

Moolgavkar SH., Luebeck EG., De Gunst M., Port RE. and Schwarz M.(1990) Quantitative analysis of enzyme-altered foci in rat hepatocarcinogenesis experiments I:Single agent regimen. Carcinogenesis 11:1271-1278.

Oberhammer F., Pavelka M., Sharma S., Tiefenbacher R., Purchio TA., Bursch W. and Schulte-Hermann R. (1992) Induction of apoptosis in cultured hepatocytes and in regressing liver by transforming growth factor β1. Proc.Natl.Acad.Sci.U.S.A. 89:5408-5412.

Oberhammer F., Bursch W., Tiefenbacher R., Froschl G., Pavelka M., Purchio T. and Schulte-Hermann R. (1993a) Apoptosis is induced by transforming growth factor β1 within 5 hours in regressing liver without significant fragmentation of the DNA. Hepatology 18:1238-1246.

Oberhammer F., Wilson JW., Dive C., Morris ID., Hickman JA., Wakeling

AE., Walker PR. and Sikorska M. (1993b) Apoptotic death in epithelial cells: cleavage of DNA to 300 and/or 50 kb fragments prior to or in the absence of internucleosomal fragmentation. The EMBO J. 12:3679-3684.

Oberhammer F., Fritsch G., Schmied M., Pavelka M., Printz D., Purchio T. and Lassmann H. (1993c) Condensation of the chromatin at the membrane of an apoptotic nucleus is not associated with activation of an endonuclease. J.Cell Sci. 104:317-326.

Ogasawara J., Watanabe_Fukunaga R., Adachi M., Matsuzawa A., Kasugai T., Kitamura Y., Itoh N., Suda T. and Nagata S. (1993) Lethal effect of the anti-Fas antibody in mice. Nature 364:806-809.

Pitot HC. and Sirica AE. (1980) The stages of initiation and promotion in hepatocarcinogenesis. Biochim.Biophys.Acta 605:191-215.

Roberts AB., Thompson NL., Heine U., Flanders C. and Sporn MB. (1988) Transforming growth factor-β: possible roles in carcinogenesis. Br.J.Cancer 57:594-600.

Russell WE., Coffex RJ.Jr., Ouellette AJ., Moses HL. (1988) Type β transforming growth factor reversibly inhibits the early proliferative response to partial hepatectomy in the rat. Proc.Natl.Acad.Sci.U.S.A. 85:5126-5130.

Sarraf CE. and Bowen ID.(1988) Proportions of mitotic and apoptotic cells in a range of untreated experimental tumors. Cell Tiss. Kin. 21:45-49.

Schulte-Hermann R. (1974) Induction of liver growth by xenobiotic compounds and other stimuli. Crit.Rev.Toxicol. 3:97-158.

Schulte-Hermann R., Ohde G., Schuppler J. and Timmermann-Trosiener I. (1983) Promotion of spontaneous preneoplastic cells in rat liver as a possible explanation of tumor production by nonmutagenic compounds. Cancer Res. 43:839-844.

Schulte-Hermann R. (1985) Tumor promotion in the liver. Arch.Toxikol. 57:147-215.

Schulte-Hermann R., Timmermann-Trosiener I., Barthel G. and Bursch W. (1990) DNA synthesis, apoptosis and phenotypic expression as determinants of growth of altered foci in rat liver during phenobarbital promotion.Cancer Res. 50:5127-5135.

Schwall RH., Robbins K., Jardieu P., Chang L., Lai C. and Terrell TG. Activin induces cell death in hepatocytes in vivo and in vitro. Hepatology 18:347-356.

Schweichel J.-U. and Merker HJ.(1973) The morphology of various types of cell death in prenatal tissues. Teratology 7:253-266.

Szende B., Zalatanai A., Schally AV. (1989) Programmed cell death (apoptosis) in pancreatic cancers of hamsters after treatment with analogs of both luteinizing hormone-releasing hormone and somatostatin. Proc.Natl.Acad.Sci.U.S.A. 83:1643-1647.

Wyllie AH. (1980) Glucocorticoid-induced thymocyte apoptosis is associated with endogenous activation. Nature 284:555-556.

Zakeri ZF., Quaglino D., Latham T. and Lockshin RA. (1993) Delayed internucleosomal DNA fragmentation in programmed cell death. FASEB J. 7:470-478.

CELL PROLIFERATION AND CELL DEATH IN THE LIVER: PATTERNS, MECHANISMS AND MEASUREMENTS

Malcolm Alison, Matthew Golding and Catherine Sarraf
Deparment of Histopathology
Royal Postgraduate Medical School
Du Cane Road, London W12 ONN
UK

INTRODUCTION

Measurements of cell proliferation and cell death in the liver are important in development, normal cell turnover, wound healing and tumour progression. A large literature exists on the subject of cell proliferation, not least because of the feeling that the rate of growth of any human or animal tumour will reflect its subsequent behaviour pattern, though as persuasively illustrated many years ago by Steel (Steel 1977) this need not be the case. Before addressing the techniques it is necessary to consider the proliferative status of the liver. Figure1 depicts the cell cycle, and the very low proliferative rate in the normal liver strongly suggests that most hepatocytes are in the G_0 phase.

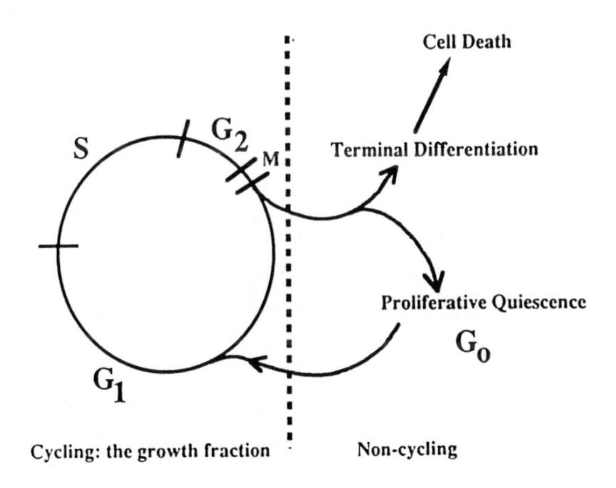

Figure 1. Possible phases in the life history of a cell.

NATO ASI Series, Vol. H 88
Liver Carcinogenesis
Edited by G. G. Skouteris

So any stimulus increasing the proliferative rate does so by triggering cells into the cell cycle rather than by shortening the cell cycle of already cycling cells. In young adults at least, there is no evidence that there are any hepatocytes which have terminally differentiated, which would render their re-entry into the cell cycle impossible (Wright and Alison 1984; Alison 1986).

The counting of mitotic figures has been the traditional method of assessing cell proliferation, and is derived by simple observation. However there are many pitfalls in the accurate measurement of the mitotic index (MI) (Quinn and Wright 1990), particularly in the liver, where tissue heterogeneity necessitates the counting of vast numbers of cells because the number of mitoses is at best moderate since mitosis occupies such a short part of the cell's life history (Wright and Alison 1984). The DNA synthetic (S) phase occupies a much longer part of the cell cycle, and the labelling index (LI) is thus a higher index than the MI and is therefore (theoretically) a more accurate measurement. The LI was first measured by tritiated thymidine, but with the widespread use of immunocytochemistry thymidine has been replaced by 5-bromo-2'-deoxyuridine (BrdUrd)-a thymidine analogue (Gratzner 1982).

Thymidine and BrdUrd labelling both suffer from the disadvantage of the DNA substrate having to be administered prior to detection, so precluding retrospective studies on archival material. This limitation was overcome by the introduction of the monoclonal antibody Ki-67 (Gerdes 1983), which recognizes a bimolecular complex of 345 and 395kD (Schluter et al 1993). The human Ki-67 antigen is expressed in all stages of the cell cycle and is thus a measure of the growth fraction, but is generally limited to human snap frozen material. The antigen termed proliferating cell nuclear antigen (PCNA) promised to replace Ki-67 because it could be detected in routinely processed tissue. PCNA, also termed cyclin, is a 36kD, non-histone nuclear protein required for DNA synthesis and repair. The prototype antibody, PC10 (Waseem and Lane 1990), labels cells in the S phase and beyond because of the long half-life of the protein (Sarraf et al 1991), but its value has recently been questioned because the antigen can be readily destroyed by prolonged fixation and equally well be expressed in 100% of cells by aggressive antigen retrieval procedures (Schwarting 1993). Whatever the relative merits and detractions of the various labelling methods they all rely on the microscopical assessment of tissue sections, a tedious and time-consuming exercise. Nevertheless they are all preferred to measurements of radiolabelled thymidine uptake, 'the DNA specific activity', which does not discriminate between cell types. This is totally inappropriate for a tissue such as the liver where up to 40% of cells in the liver are not hepatocytes.

In tissues such as the epidermis and gastrointestinal mucosa, cells proliferate, differentiate and die in a highly ordered and predictable fashion. In these tissues cells destined for terminal differentiation proceed through a short series of cell divisions to amplify their numbers, before permanently decycling from the proliferative cell cycle and migrating into an anatomically distinct differentiated compartment which heralds their imminent demise. Such a tissue hierarchy is a feature of the so-called **continually renewing** tissues such as gut, epidermis and bone marrow, while liver with its almost negligible rate of cell proliferation is classified as a **conditionally renewing** population - a population of essentially proliferatively quiescent cells which nevertheless retains the ability to re-enter the proliferative cell cycle after a stimulus such as cell damage. However, the wisdom of the emplacement of the liver in the conditionally renewal category has recently been challenged by Zajicek et al (1985) who suggested that hepatocytes migrated from the periportal area towards the terminal hepatic veins where they underwent apoptosis. However hard evidence for the apoptosis of perivenular hepatocytes as a crucial component in the maintenance of the steady state in a hepatic renewal system is still lacking. Though hepatic cell death is not a prominent feature of normal adulthood, it is certainly a major component in many hepatic diseases and thus needs to be correctly identified and quantified. In 1980, Wyllie et al proposed a new classification of cell death based on morphological criteria which separated the degradative reactions which occur after cell death due to generally severe perturbations (necrosis), from a gene-directed programme of cell death (apoptosis) caused by both physiological and pathological conditions. Generally speaking, necrosis follows from severe environmental trauma which either directly damages the plasma membrane (complement-induced cytolysis) or interferes with the generation of energy by blocking the synthesis of ATP (anoxia, ischaemia). As a consequence, energy-dependent ion-pumping mechanisms are impaired causing various ions to move down their concentration gradients across the plasma membrane; notably an entry of sodium and calcium and a loss of potassium. These ion movements result in a loss of plasma membrane volume control, causing an influx of water into the cell (cell and organelle oedema, formerly called hydropic degeneration), manifest as acute cell swelling or "cell ballooning". Initially such changes are reversible, but prevailing adverse conditions send the cell on a downward spiral particularly with the sustained increased levels of cytosolic calcium causing disruption of the cytoskeleton and activating membrane located degradative phospholipases and proteases. Together with a switch to anaerobic glycolysis, a decrease in intracellular pH, and a reduction of macromolecular synthesis, the affected cell dies with the accompanying rupture of organelles and the plasma membrane, manifest by light microscopy as coagulative necrosis. On the other hand, apoptosis (apo'-pto'-sis: Gr 'dropping off', as leaves from trees [Kerr et al 1972]) is not a passive phenomenon but is gene directed, usually requiring ongoing protein synthesis. Apoptosis is not new to liver pathology, but terms such as 'shrinkage necrosis', 'acidophil bodies' and 'Councilman bodies' have undoubtedly been used to describe what the modern histopathologist would regard as apoptotic hepatocytes (Alison and Sarraf 1992). The apoptotic cell death pathway has been conserved through much of evolution (Vaux et al 1993), and though the activation of a non-lysosomal endonuclease which cleaves double-stranded DNA into oligonucleosome length fragments is considered a key event in many apoptotic cells (Arends et al 1990), degradation of chromatin at a higher level of organisation may be a more proximate event (Tomei et al 1993).

Viewed by light microscopy, the difference between apoptotic cell death and necrosis is not always apparent. The large lakes of centrilobular necrosis caused by the archetypal hepatotoxin carbon tetrachloride present no problem in recognition, but when cell death occurs discretely, affecting single cells, how do we know if it is single cell necrosis or apoptosis? In the classical *in vitro* model of apoptosis in which thymocytes from young rats are exposed to glucocorticoids, large numbers of apoptotic cells can be harvested, and the 'ladder pattern' of degraded DNA products can be readily demonstrated (Arends et al 1990). To isolate apoptotic hepatocytes from an intact liver, and in sufficient numbers for DNA analysis would be a difficult task. The most practicable and reliable way of unequivocally distinguishing between apoptosis and necrosis is by transmission electron microscopy. Cell blebbing and organelle disruption are features of necrotic cells, while cell condensation and fragmentation into a number of membrane-bound 'apoptotic bodies', initially containing well preserved organelles and often, condensed chromatin, are hallmarks of apoptotic cell death. Much effort is being spent on developing a histological 'marker' of apoptosis, and Fesus and colleagues (Piacentini et al 1991; Fesus et al 1991) describe how tissue transglutaminase activity is enhanced in hepatocytes *in vitro*; such activity can be detected immunohistochemically using an anti-human tissue transglutaminase antibody. Very recently the task of identifying isolated apoptotic cells has been made easier by the development of a technique in which tissue sections are incubated with an appropriately labelled nucleotide triphosphate in the presence of DNA polymerase - *in situ* end-labelling (Wijsman et al 1993; Ansari et al. 1993). The technique relies on the presence of DNA strand breaks in apoptotic cells, and the labelled DNA is identified immunohistochemically. Thus studies of cell death require the correct recognition of the process involved as well as the appropriate quantification of the volume of tissue or percentage of cells affected. This paper describes steps for the accurate assessment of cell proliferation and cell death in liver tissue.

MATERIALS AND METHODS

Experimental Animals-Male albino Wistar rats weighing 190-220g were used for the partial hepatectomy study. This is the classical model of liver cell proliferation and thus the appropriate model to assess a proliferation marker. Cell proliferation in the liver was stimulated by the standard two-thirds partial hepatectomy involving removal of the two anterior liver lobes. Animals were killed in groups of three at 12, 18, 24, 48 and 72 hr after the operation, with each animal receiving a single ip injection of BrdUrd (Sigma Chemical Co, Poole, Dorset) at a dosage of 50mg/kg body weight 1 hr before death.

Tissue Processing-Samples of liver tissue 3-4 mm in thickness were taken immediately after death and fixed for either 24 hr in formal saline or 4 hr in Carnoy's solution (ideal for BrdUrd immunostaining) and then processed. All tissues were embedded in paraffin wax. Archival human material from Hammersmith Hospital had been routinely fixed in formal saline.

Immunocytochemistry-To visualize BrdUrd incorporation immunostaining was performed on sections of 5μm thickness which were mounted on poly L-lysine-coated glass slides. Immunostaining was carried out as previously described (Sarraf et al 1991). Briefly, sections were deparaffinized with CNP, hydrated in 100% ethanol, and endogenous peroxidase inactivated by immersing the slides in 98.4% methanol/1.6% hydrogen peroxide for 45min. DNA was denatured by immersing the slides in 2 N HCl for 1 hr, followed by rinsing in PBS. If the tissue had been fixed in formal saline, sections were then treated with 0.2% porcine pancreas trypsin (ICN Biochemical, ICl, U.K.) as recommended by Hayashi et al (1988). The optimum trypsinization time for clear subsequent immunostaining was 15min at 37°C; this step was not necessary if the tissue had been fixed in Carnoy's solution. Immunostaining was performed using a mouse monoclonal antibody to BrdUrd (Dako, U.K. Ltd), applied in a 1:20 dilution to the sections and left overnight at 4°C. Negative controls used PBS in place of the the primary antibody. All sections were then stained by the indirect immunoperoxidase method. The sections were covered with a 1:50 dilution of peroxidase conjugated rabbit, anti-mouse antibody (Dako) and were incubated for 1 hr at room temperature. The sections were washed three times in PBS for 5 min each. The peroxidase reaction was developed using the diaminobenzidine method, and sections were counterstained with Cole's haematoxylin for 1 min. To visualize PCNA, PC10 (NCL-PCNA, Novocastra Labs, Newcastle upon Tyne) was applied at a 1:20 dilution in PBS on the formalin fixed tissue, either for 2 hr at room temperature or overnight at 4°C. This was followed by a 1 hr incubation with a 1:200 dilution of biotinylated goat anti mouse immunoglobulin. After washing with PBS, a streptavidin complex (Dako Ltd.) was applied for 30 min before further washing and development with diaminobenzidine: all counterstaining was with Cole's haematoxylin.

Measurement of Labelling Indices- Labelled nuclei were counted in a uniform pattern using a square ocular eyepiece graticule and the LI for either BrdUrd incorporation or PCNA immunoreactivity was expressed as the percentage of labelled nuclei per 2000 nuclei; all cell counting was performed using a x40 magnification objective lens.

In addition to evaluation of routinely fixed and stained tissue sections for analysis of cell death, electron microscopy and *in situ* end labelling are useful adjuncts.

Electron microscopy-Tissue samples, not exceeding $1mm^3$ in volume were fixed in 2% glutaraldehyde for 2 hr. After washing in phosphate buffer, tissues were osmicated and dehydrated in acidified DMP before routine embedding in TAAB resin. 1µm sections were cut and stained with toluidine blue for observation at light microscope level and selection of relevant blocks, followed by ultrathin sections of approximately 100nm, collected on nickel grids and stained with uranyl acetate and lead citrate, for observation on a Philips CM-10 electron microscope.

In Situ End Labeling (ISEL)-This is a modification of the protocol from Wijsman et al (1993). Tissues were prepared as above and 3 µm sections cut and mounted onto silane coated slides. After blocking for endogenous peroxidase, sections were incubated for 1 hr at 37°C with a reaction buffer (50 mM Tris, 5 mM $MgCl_2$, 10 mM b- Mercaptoethanol, 0.005% BSA [Fraction V]) containing a final concentration of 0.01 mM each of dATP, dGTP, dCTP (Pharmacia); 0.3 μM Biotin-16-dUTP (Boehringer Mannheim) and 2.5 Units/ml of Klenow fragment of DNA polymerase 1 (Pharmacia). Sections were rinsed well with distilled water followed by PBS and then incubated with Strep.ABC/ HRP (Dako) for 30 min at room temperature. Sections were developed with DAB, conterstained and mounted as above.

Results

BrdUrd Labelling After Partial Hepatectomy

Figure 2 shows the changes in labelling in the hepatocytes (parenchymal cells) and littoral cells with time after resection. As expected there was little evidence of BrdUrd incorporation at 12 hr, but the subsequent customary wave of proliferative activity was confirmed. As expected the kinetics of the littoral cell response was different from that of hepatocytes, with an approximate time delay of 24 hr in terms of the initiation of cell proliferation in the littoral cells. No attempt was made to discriminate between the various types of littoral cells in this study, though the cells which line the sinusoids would principally be endothelial cells, Kupffer cells and Ito (fat-storing) cells. The precise temporal pattern of cell proliferation after hepatectomy has been well defined (Fabrikant 1968), but the clearly ephemeral nature of the induced response make it mandatory that one takes readings over a broad span of time if claims regarding alterations to the induced response are made. Another point to consider is that there is a distinct acinar heterogeneity in terms of proliferative activity, at least in the first 36 hr after resection, with most proliferative cells congregated around the points of entry of the afferent blood supply-in zone 1 of the acinus (Figure 3). Thus considerable care needs to be exercised in order not to bias the count to a particular region, in effect a large sample of cells needs to be counted!

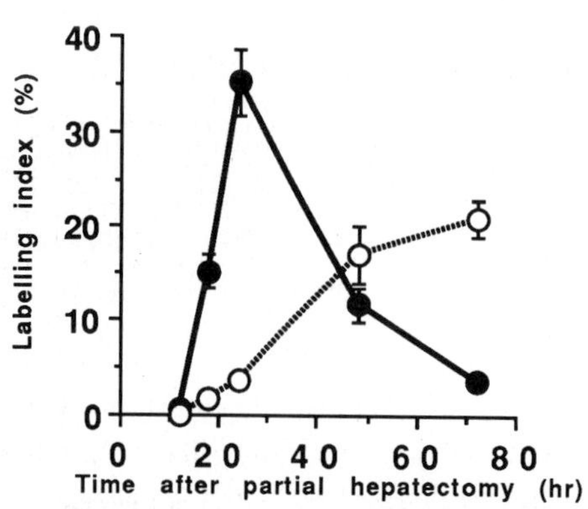

Figure 2. Changes in the labelling index in hepatocytes (O) and littoral cells (O) with time after resection in the rat liver.

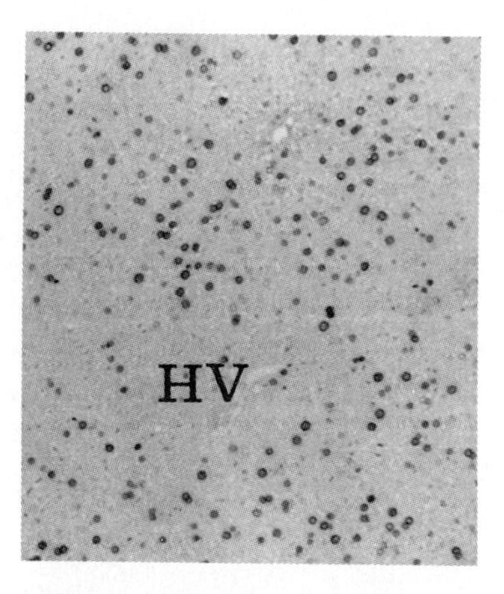

Figure 3. Photomicrograph of BrdUrd labelling at 24 hr resection, note relative lack of labelling around hepatic veins (HV).

When PCNA immunoreactivity is scored (Figure 4) the labelling index is higher than the BrdUrd LI because PCNA is still detected in post-DNA synthetic cells. This in itself does not negate the use of this marker, but perhaps more useful will be markers of the growth fraction such as the MIB 1 antibody (Gerdes et al 1992).

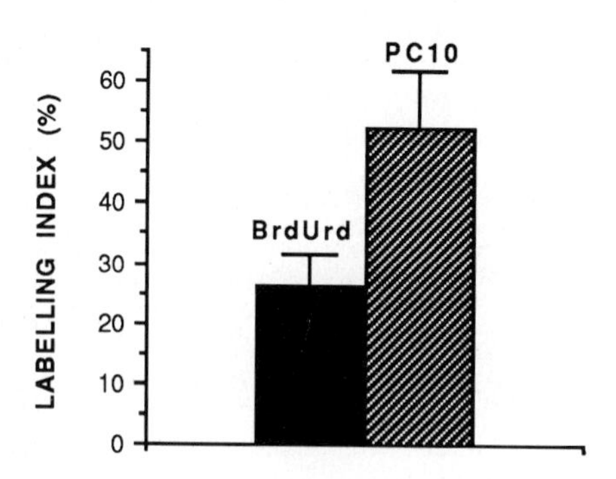

Figure 4. PCNA (PC10) and BrdUrd labelling indices in the rat liver at 24 hr after two-thirds partial hepatectomy. Mean $\pm$ SD, n=4.

Cell Death in the Liver

The quantification of cell death in the liver poses some of the same problems encountered in measuring proliferation, namely how many cells do we scan for an accurate estimate, and how do we compensate for tissue heterogeneity? In addition we have to be able to distinguish between necrosis and apoptosis. If we have large 'lakes' of necrosis, seen for example after carbon tetrachloride, then we have to able to estimate the volume of tissue affected, and this is normally done by point counting or image analysis. On the other hand, if apoptosis is involved then it may be more feasible to count the apoptotic index in much the same way we would count the labelling or mitotic index. At light microscope level it is not always possible to be unequivocal about recognising necrotic and apoptotic hepatocytes, it is not uncommon to find cell ballooning (a forerunner of necrosis) and small acidophilic hepatocytes together (Figure 5); in response to the same stimulus do some cells undergo necrosis while others succumb to an apoptotic death? A further point to consider is that apoptotic hepatocytes can fragment into a number of apoptotic bodies (Figure 6) making it difficult to count the number of apoptotic cells, particularly if adjacent cells have died by this means.

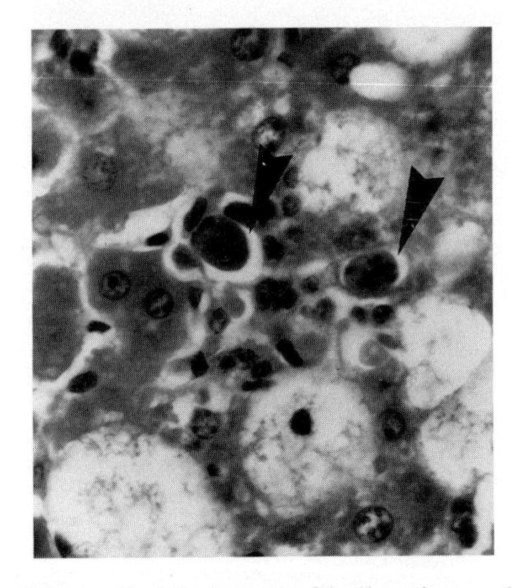

Figure 5. Admixture of ballooning and small acidophilic (?apoptotic) liver cells (arrows).

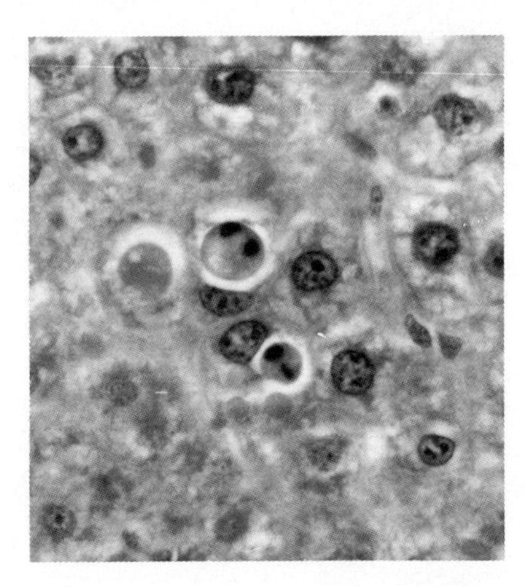

Figure 6. Three apoptotic bodies, probably phagocytosed by nearby hepatocytes, two contain chromatin

At light microscope level apoptotic cells are easily recognisable; they occur individually or in very small groups surrounded by normal healthy tissue. They have condensed, basophilic chromatin, a small amount of eosinophilic cytoplasm and are typically surrounded by a 'halo' (Figure 6): larger apoptotic cells break up and resultant fragments, apoptotic bodies, may or may not have nuclear components. At these magnifications it is impossible to distinguish with any certainty whether the halo is a result of retraction of the apoptotic cell or body from its neighbours or, as is more likely, that it is a phagocytic vacuole whose cytoplasmic boundaries are too fine to be observed. Apoptotic bodies are rapidly phagocytosed by macrophages or neighbouring epithelial cells (neutrophils play no part) and it seems that the vast majority are already within phagocytosing cells when they become apparent. Where discernible, the only clue to previous phagocytosis of an apoptotic body may be the distortion of the nucleus of the ingesting cell. Interestingly the apoptotic (Councilman) bodies seen in viral hepatitis do not seem to be readily phagocytosed (Figure 7), and ISEL can highlight apoptotic cells (Figure 8) though even necrotic cells have fragmented DNA which can incorporate added nucleotides.

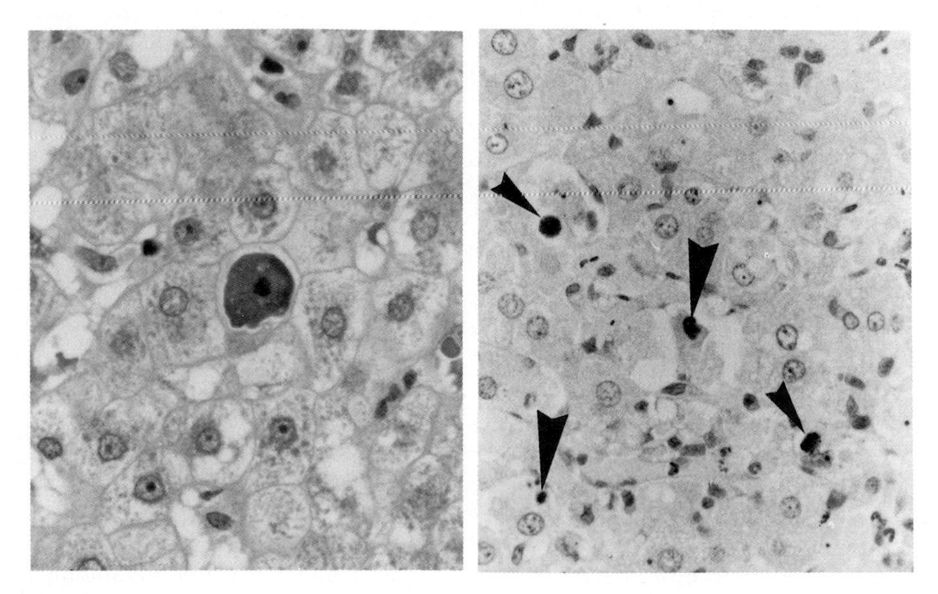

Figure 7. A clearly visible Councilman body with a well defined halo in human viral hepatitis

Figure 8. Apoptotic cells (arrows) visualised by ISEL seen by the peroxidase-DAB reaction product

When ultrastructural features are examined, apoptotic cells are found to have a remarkably stereotyped morphology. Whether free or within a phagocytic vacuole, the earliest intracellular changes are condensation of the cytoplasm and pushing of one or more dense crescents of chromatin to the periphery of the nucleus (Figure 9).

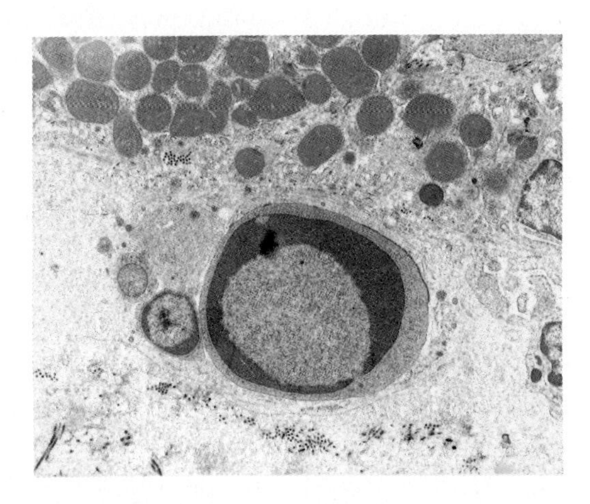

Figure 9. An apoptotic liver stem (oval) cell in the rat liver in a model of oval cell proliferation. Note the organelle-poor cytoplasm of the oval cell, a feature of an undifferentiated cell, and the prominent margination of the chromatin under the nuclear membrane, typical of an apoptotic cell.

The increased electron density of the cytoplasm is a result of cramming together of organelles and free ribosomes (Figure 10) due to the reduction in water content of the apoptotic hepatocyte. Integrity of organelles, however, is largely maintained and there is no initial swelling of the cisternae of endoplasmic reticulum. Mitochondria have none of the distension that is characteristic of necrotic cells, and even in the early period after the phagocytosis of apoptotic bodies by neighbouring cells they can look remarkably intact (Figure 10). The final stages of apoptotic disintegration within phagosomes are similar to degradation of any cellular material within secondary lysosomes, a process known as secondary necrosis (Figure 11): the ingested apoptotic fragments are only distinguishable from other heterophagic vacuoles by the presence of chromatin

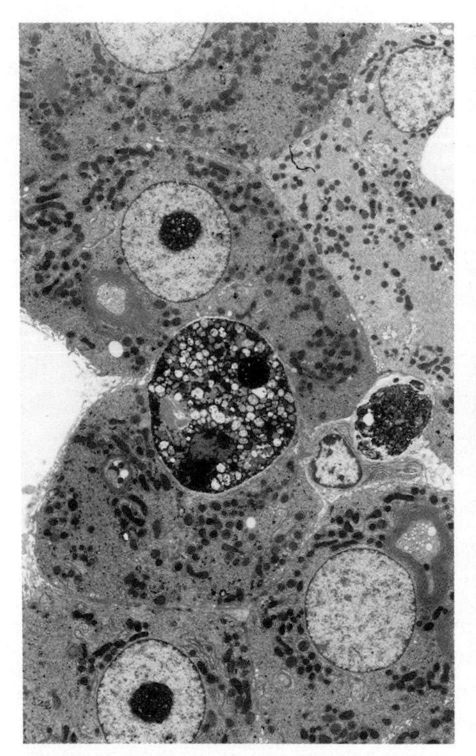

Figure 10. An apoptotic body being phagocytosed by a hepatocyte - note cramming together of organelles.

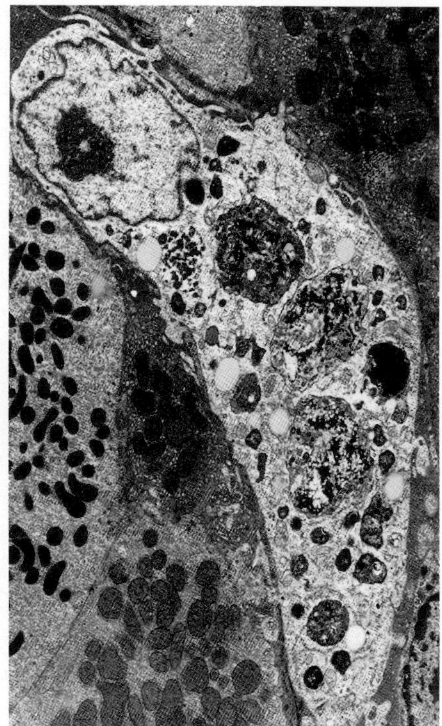

Figure 11. A Kupffer cell with many heterophagic vacuoles, some of which contain chromatin confirming their apoptotic ancestry.

DISCUSSION

The widespread use of immunocytochemistry has resulted in an explosion of renewed interest in assessing cell proliferation, particularly in clinical material, and reviews on the subject abound (Hall and Levison 1990; Quinn and Wright 1990; Linden et al.1992; Hall et al. 1992; Schwarting 1993). Methods based on thymidine and BrdUrd detection suffer from the limitation that these substances need to be first administered. The Ki-67 antigen originally could only be detected in frozen material, though a new monoclonal antibody, designated MIB 1, can detect the Ki-67 antigen in paraffin sections (Gerdes et al 1992). However the Ki-67 antigen is expressed throughout the cell cycle and a roughly similar distribution is claimed for PCNA. Additionally, PCNA staining needs to be carefully evaluated; different commercially available antibodies to PCNA yield different results, the antigen is very prone to destruction upon prolonged fixation, and antigen retrieval can cause staining in all cells (Hall et al. 1990; Leong et al 1993; Schwarting 1993). Thus there are a number of proliferation markers commercially available, either directed against S phase cells or even all cycling cells (the growth fraction, see Figure 1), though with the caveat that all immunocytochemical procedures are subject to the vaguaries of appropriate fixation, fixation time etc.

Cell proliferation is not only an essential component in any regenerative response to injury, as enhanced proliferation also plays a role in the expansion of premalignant cell populations. In addition, cell proliferation is considered pivotal to the initiation process, since genetic damage must be replicated prior to DNA repair. Hence in most models of liver carcinogenesis there is first damage to the liver to stimulate a regenerative response, before exposing the liver to a genotoxic agent. In man also, the processes of cell damage, regeneration and neoplasia seem to be inextricably linked since hepatocellular carcinoma is generally associated with agents which damage liver cells either directly (alcohol) or indirectly (hepatitis B virus). Thus cell damage, like cell proliferation features prominently in liver disease.

Liver cell damage features strongly in many liver diseases including those that are immunologically mediated. This type of injury results from cytotoxic lymphocytes attacking antigens exposed on the surface of cells in the liver, and would include diseases such as acute hepatitis (viral and drug-induced), chronic active hepatitis, primary biliary cirrhosis and possibly alcoholic liver disease. Viral infections of the liver may be due to the hepatotropic viruses (the hepatitis viruses) or to viruses which affect many organs including the liver. The fundamental lesion of acute viral hepatitis is illustrated in Figure 7. These, the Councilman bodies first described in yellow fever are undoubtedly apoptotic hepatocytes. Certain drugs can also lead to a liver picture mimicing viral hepatitis, often with an intermediate metabolite acting as a hapten, combining with a normal membrane antigen, rendering it antigenic. Sensitized lymphocytes are produced and, on re-exposure a delayed hypersensitivity reaction ensues and hepatocyte cell death occurs. Drugs falling into this category include probably phenytoin, chlorpromazine and β-aminosalicyclic acid.

Many hepatotoxic chemicals can be classified as either Type I (their effects are predictable, dose dependent, occur in most individuals, readily reproducible in animals), or Type II, where the reactions are not predictable, only seen after extensive screening and can rarely if at all be reproduced in experimental animals. Hepatotoxic compounds can be further divided into direct hepatotoxins (eg CCl_4) which primarily injure the structural basis of hepatic metabolism, and indirect hepatotoxins (eg paracetamol and galactosamine) which interfere with a specific metabolic pathway; thus structural injury is secondary to the metabolic lesion. Furthermore most intrinsic hepatotoxins cause zonal cell death, most commonly in zone 3 where this reflects the concentration of the enzyme system responsible for the conversion of the agent to its hepatotoxic metabolite. For example, CCl_4 is a well known hepatotoxic solvent whose toxicity crucially depends on a cytochrome P450-dependent monooxygenase which is located in the smooth endoplasmic reticulum of perivenular hepatocytes. A free radical ($CCl_3^{\bullet}$) may be responsible for the peroxidation of microsomal lipids leading to structural damage, though this may be secondary to the cells suffering from 'oxidative stress' whereby glutathione (GSH) is overwhelmingly oxidized to its disulphide (GSSG) and GSH is not available to act as a reductant in the metabolism of free radicals (Fawthrop et al 1991). Paracetamol is a widely used analgesic which is hepatotoxic at high dose, again producing centrilobular necrosis because it is oxidised to its toxic metabolite NABQI by the cytochrome P450 system; NABQI may exert much of its cytotoxicity through covalent binding to proteins (Boobis et al 1992). Thioacetamide causes centrilobular necrosis in the rat liver (Ledda-Columbano et al 1991), but very acutely (within 3 hours) some hepatocytes may undergo apoptosis. Galactosamine depresses the uracil nucleotide-dependent biosynthesis of macromolecules, and causes diffuse cell death (probably apoptosis, see Figure 8) throughout the liver parenchyma. The degree of injury is very much dose-dependent, and so galactosamine is widely used in models of fulminant hepatic failure.

Chronic liver injury can lead to cirrhosis and the commonest cause of cirrhosis in Europe and the USA is long-term alcohol abuse. Ethanol is oxidised in the liver to acetaldehyde and acetate, and the earliest and most common pathological manifestation is alcoholic steatosis (fatty change), which can occur throughout the liver but which is often predominantly in zone 3. This leads on to alcoholic hepatitis with ballooning degeneration, liver cell necrosis and an influx of inflammatory cells into the affected area. Of course, hepatocellular carcinoma usually develops in a cirrhotic setting. A wide range of drugs and environmental pollutants induce hyperplasia and/or hypertrophy of hepatocytes, at least in rats. This growth can be purely 'additive' in that it is not preceded (initiated) by a necrogenic event. However, once the stimulus for hyperplasia has been removed, the liver rapidly reverts to its normal size. In almost all cases this is achieved by an increased incidence of apoptosis, and this occurs in rat livers after lead nitrate injection (Columbano et al 1985) or cyproterone acetate treatment (Alison et al 1987). In the latter case, cells which don't participate in the induced hyperplasia appear to be selectively deleted and TGF-β is believed to have a role in the elimination process (Bursch et al 1985 ; 1993).

There is no doubt that many carcinogens, particularly the nitrosamines, cause both apoptosis and necrosis, predominantly in the centrilobular region (Pritchard and Butler 1989; Ray et al 1992). In some experimental models of liver carcinogenesis this induced necrosis is pivotal to the generation of dysplastic foci, since regeneration is initiated at the time of carcinogen exposure (Alison and Hully 1991). Elevated levels of apoptosis are often found in these foci (Columbano et al 1984) and indeed some tumour promoters appear to stimulate the growth of these foci by preventing apoptosis (Gerbracht et al 1990). In conclusion there is no doubt that cell proliferation and cell death play vital roles in liver disease, particularly neoplasia, making it imperative that these two reactions are correctly identified and quantified.

REFERENCES

Alison MR (1986) Regulation of liver growth. Physiol Rev 66: 499-541

Alison MR, Hully JR (1991) Stimulus dependent phenotypic diversity in the resistant hepatocyte model. In: Columbano A, Feo F, Pascale R, Pani P (eds). Chemical carcinogenesis 2. Modulating factors. New York: Plenum, pp563-577

Alison MR, Sarraf CE (1992) Apoptosis: - a gene-directed programme of cell death. J R Coll Phys 26: 25-35

Alison MR, Wilkins MJE, Walker SM, Hully JR (1987) Cell population size control in the rat liver: the response of hepatocytes in various proliferative states to a mitogenic stimulus. Epithelia 1: 53-64

Ansari B, Coates PJ, Greenstein BD, Hall PA (1993)*In situ* end-labelling detects DNA strand breaks in apoptosis and other physiological and pathological states. J Pathol 170: 1-8

Arends MJ, Morris RG, Wyllie AH (1990) Apoptosis. The role of the endonuclease. Am J Pathol 136: 593-608

Boobis AR, Fawthrop DJ, Davies DS (1992) Mechanisms of cell injury. In: Oxford Textbook of Pathology. Vol 1 (eds JOD McGee, PG Isaacson, NA Wright) Oxford University Press, Oxford. pp181-193

Bursch W, Oberhammer F, Jirtle RL et al (1993) Transforming growth factor-β1 as a signal for induction of cell death by apoptosis. Br J Cancer 67: 531-536

Bursch W, Taper HS, Lauer B, Schulte-Hermann R (1985) Quantitative histological and histochemical studies on the occurrence and stages of controlled cell death (apoptosis) during regression of rat liver hyperplasia. Virch Arch [Cell Pathol] 50: 153-166

Columbano A, Ledda-Columbano GM, Coni PP et al (1985) Occurrence of cell death (apoptosis) during the involution of rat liver hyperplasia. Lab Invest 52: 670-673

Columbano A, Ledda-Columbano GM, Rao PM, Rajalakshmi S, Sarma DSR (1984) Occurrence of cell death (apoptosis) in preneoplastic and neoplastic liver cells. Am J Pathol 116: 441-446

Fabrikant JI (1968) The kinetics of cellular proliferation in the regenerating liver. J Cell Biol 36:551-565

Fawthrop DJ, Boobis AR, Davies DS (1991) Mechanisms of cell death. Arch Toxicol 65: 437-444

Fesus L, Davies JA, Piacentini M (1991) Apoptosis: molecular mechanisms in programmed cell death. Eur J Cell Biol 56:170-177

Gerbracht U, Bursch W, Kraus P et al (1990) Effects of hypolipidaemic drugs nafenopin and clofibrate on phenotypic expression and cell death (apoptosis) in altered foci of rat liver. Carcinogenesis 11: 617-624

Gerdes J, Becker MHG, Key G (1992) Immunohistological detection of tumour growth fraction (Ki-67 antigen) in formalin fixed and routinely processed tissues. J Pathol 168: 85-86

Gerdes J, Schwab U, Lemke H, Stein H (1983) Production of a monoclonal antibody reactive with a human nuclear antigen associated with cell proliferation. Int J Cancer 31:13-20

Gratzner HG (1982) Monoclonal antibody to 5-bromo and 5-iododeoxyuridine: a new reagent for detection of DNA replication. Science 218: 474-475

Hall PA, Levison DA (1990) Review: Assessment of cell proliferation in histological material. J Clin Path 43:184-191

Hall PA, Levison DA, Woods AL, Yu CCW, Kellock DB, Watkins JA, Barnes DM, Gillett CE, Camplejohn R, Dover R, Waseem NH, Lane DP (1990) Proliferating cell nuclear antigen (PCNA) immunolocalization in paraffin sections: an index of cell proliferation with evidence of deregulated expression in some neoplasms. J Pathol 162: 285-294

Hall PA, Levison DA, Wright NA (eds) (1992) Assessment of cell proliferation in clinical practice. Springer-Verlag, London UK

Hayashi Y, Koike M, Matsutani M, Hoshino T (1988) Effects of fixation time and enzymatic digestion on immunohistochemical demonstration of bromodeoxyuridine in formalin-fixed, paraffin embedded tissue. J Histochem Cytochem 36: 511-514

Kerr JFR, Wyllie AH, Currie AR (1972) Apoptosis: a basic biological phenomenon with wide-ranging implications in tissue kinetics. Br J Cancer 68: 239-257

Ledda-Columbano GM, Coni P, Curto M et al (1991) Induction of two different modes of cell death, apoptosis and necrosis, in rat liver after a single dose of thioacetamide. Am J Pathol 139: 1099-1109

Leong ASY, Milos J, Tang SK (1993) Is immunolocalisation of proliferating cell nuclear antigen (PCNA) in paraffin sections a valid index of cell proliferation? Appl Immunohistochem 1:127-135

Linden MD, Torres FX, Kubus J, Zarbo RJ (1992) Clinical application of morphologic and immunocytochemical assessments of cell proliferation. Am J Clin Pathol 97(Suppl1): S4-S13

Piacentini M, Autuori F, Dini L et al (1991) 'Tissue' transglutaminase is specifically expressed in neonatal rat liver cells undergoing apoptosis upon epidermal growth factor stimulation. Cell Tissue Res 263: 227-235

Pritchard DJ , Butler WH (1989) Apoptosis- the mechanism of cell death in dimethylnitrosamine-induced hepatotoxicity. J Pathol 158: 253-260

Quinn CM, Wright NA (1990) The clinical assessment of proliferation and growth in human tumours: evaluation of methods and applications as prognostic variables. J Pathol 160:93-102

Ray SD, Sorge CL, Kamendulis LM, Corcoran GB (1992) Ca^{2+} activated DNA fragmentation and dimethylnitrosamine-induced hepatic necrosis: effects of Ca^{2+} endonuclease and poly (ADP-ribose) polymerase inhibitors in mice. J Pharmacol Exp Ther 263: 387-394

Sarraf CE, McCormick CSF, Brown G, Price YE, Hall PA, Lane DP, Alison MR (1991) Proliferating cell nuclear antigen immunolocalization in gastro-intestinal epithelia. Digestion 50: 85-91

Schluter C, Duchrow M, Wohlenberg C, Becker MHG, Key G, Flad H-D, Gerdes J (1993) The cell proliferation-associated antigen of antibody Ki-67: a very large, ubiquitous nuclear protein with numerous repeated elements, representing a new kind of cell cycle-maintaining proteins. J Cell Biol 123:513-522

Schwarting R (1993) Little missed markers and Ki-67. Lab Invest 68:597-599

Steel GG (1977) Growth kinetics of tumours. Clarendon Press, Oxford

Tomei LD, Shapiro JP, Cope FO (1993) Apoptosis in C3H/10$^1/_2$ mouse embryonic cells: evidence for internucleosomal DNA modification in the absence of double-strand DNA cleavage. Proc Natl Acad Sci 90: 853-857

Vaux DL (1993) Toward an understanding of the molecular mechanisms of physiological cell death. Proc Natl Acad Sci 90: 786-789

Waseem NH, Lane DP (1990) Monoclonal antibody analysis of the proliferating nuclear antigen (PCNA). J Cell Sci 96:121-129

Wijsman JH, Jonker RR, Keijzer R, Van de Velde CJH, Cornlisse CJ, Van Dierendonck JH (1993) A new method to detect apoptosis in paraffin sections: in situ end-labelling of fragmented DNA. J Histochem Cytochem 41: 7-12

Wright NA, Alison MR (1984) The biology of epithelial cell populations. Clarendon Press, Oxford

Wyllie AH, Kerr JFR, Currie AR (1980) Cell death: the significance of apoptosis. Int. Rev Cytol 68: 251-306

Zajicek G, Oren R, Weinreb JR (1985) The streaming liver. Liver 5: 293-300

PRENEOPLASTIC CHANGES DURING NON-GENOTOXIC HEPATOCARCINOGENESIS

Christopher J.Powell[1], Beatrice Secretan and Suzanne Cottrell
DH Department of Toxicology
St.Bartholomew's Hospital Medical College
Dominion House
London EC1 7ED
United Kingdom

INTRODUCTION

Potential new drugs undergo *in vitro* genotoxicity tests early in their development program and those found to be positive are usually screened out.Hence, compounds tested in preclinical carcinogenicity studies are usually non-genotoxic. The non-genotoxicity of a compound is operationally defined by 1)its inability to cause mutations in several bacterial chromatid exchange and 2) its inability to produce covalent DNA adducts or induce DNA repair (measured as unscheduled DNA synthesis). Genotoxic agents are believed to directly affect the genes controlling cell growth and differentiation or the regulatory sequences that control them, and pose a significant risk at any level of exposure. Non-genotoxic carcinogens on the other hand act via indirect or epigenetic mechanisms related to alterations of gene expression.Their effects are believed to be thresholded and to occur only at high or cytotoxic concentrations. Thus, the significance of non-genotoxic carcinogens in terms

[1]to whom correspondence should be addressed

NATO ASI Series, Vol. H 88
Liver Carcinogenesis
Edited by G. G. Skouteris

of human safety is highly disputed and there is real need for a better understanding of the underlying mechanisms. Non-genotoxic carcinogens include mitogenic agents such as TSH, peroxisome proliferators such as fibrate and hypolipaedemic drugs, synthetic estrogens, phorbol esters and phenobarbitone. This paper reports the localisation and characterisation of early changes occurring in the livers of Han-Wistar rats given methapyrilene from 6 weeks to 6 months, includes a comparison with an hepatotoxin (paracetamol) and a genotoxic hepatocarcinogen (aflatoxin B1), and reviews the toxic and carcinogenic effects of methapyrilene and the possible underlying mechanisms.

Toxicity and Carcinogenicity of methapyrilene

Methpyrilene, a competitive H_1 histamine antagonist and sleep aid, was widely sold over-the-counter for over 20 years until Lijinsky discovered its carcinogenic potential in F344 rats. The highest tolerated dose.in rats was established as 1.000 ppm (Lijinsky et al.,1980). At this dose, it induced hepatocellular carcinomas with almost 100% incidence within 43 to 64 weeks, as well as cholangioadenomas and carcinomas (Ohshima et al., 1984). At a dose of 250 ppm, 98% of the rats developed hepatocellular carcinomas and/or hepatic nodules within 110 weeks, whereas at 125 ppm, 40% of the rats had nodules but no carcinomas were found (Lijinsky, 1894). It was thus proposed that the threshold level of carcinogenicity was 200 ppm.

More interestingly, methapyrilene was not carcinogenic in mice (Brennan & Creasia, 1982), Syrian hamsters or guinea-pigs at the highest tolerated dose (Lijinsky et al.,1983). Neither were any of the structurally similar or biologically related compounds when tested at similar or higher doses (Lijinsky and Kovatch, 1986; Habs et al., 1986; Lijinsky, 1984).

Before Lijinsky discovered its carcinogenicity, other groups had shown that methapyrilene was not mutagenic in Ames' test (Andrew et al., 1980) and was unable to transform Syrian hamster embryo cells in culture (Pienta, 1980). Its non-genotoxicity was then confirmed by many *in vitro* and *in vivo* short-term assays (Oberly et al., 1993; Iype et al., 1982; Lijinsky & Yamashita, 1988; Probst & Neal, 1980; McQueen & Williams 1981; Budroe et al., 1984). Several groups have studied the combined effects of other hepatocarcinogens with methapyrilene. Its potential as initiator and promoter of hepatocarcinogenicity was assessed using the detection of GST-P positive foci as an end point. After partial hepatectomy, methapyrilene was given as initiator (followed by phenobarbitone promotion) or as promoter (preceded by the genotoxic carcinogen nitrosodiethylamine) and appeared to be a weak initiator when given at doses higher than 200 ppm, confirming the threshold level of carcinogenicity. A choline-deficient diet enhanced both the initiating and promoting effects of methapyrilene (Perera et al., 1987).Lijinsky and co-workers (1992) showed an interesting triple cooperative or synergistic effect between nitrosodiethylamine (NDEA), phenobarbital (PB) and methapyrilene (MP) (Table I). The biological basis of this intriguing finding is not presently understood.

Table I. Tumors in rats given NDEA, MP, PB or PH (no. of rats with tumors)

Group		Median week of death	Effective no. of rats	Liver			
				Adenoma	Carcinoma	Carc or aden	Hemangiosarcoma
1	NDEA	94	21	2	–	2	–
2	MP	103	21	1	–	1	–
3	NDEA + MP + PB	88	21	13	10**	17**	4
4	NDEA + MP	88	24	8	2	10**	1
5	NDEA + PB	94	24	3	–	3	1
6	MP + PB	107	24	1	–	1	–
7	PB	102	24	–	–	–	–

**$P < 0.01$

On balance, methapyrilene is widely considered to be an epigenetic or non-genotoxic carcinogen. Both the species and structural specificities give the compound a particular interest, but the mechanism by which the progressive genetic change that characteristics neoplasia occurs remains unexplained.

Possible mechanisms involved

The metabolism of methapyrilene is incompletely reported and very complex. Many metabolites have been detected in both *in vivo* and *in vitro* systems, but the complete pathways are yet to be elucidated and a potentially highly reactive intermediate to be identified. The metabolism is NADPH-dependent, implicating mixed functions oxidases and/or FAD monooxygenases (Lampe & Kammerer, 1987; Ziegler et al.,1981). The specific enzyme systems involved remain uncharacterised, as do the effects of enzyme-modulation on methapyrilene's toxicity and carcinogenicity. Studies have shown both qualitative and quantitative differences between rat and other species in the metabolism of methapyrilene (Lampe & Kammerer, 1990; Kammerer & Schmitz, 1987; Kelly et al., 1992). These differences are likely to be at least part of the reason for the species-specificity of the compound as a carcinogen, but none are yet sufficient to exclusively account for this.

As with other liver damaging agents, methapyrilene is found to decrease membrane-bound enzyme levels, including cytochrome P450, in liver microsomes (Wrighton et al.,1991). However, the IA1 isoform of cytochrome P450 (Wrighton et al., 1991) and the microsomal epoxide hydrolase (Graichen et al., 1985) are increased, and this has sometimes been associated with increased formation of reactive epoxides. In the same study, an increase in cytosolic DT-diaphorase was observed, which could reflect the formation of a pyridol metabolite of methapyrilene. However, all these mechanisms are hypothetical.

Methapyrilene strongly induces proliferation of short and bizarrely shaped mitochondria in periportal hepatocytes *in vitro* and *in vivo* (Reznik-Schuller & Gregg, 1983; Iype et al., 1985). The compound binds to cytosolic liver proteins (Lijinsky & Muschlik, 1982) with a maximum 6 h after exposure (Reznik-Schuller & Lijinsky, 1981), which results in charged-shift alterations of more than 100 proteins by phosphorylation, glycosylation or carbamylation (Anderson et al., 1992; Zeindl-Eberhart & Rabes, 1992), only some of which are of mitochondrial origin (Richardson et al., 1994). Methapyrilene-induced mitochondrial proliferation does not appear to have been observed in other species (Lijinsky et al., 1983) or with structurally similar compounds (Reznik-Schuller, 1982), suggesting that it might play a role in the carcinogenic process. The cause of this unusual phenomenon is inadequately explored and comparatively few compounds have this potential. One distinct possibility is that mitochondrial proliferation is a secondary adaptative response to inhibition or normal intracellular energy states, perhaps as a consequence of uncoupled oxidative phosphorylation (Copple et al., 1992).

Methapyrilene also binds to mitochondrial membranes and induces lipid peroxidation in rat livers (Perera et al., 1985). However, lipid peroxidation was also induced with analog compounds and is therefore not relevant (Hernandez & Lijinsky, 1989). Hernandez et al. (1989) have observed an increase in liver

DNA methylation after 20 weeks of methapyrilene treatment, but they were not able to show any higher gene expression (Hernandez et al., 1991). More recently, Copple et al. (1992) have shown that methapyrilene decreases mitochondrial DNA and protein synthesis both in rats and mice, thus appearing to exclude a possible involvement in the carcinogenicity of methapyrilene.

Preneoplastic Effects of methapyrilene

A short-term experiment was conducted in which Han-Wistar rats were given 1.000 ppm methapyrilene in the diet for up to 12 weeks, in order to examine the early preneoplastic changes caused by a non-genotoxic compound. Histology and immunocytochemistry showed that methapyrilene caused marked proliferation of oval cells and bile ducts with interlobular bridging, or in the most severe cases, effacement of normal liver architecture by ductal expansion (Fig.1a). A marked increase in reticulin accompanied oval cell proliferation (Fig.1b). Within the areas of oval cells, distinctive "trapped" basophilic hepatocytes were detectable. Active periportal necrosis and compensatory regeneration of hepatocytes were also observed, and in some case the regenerative hyperplasia was clearly nodular. Treatment with methapyrilene also affected the nucleus:karyomegaly, often accompanied by cytomegaly and an increased number of binucleated cells, were noted (Fig.1c).Immunochemistry for GST-P revealed a strong positivity in foci of altered hepatocytes with a weaker but regular zonal pattern of enzyme induction in periportal areas (Fig.1d). This zonal specificity of GST-P is

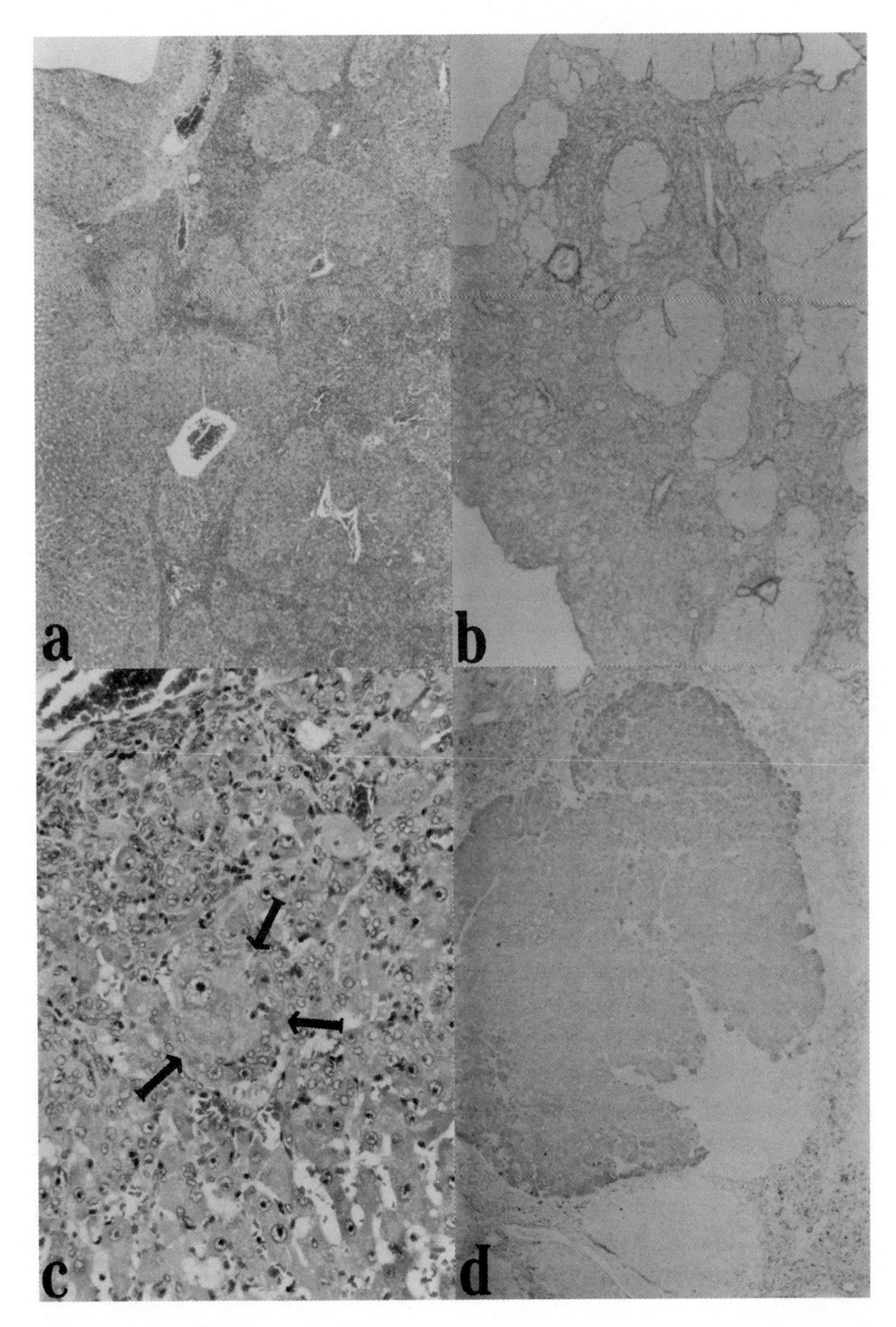

Fig. 1. (see page 223 for caption)

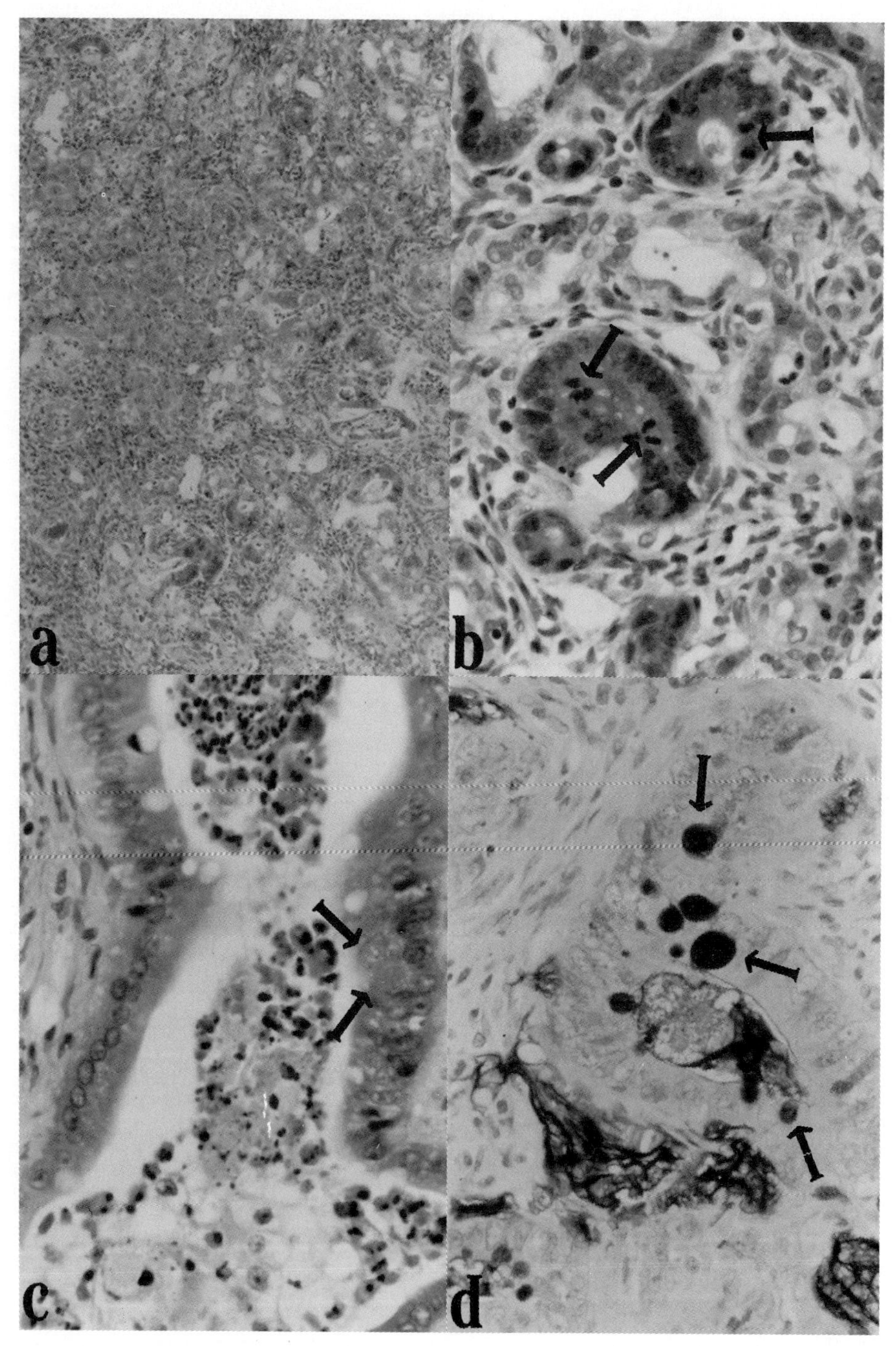

Fig. 2. (see page 223 for caption)

Fig. 1: Liver following 12 weeks administration of 1,000 ppm methapyrilene. A and B (x13): nodular-like regenerative hepatocyte areas are separated by solid areas and dissecting strands of proliferating bile ducts, oval cells and fibrous stroma;C (x66):Hepatocyte cytomegaly and karyomegaly (arrowed) in a periportal area infiltrated by proliferating bile ducts and hepatocytes; D (x13): Multilobular focus of "altered" i.e. enlarged hyperproliferative, GST-P expressing hepatocytes.
(A+C:H&E staining; reticulin stain; D:immunocytochemical staining for glutathione S-transferase P).

Fig. 2: hepatic cholangiofibrosis following 12 weeks administration of 1,000 ppm methapyrilene.
A (x13): profuse hyperproliferating and slightly dysplastic bile ducts are accompanied by a highly cellular connective stroma containing inflammatory cells, siderophages, occasional "entrapped" hepatocytes and blood vessels. B-D (x130) :basophilic bile ducts are lined by tall columnar epithelium,show frequent mitotic figures (arrowed in B) and contain cytoplasmic and inflammatory cell debris in the lumen (C); mucin secreting goblet cells are identified as eosinophilic apical granules (arrowed in C) and stain positively with periodic acid Schiff reagent (arrowed in D).
(A-C:H&E staining)

uncommon, since most carcinogens only induce the formation of focal enzyme expression. In some lobes, cholangiofibrosis and bile duct goblet cell metaplasia were observed (Fig.2). Methapyrilene therefore causes substantial cytotoxicity at this potently carcinogenic dose, a point often ignored in other reports, despite the liver being quite abnormal prior to the appearance of altered hepatocyte foci. The regenerative proliferation of hepatocytes is likely to exert a promoting effect on any epigenetic changes. Moreover, the mild dysplasia and cholangiofibrosis are potential preneoplastic steps towards cholangiocarcinoma.

In further studies of up to 6 months duration, groups of Han-Wistar rats were given 400 ppm methapyrilene in the diet. Histological examination included dual immunocytochemical staining for detection of GST-P positive foci and determination of cell proliferation rate with bromodeoxyuridine.

The effects of methapyrilene were dose-related and generally showed a lower incidence of the abnormalities noted in the shorter-term higher dose experiment reported here.

Induction of foci of altered hepatocytes

After 6 months, methapyrilene caused a 5fold increase in the number of GST-P positive foci compared to the control, located specifically in the periportal zones. By comparison, the genotoxic carcinogen aflatoxin B1 induced a 20fold increase in GST-P hepatocyte foci, but the hepatotoxin paracetamol had no effect.

At this stage, the altered foci were extremely difficult or even impossible to detect by routine histological staining and examination methods. Foci, since they are derived from phenotypically distinct cells, form an heterogeneous population, and themselves often develop into an heterogeneous population of cells. Nonetheless, they are usually histologically divided into basophilic, eosinophilic, and then basophilic, foci and adenomas (Ohshima et al., 1984), and finally hepatocellular carcinomas, prompting the suggestion that each type of focus corresponds to a particular stage in tumorigenesis. Thus basophilic foci would represent a more advanced form. The fact that we were not able to localise the foci by routine histology suggests that they were of a very early, still highly differentiated type.

Cell proliferation rates

Chronic exposure to methapyrilene, like other non-genotoxic hepatocarcinogens, increases periportal hepatocyte proliferation, whereas paracetamol, like most hepatotoxins, increased hepatocyte proliferation in centrilobular zones. Minor and more sporadic cell proliferation was observed with aflatoxin B1 treatment.

The relationship between cell proliferation in normally differentiated hepatocytes and in those with an altered phenotype is a crucial component of hepatocarcinogenicity. The question as to whether foci have a higher proliferation rate than the surrounding normal hepatocytes is confounded by focus heterogenicity and the potential for proliferation rate while the majority of them were in a relatively quiescent state. These proliferative subtypes may ultimately correlate with the various histologically subtypes, although no such relationship is yet apparent.

Conclusion

The preneoplastic effects of this potent non-genotoxic carcinogen appear to be dose related. The relatively high potency and thus correspondingly short time frame in which the preneoplastic changes develop, identify this as a suitable model for examining mechanisms of epigenetic carcinogenesis. The sequential changes in the control of cell proliferation leading to autonomous growth, and in particular the point at which changes in growth control become irreversible, are important in determining mechanisms of carcinogenicity. Even though these mechanisms are not presently understood, the exposure threshold for their occurrence and the sequence that they follow are relevant for the human safety evaluation of non-genotoxic carcinogens.

REFERENCES

Anderson NL, Copple DC, Bendele RA, Probst GS, Richardson FC (1992) Covalent protein modifications and gene expression changes in rodent liver following administration of methapyrilene: a study using two-dimensional electrophoresis.Fund Appl Toxicol 18,570-580.
Andrew AW, Fornwald JA, Lijinsky W (1980) Nitrosation and mutagenicity of some amine drugs. Toxicol Appl Phrarmacol 52,237-244.
Brennan LM & Creasia DA (1982) The effects of methapyrilene hydrochloride on hepatocarcinogenicity and phenobarbital-induced sleeping time in rats and mice. Toxicol Appl Phramacol 66,252-258.
Brudoe JD, Shaddock JG, Casciano DA (1984) A study of the potential genotoxicity of methapyrilene and related antihistamines using the hepatocyte/DNA repair assay. Mutat Res 135,131-137.
Copple DM, Rush GF, Richardson FC (1992) Effects of methapyrilene

measured in mitochondria isolated from naive and methapyrilene-treated rats and mouse hepatocytes. Toxicol Appl Pharmacol 116,10-16.

Couri D, Wilt SR, Milks NM (1982) Methapyrilene effects on initiation and promotion of gamma-glutamyl-transpeptidase positive foci in rat liver. Res Communic Chem Path Pharmacol 35(1), 51-61.

Glauert HP & Pitot HC (1989) Effects of the antihistamine methapyrilene as an initiator of hepatocarcinogenesis in female rats. Cancer Lett 46, 189-194.

Graichen ME, Neptun DA, Dent JG, Popp JA, Leonard TB (1985) Effects of methapyrilene on rat hepatic xenobiotic metabolising enzymes and liver morphology. Fund Appl Toxicol 5, 165-174.

Habs M, Shubik P, Eisenbrand G (1986) Carcinogenicity of methapyrilene hydrochloride, mepyramine hydrochloride, thenyldiamine hydrochloride and pyribenzamine hydrochloride in Sprague-Dawley rats. J Cancer Res Clin Oncol 111, 71-74.

Hernandez L & Lijinsky W (1989) Glutathione and lipid peroxide levels in rat liver following administration of pethapyrilene and analogs. Chem-Biol Interact 69, 217-224.

Hernandez L, Allen PT, Poirier LA, Lijinksy W (1989) S-adenosylmethionine, S-edenosylhomocysteine and DNA methylation levels in the liver of rats fed methapyrilene and analogs. Carcinogenesis 10,557-561.

Hernandez L, Petropoulos CJ, Hughes SH, Lijinsky W (1991) DNA methylation and oncogene expression in methapyrilene-induced rat liver t tumors and in treated hepatocytes in culture. Mol Carcinogenesis 4, 203-209.

Iype PT, Bucana CD, Kelley SP (1985) Carcinogenesis by nonmutagenic chemicals: early response of rat liver cells induced by methapyrilene. Cancer Res 45,2184-2191.

Iype PT, Ray-Chaudhuri R, Lijinsky W.,Kelley SP (1982) Inability of methapyrilene to induce sister-chromatid exchanges *in vitro* and *in vivo*. Cancer Res 42,4614-4618.

Kammerer RC & Schmitz DA (1987) Species differences in the *in vitro* metabolism of methapyrilene. Xenobiotica 17(9), 1121-1130.

Kelly DW, Holder CL, Korfmacher WA, Getek TA, Lay JOjr., Casciano DA, Shaddock JG, Duhart HM, Slikker W Jr. (1992) Metabolism of methapyrilene by Fisher-344 rats and B6C3F1 mouse hepatocytes. Xenobiotica 22(12), 1367-1381.

Lampe MA & Kammerer RG (1987) The effect of chronic methapyrilene treatment on methapyrilene metabolism *in vitro*. Carcinogenesis 8(2) 221-226.

Lampe MA & Kammerer RC (1990) Species differences in the metabolism and macromolecular binding of methapyrilene: a comparison of rat, mouse, hamster. Xenobiotica 20(12),1269-1280.

Lijinsky W & Kovatch RM (1986) Carcinogenicity studies of some analogs of the carcinogen methapyrilene in F344 rats. J Cancer Res Clin Oncol 112,57-60.

Lijinsky W & Muschik GM (1982) Distribution of the liver carcinogen methapyrilene in Fisher rats and its interaction with macromolecules. J Cancer Res Clin Oncol 103,69-73.

Lijinsky W & Yamashita K (1988) The binding of methapyrilene and similar antihistamines to rat liver DNA examined by 32P-postlabeling.Cancer Res 48,6475-6477.

Lijinsky W (1984) Chronic toxicity of pyrilamine maleate and methapyrilene hydrochloride in F344 rats. Food Chem Toxicol 22, 27-30.

Lijinsky W, Knutsen G, Reuber MD (1983) Failure of methapyrilene to induce tumors in hamsters or guinea-pigs. J Toxicol Envir Health 12,653-657.

Lijinksy W, Kovatch RH, Thomas BJ (1992) The carcinogenic effect of methapyrilene combined with nitrosodiethylamine given to rats in low doses. Carcinogenesis 13(7),1293-1297.

Lijinsky W, Reuber MD, Blackwell B-N (1980) Liver tumors induced in rats by oral administration of the antihistaminic methapyrilene hydrochloride. Science 205,817-819.

McQueen CA & Williams GM (1981) Characterisation of DNA repair elicited by carcinogens and drugs in the hepatocyte primary culture/DNA repair test. J Toxicol Envir Health 8,463-477.

Oberly TJ, Scheuring JC, Richardson KA, Richardson FC, Garriott ML (1993) The evaluation of methapyrilene bacterial mutation with metabolic activation by Aroclor-induced, methapyrilene-induced and noninduced rat liver S9. Mut Res 299, 77-84.

Ohshima M, Ward JM, Brennan LM, Creasia DA (1984) A sequential study of methapyrilene hydrochloride-induced liver carcinogenesis in male F344 rats. JNCI 72(3), 759-765.

Perera MIR, Ktyal SL, Shinozuka H (1985) Methapyrilene induces lipid peroxidation of rat liver cells. Carcinogenesis 6(6),925-927.

Perera MIR, Katyal SL, Shinozuka H (1987) Choline deficient diet enhances the initiating and promoting effects of methapyrilene hydrochloride in rat liver as assayed by the induction of gamma-glutamyltranspeptidase-positive

hepatocyte foci. Br J Cancer 56,774-778.
Pienta RJ (1980) Evaluation and relevance of the Syrian hamster embryo cell system.In: G.M.Williams et al. (eds.) The predictive value of short screening tests in carcinogenicity evaluation, pp. 149-169. Amsterda: Elsevier/M.Holland.
Probst GS & Neal SB (1980) The induction of unscheduled DNA synthesis by antihistamines in primary hepatocyte cultures. Cancer Lett 10, 67-73.
Reznik-Schuller H & Lijinsky W (1981) Morphology of early changes in liver carcinogenesis induced by methapyrilene. Arch. Toxicol. 49, 79-83.
Reznik-Schuller H & Lijinsky W (1982) Ultrastructural changes following treatment with methapyrilene and some analogs. Ecotoxicol Envir Safety 6,328-335.
Reznik-Schuller HM & Gregg M (1983) Sequential morphologic changes during methapyrilene-induced hepatocellular carcinogenesis in rats. JNCI 71(5),1021-1027.
Richardson FC, Horn DM, Anderson NL (1994) Dose-response in rat hepatic protein modification and expression following exposure to the rat hepatocarcinogen methapyrilene. Carcinogenesis 15(2),325-329.
Wrighton SA, Van den Branden M, Brown TJ, van Pelt CS, Thomas PE, Shirley LA (1991) Modulation of rat hepatic cytochromes P-450 by chronic methapyrilene treatment.Biochem Pharmacol 42(5_,1093-1097.
Zeindl-Eberhart E & Rabes HM (1992) Variant protein patterns in hepatomas and transformed liver cell lines determined by high resolution two-dimesional gel electrophoresis (2-DE) Carcinogenesis 13(7),1177-1182.
Ziegler R, Ho B, Castagnoli N Jr. (1981) Trapping of metabolically generated electrophilic species with cyanide ion: metabolism of methapyrilene. J Med Chem 24, 1133-1138.

mRNA COMPOSITION OF RAT LIVER TUMORS INITIATED BY AROMATIC AMINES

Annette Bitsch, Martina Jost, Hedwig Richter
Institute of Pharmacology and Toxicology
University of Würzburg
Versbacherstr. 9
97078 Würzburg
Germany

INTRODUCTION

Many polycyclic aromatic amines are mutagenic. Their genotoxic properties are explained by the formation of metabolites that react with DNA by forming adducts (Kriek 1969, Franz et al. 1986). These DNA adducts may cause mutations which are responsible for tumorigenic effects. The carcinogenesis induced by aromatic amines is very often characterized by high tissue specificity. Chronic feeding of 2-acetylaminofluorene (AAF) typically produces liver tumors in rodents, 2-acetylaminophenanthrene (AAP) mammary tumors and *trans-4*-acetylaminostilbene (AAS) Zymbal's gland tumors (Neumann et al. 1970). Many attempts have been made to correlate DNA binding or the resulting lesions with tissue specific tumor formation. However, this phenomenon cannot be readily explained by the extent of DNA modifications. All three amines generate comparable DNA adduct levels in rat liver and moreover in liver more extensively than in any other tissue (Neumann 1983, Ruthsatz and Neumann 1988, Gupta et al. 1989). Therefore, DNA binding may reflect the formation of some critical lesions related to

NATO ASI Series, Vol. H 88
Liver Carcinogenesis
Edited by G. G. Skouteris

tumor initiation but is not sufficient to explain carcinogenic effects. AAS and AAP are able to produce liver tumors in rats, but only if initiation is followed by some promotion treatment (Hammerl 1989). AAF in contrast is a complete carcinogen for this tissue. This raises the question whether organ specificity is a consequence of different genotoxic properties and whether complete carcinogens produce lesions in different genes of target tissues.

Multiple structural and functional cellular alterations, like expression of oncogenes and fetal proteins as well as changes in enzyme patterns, take place during chemically induced liver carcinogenesis. Some of these changes are supposed to be attributed to the irreversible events during initiation. The cooperation and subsequent activation of oncogenes during tumor formation has been demonstrated in several studies (Land et al. 1983, Hinds et al. 1990, Hsieh et al. 1991, Hunter 1991). Recently we have shown that in contrast to mice rat liver tumors induced by aromatic amines do not have mutated or overexpressed H-ras genes (Bitsch et al. 1993). Now we attempted to investigate changes in the expression for a set of activated genes in tumors. Using the tumor as the biological endpoint it should be possible to investigate changes in the gene expression that may directly or indirectly be associated with the genotoxic effects of the initiators.

Several approaches have been pursued to identify relevant changes in gene expression during carcinogenesis. Many of them used the high resolution two-dimensional el electrophoresis to demonstrate alterations in the amounts of cellular proteins. Novikoff (Takami and Busch 1979) and Morris hepatomas (Takami et al. 1979) as well as regenerating rat liver (Kadofuku and Sato 1985) were compared to normal liver. Wirth et al. (1987) detected numerous quantitative changes but only a few qualitative alterations in tumors induced by the Solt-Farber model. We have chosen the initiation-promotion model according to Peraino et al. (1981) with new-born Wistar rats. Tumors were initiated with the three amines mentioned before, phenobarbital was used for promotion. mRNA was isolated from these tumors and *in vitro* translated into protein. Following the separation on SDS-PAGE the pattern of translation products was compared with that from control tissue. With this approach we demonstrated several quantitative and a few qualitative alterations in the protein pattern in tumors, indicating modifications in gene expression in the tumors. Differences in the pattern of differently initiated tumors led us to assume, that some of these effects may be specific for the initiators. But the overall changes in RNA content in tumor tissue as compared to control liver were less than expected.

MATERIALS AND METHODS

Chemicals: 2-Acetylaminofluorene (AAF) was obtained from Merck (Darmstadt, FRG), and phenobarbital (PB) from Serva (Heidelberg, FRG). 2-Acetylaminophenanthrene (AAP) and trans-4-acetylaminostilbene (AAS) were synthesized and characterized according to Calder and Williams (1974) and Metzler and Neumann (1971). ^{35}S-Methionine was received from Amersham Buchler (Braunschweig, FRG).

Animals and Treatment: As a model system we used new-born Wistar rats. Pregnant female rats were purchased from Deutsche Versuchstieranstalt (Hannover, FRG). New-borns were randomly divided into five treatment groups. The three aromatic amines were dissolved in tinned milk and orally administered at days 5, 7, 9, 11 postnatally as the initiation treatment: AAF (1.5 mmol/kg body weight), AAP (1.5 mmol/kg) and AAS (0.15 mmol/kg). Two control groups were treated with the solvent only. The rats were weaned at day 26 and maintained on normal diet with Altromin 1324 (Altrogge, Lage, FRG) and water ad libitum in an air-conditioned room with a 12 h light/dark cycle. Promotion with phenobarbital (500ppm added to the drinking water) was started at day 35. One of the control groups received the promoter only. Groups of 3-5 rats of both sexes were sacrificed after 26, 52, 78 and 104 weeks. The rats were anaesthetized with ether, the livers were perfused with isotonic NaCl-solution, resected and weighed. Tumors - if present - were separated from the surrounding tissue. For further investigations this material was frozen immediately in liquid nitrogen and stored.

RNA-Isolation: RNA isolated from pulverized frozen tissue by lysis in guanidinium isothiocyanate solution, chloroform/phenol extraction and two precipitation steps as described by Chirgwin et al. (1971) Poly A+RNA was purified by affinity chromatography to oligo-dT-cellulose according to Aviv & Lederer (1972) and Pemberton et al. (1975).

***In vitro* Translation:** Limiting amounts of isolated poly A+RNA were translated *in vitro* in an amino acid depleted cell free wheat germ system as described by the supplier (Promega). The added mixture of amino acids included ^{35}S-methionine to label the translation products. In vitro translation was carried out at 27°C for 1 hour. Incorporation of radioactivity was determined and compared with that of a control RNA. Translated proteins were precipitated with 10% trichloroacetic acid. The wheat germ system preferentially translates RNA species in the lower molecular weight range (Roberts and Paterson 1973).

Protein Analysis: Precipitated translation products were resolved in a gel-loading buffer (Laemmli 1970). Equal amounts of radioactivity from each sample were separated on a discontinuous polyacrylamide gel system. SDS-PAGE was performed using 0.75 mm thick 10% gels. Gels were run without external cooling at 15 mA constant current using the buffer system described by Laemmli (1970) until the bromophenol blue dye reached the bottom. Fixation and staining of proteins in the gel was done in 0.25% coomassie blue (Serva) in 45% methanol, 9% acetic acid, destaining in 10% methanol, 7% acetic acid. Molecular weights were estimated using a protein standard (carboanhydrase 19kD, ovalbumin 45kD, glutamic dehydrogenase 55kD, albumin 66kD and phosphorylase b 97kD) was used. Protein gels were dried and exposed to a phosphorimager screen (Molecular Dynamics) to determine relative amounts of radioactivity for each protein band. Translation products were documented by autoradiography on hyperfilm β-max (Amersham Buchler).

RESULTS

The formation of numerous liver tumors after 105 weeks as well as the preneoplastic and neoplastic lesions 26 and 52 weeks after initiation treatment proved the initiating properties of the used aromatic amines in our animal system (Hammerl 1989, Bitsch et al. 1993). The initiation potency decreased in the order AAS>AAP>AAF.

Poly A+RNA was isolated from the obtained rat liver tumors and compared with poly A+RNA in control liver by using the *in vitro* translation in a cell free system as a method to identify translatable RNAs. Following separation in a 10% SDS-PAGE, the translated proteins, each of them corresponding to one RNA species, were quantified by the incorporated radioactivity. This incorporation might be different for different proteins and depends on the methionine content of the protein. The comparison of two proteins or groups of proteins with the same electrophoretic migration includes the assumption that they are identical. Even when preparations of translation products and gel electrophoreses are performed under almost identical conditions, gels come out a little different. Therefore, translated proteins from control and tumor tissue have to be separated in parallel in the same run for comparison. Experiments were carried out with tumor RNA from three and more different animals of each sex. Prominent protein bands and proteins, that showed distinct alterations in at least more than two gels were compared and only those differences which were almost identical in three independent electrophoreses were considered further. In order to facilitate a comparison of results, corresponding protein bands in different gels received an arbitrary number.

First, mRNA activity in livers from controls and phenobarbital treated animals was compared to account for effects resulting from enhanced proliferation. Except a few changes of proteins with molecular weights in the range of 45kD to 55kD, proliferation did not change the pattern of translation products significantly (fig. 1).

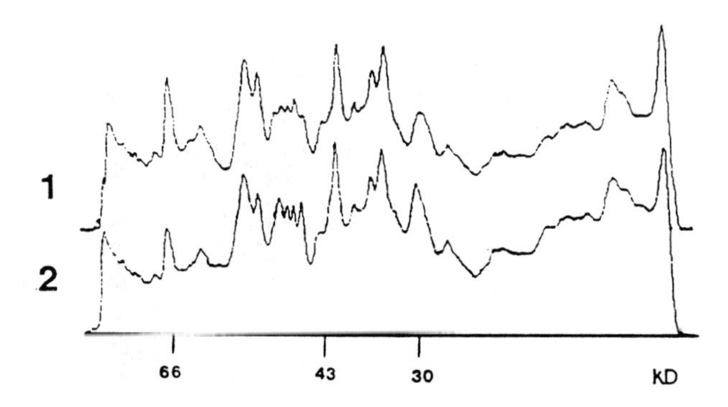

Fig. 1 Radioactivity scan of *in vitro* translation products from control and phenobarbital treated liver tissue.
mRNA was isolated from control (1) and phenobarbital treated (2) liver, *in vitro* translation products were separated on SDS-PAGE and scanned for radioactivity.

In tumor tissue of female AAF and AAP treated animals, one translation product of about 32kD (protein no.14), which was present in control liver, disappeared - a fact, that may indicate the down-regulation or disappearance of the corresponding mRNA species (Fig.2, Fig.3). This protein did not disappear in all AAS initiated tumors in female rats. A protein with the same molecular weight and nearly the same radioactivity was detected in one preparation.

Further changes in the translation pattern could be demonstrated in the molecular weight range of 43-55kD (Fig. 3). But in contrast to the above change, they were only of limited quantitative nature. The amount of protein no. 15, which has an apparent molecular weight of 31kD, was doubled in AAF and AAP initiated tumors and protein no. 2 (55kD) was 160% in AAS tumor tissue as compared to the control.

Fig. 2

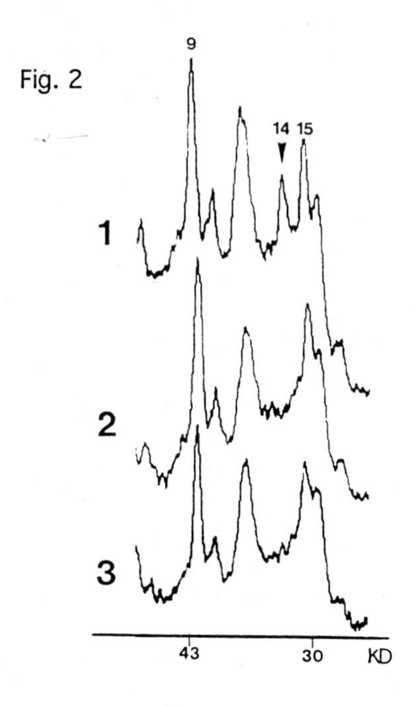

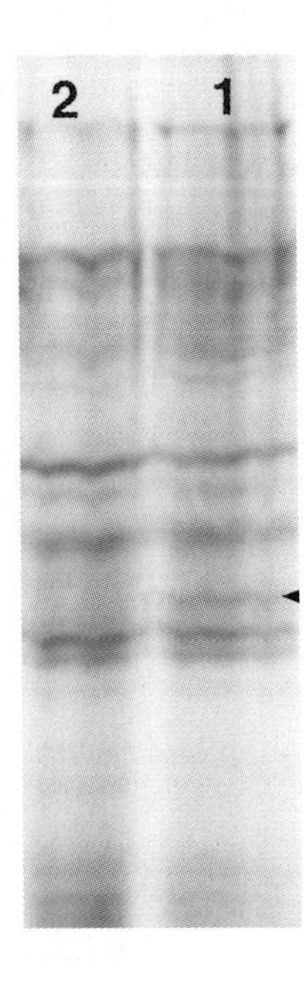

Fig. 2 Radioactivity scan of *in vitro* translation products from control liver and AAF and AAP initiated tumors.
mRNA was isolated from control (1), AAF (2) and AAP (3) initiated tumors from female animals, *in vitro* translation products were separated on SDS-PAGE and scanned for radioactivity.

Fig. 3 Autoradiography of *in vitro* translation products from control liver and AAF initiated tumors.
mRNA was isolated from control (1) and AAF (2) initiated tumors from female animals, *in vitro* translation products were separated on SDS-PAGE and exposed to a film.

Differences in the translation pattern between tumors from treated male rats and control tissue occurred in the molecular weight range of 52-57kD, 33-43kD and around 30kD (Fig. 4). No significant changes could be detected at 32kD. An increased expression of RNA corresponding to protein no. 15 (31kD) was obvious in all tumors.

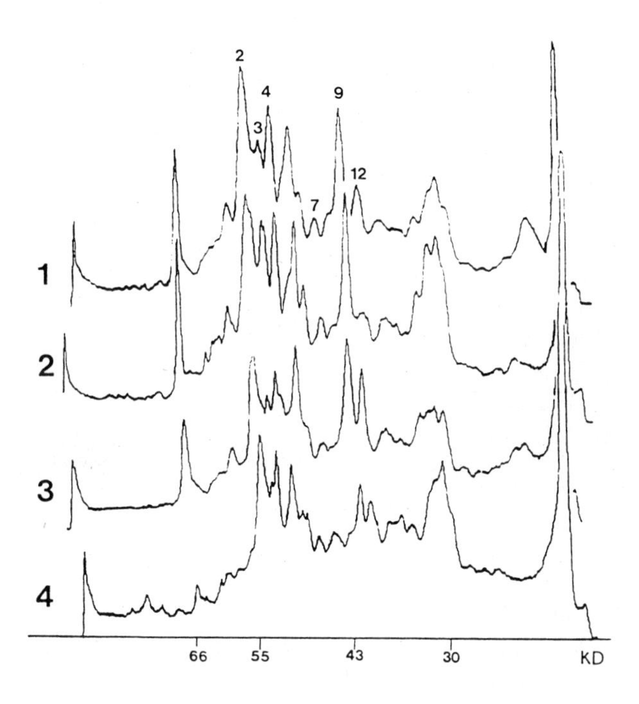

Fig. 4 Radioactivity scan on *in vitro* translation products from control liver and differently initiated tumors.
mRNA was isolated from control (1), AAF (2), AAS (3) and AAP initiated (4) tumors in the liver of male rats. *In vitro* translation products were separated on SDS-PAGE and scanned for radioactivity.

Protein no. 2 (55kD) was increased in tumors from AAS treated males as already described for females (Fig. 5 A).

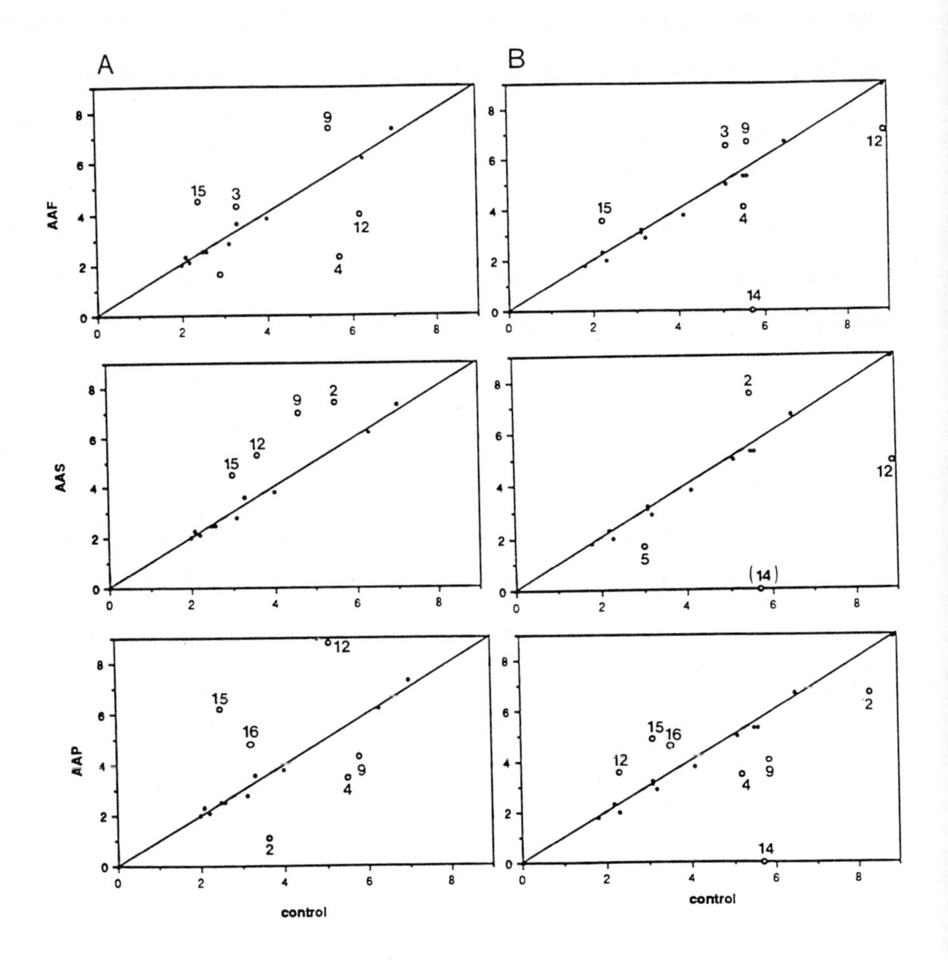

Fig. 5 Correlation of the relative fraction of *in vitro* translated proteins directed by mRNA obtained from tumors and control liver tissue. Proteins near the diagonal line represent constitutive expressed proteins, the numbered spots up- or down-regulated proteins. A: tissues of male animals, B: tissues of female animals.

The translated proteins for those RNAs which are constitutively expressed give a value closed to the diagonal line. Those proteins with values below the line are considered down-regulated, those above the line are up-regulated. For example, protein no. 15 (31kD) appears to be up-regulated in all tumors, protein no. 16 (30kD) seems to be up-regulated only in AAP initiated tumors in male and female rats. Protein no. 9 (43kD) is high in AAF initiated but low in AAP initiated tumors. Protein no. 12 (38kD) is down-regulated in AAF initiated, protein no. 9 (43kD) in AAP initiated and protein no. 12 in both AAF and AAS initiated tumors.

A sex specific difference could regard protein no. 14 (32kD) as well as proteins no. 5 and no. 12 (50kD, 38kD), which are low in females only in AAS initiated tumors (Fig. 5B).

In order to enable an easier comparison of mRNA expression in tumors initiated by different arylamines, the date were processed in the following way: the radioactivity contained in individual protein bands was expressed as the percentage of total radioactivity in the corresponding lane of the gel. The obtained relative fraction was then compared pairwise between tumor and control tissue in a plot (Fig. 5).

DISCUSSION

AAF induced tumors and neoplastic lesions have been analyzed for gene expression and enzyme modifications n several other laboratories. A tumor specific increase in aldehyde dehydrogenase was shown in AAF induced tumors by Lindahl et al. (1982). Sell et al. (1987) demonstrated the increased expression of alpha-fetoprotein and albumin. Changes in the amount of glutathione-S-transferases could be detected in different AAF induced rat liver tumors (Kitahara et al. 1984, Sugioka et al. 1985, Wirth et al. 1987). In these studies usually the level of the investigated protein was determined in the tumor.

For the comparison of tumors initiated by different arlyamines, a new approach was attempted by assessing gene expression on the basis of the *in vitro* translation products of mRNA. This system is independent of post-translational changes like protein stabilization and accumulation, that may be caused by reduced degradation in nodules and hepatomas (Canuto et al. 1993). But post-transcriptional modifications like RNA stabilization or degradation, which may be a consequence of genotoxic events, can be detected.

Alterations in the relative amount of translated proteins are considered as up- or down-regulation of gene transcription into the corresponding RNA. Several changes in the protein pattern were detected in tumor tissues as compared to controls. Most of them were quantitative in nature and some like the up-regulation of a protein with an apparent molecular weight of 31kD were observed in nearly all tumor tissues. They may be involved in tumor formation in general. Others were expressed to a higher extent only in tumors initiated by a particular arylamine (like protein no. 16 (30kD) by AAP and protein no. 2 (55kD) by AAS). Another protein was up-regulated in AAF tumors and down-regulated in AAP tumors (protein no. 9, 43kD) (Fig. 4, 5). These findings indicate that some alterations in gene expression could be related to the initiation treatment.

Other investigators have also tried to find a correlation of protein expression and carcinogenesis. Most of the used 2-dimensional separation of total liver and tumor protein. Anderson et al. (1991) identified in total liver protein a spot at 55kD that correlates with a mitochondrial stress protein. A protein (protein no. 2) with an apparent molecular eight of 55kD was up-regulated in AAS initiated liver tumors. Wirth et al. (1986) demonstrated an up-regulation of a 41kD membrane bound protein in tumors of male rats that were induced according to the Solt-Farber model. They further demonstrated an increased expression of several isoforms of glutathione-S-transferase at a molecular weight of 26kD, 28kD and 29kD in this system. Chronical feeding of AAF to male rats also resulted in increased levels of glutathione-S-transferases and furthermore in a new protein at 35kD (Sugioka et al. 1985). In another study, Wirth et al. (1992) detected the down- regulation of a cytosolic protein at 30kD induced by aflatoxin treatment of cells. In AAP initiated tumors, a protein of this l\molecular weight was up-regulated. In aflatoxin transformed rat liver cell lines, Wirth et al. (1992) observed alterations in the level of a protein at a molecular weight of 43kD, that may correlate with protein no. 9. Husmans et al. (1989) identified a growth hormone receptor at 43kD. Looking for alterations of the translation products as a consequence of cell proliferation so far did not result indistinct changes of the protein pattern (Fig. 1).

In the present experiments, some sex-specific alterations of gene expression were found in tumors. The most pronounced of them was the disappearance of a protein at 32kD in tumors of AAF and AAP treated female animals (Fig. 2, 3). Another difference was the down-regulation of a protein of about 38kD in all tumors of females. Wirth et al. (1992) demonstrated the loss of a 33kD protein in oncogene transformed cell lines.

The expression of several proteins of the cytosceleton like actins or tubulins was shown to be changed in chemical induced carcinogenesis. Others are involved in glucose metabolism. Zeindl-Eberhart and Rabes (1992) found in N-methyl-N-nitrosourea induced liver tumors a protein at 35kD up-regulated and identified it as an aldose reductase (Zeindl-Eberhart et al. 1993). These findings indicate, that many proteins related to metabolism and structure of tumor cells may be involved in carcinogenesis.

It is therefore surprising, that only a few characteristic changes became apparent in the present experiments. This finding encourages further work and an attempt to identify the mRNAs, particularly those which seem to be initiator-related. Other changes may reflect adaptive responses in tumor tissue and should be associated with some of the known phenotypical alterations.

REFERENCES

Anderson, N.L., Esquer-Blasco, R. Hofmann, J.P., Anderson N.G. (1991) A two-dimensional gel database of rat liver proteins useful in gene regulation and drug effect studies. Electrophoresis 12, 907-930.

Aviv, H. and Leder, P. (1972) Purification of biologically active globin messenger RNA by chromatography on oligothymidylic acid-cellulose. Proc. Nat. Acad. Sci. USA 69, 1408.

Bitsch, A., Roschlau, H., Deubelbeiss, C., Neumann, H.-G. (1993) The structure and function of the H-ras-protooncogene are not altered in rat liver tumors initiated by 2-acetylaminofluorene, 2-acetylaminophenanthrene and trans-4-acetylaminostilbene. Toxicology Letters 67, 173-186.

Calder, I.C. and Williams, P.J. (1974) The synthesis and reactions of some carcinogenic N-(2-phenanthryl)hydroxylamine derivatives. Aust. J. Chem. 27, 1791-1795.

Canuto, R.A., Tessitore, L., Muzio, G., Autelli, R., Baccino, F.M. (1993) Tissue protein turnover during liver carcinogenesis. Carcinogenesis 14 (12), 2581-1583.

Chirgwin, J.M., Przybyla, A.E., MacDonald, R.J., Rugger, W.J. (1979) Isolation of biologically active ribonucleic acid from sources enriched in ribonuclease. Biochem. 18, 5294-5296.

Franz, R. Schulten, H.R., Neumann, H.G. (18\986) Identification of nucleic acid adducts from trans-4-acetylaminostilbene. Chem. Biol. Interact. 59, 281-293.

Gupta, R.C., Earley, K., Fullerton, N.F., Beland, F.A. (1989) Formation and removal of DNA adducts in target and nontarget tissues of rats administered multiple doses of 2-acetylaminophenanthrene. Carcinogenesis 10, 2025-2033.

Hammerl, R. (1989) Zur synergistischen Wirkung aromatischer Amine bei Der Initiierung von Tumoren in der Rattenleber. Dissertation, Tiermedizinische Fakultat, Universitat Munchen.

Hinds, P.W., Finlay, C.A., Quartin, R.S., Baker S.J., Fearon, E.R. Vogelstein, B., Levine, A.J. (1990) Mutant p53 DNA clones from human colon carcinomas cooperate with ras in transforming primary rat cells: a comparison of the "hot spot" mutant phenotypes. Cell Growth Different. 1, 571-580.

Hsieh, L.L., Shinozuka, H., Weinstein, I.B. (1991) Changes in expression of cellular oncogenes and endogenous retrovirus-like sequences during hepatocarcinogenesis induced by a peroxisome proliferator. Br. J. Cancer 64(5), 815-820.

Hunter, T. (1991) Cooperation between oncogenes. Cell 64(2), 249-270.

Husmans, B., Gustafsson, J.A., Andersson, G. (1989) Biogenesis of the somatogenic receptor in rat liver. J. Biol. Chem. 264, 690.

Kadofoku, T., Sato, T. (1985) Detection of the changes in cellular proteins n regenerating rat liver by high-resolution two-dimensional electrophoresis. J. Chromtogr. 343, 51-58.

Kitahara, A., Satoh, K., Nishimura, K., Ishikawa, T., Ruike, K., Sato, K., Glutathione-S-transferase during chemical hepatocarcinogenesis. Cancer Res. 44, 2698-2703.

Kriek, E. (1969) On the mechanism of action of carcinogenic aromatic amines. I. Binding of 2-acetylaminofluorene and N-hydroxy-2-acetylaminofluorene to rat liver nucleic acids in vivo. Chem. Biol. Interact. 1, 3-17.

Laemmli, U.K. (1970) Cleavage of structural proteins during the assembly of the head of bacteriophage T4. Nature 227, 680-685.

Land, H., Parada, L.F., Weinberg, R.A. (1983) Tumorigenic conversion of primary embryo fibroblasts requires at least two cooperating oncogenes. Nature 304, 596-602.

Lindahl, R., Clark, R., Evces, S. (1982) Histochemical localization of aldehyde dehydrogenase during rat hepatocarcinogenesis. Proc. Am. Assoc. Cancer Res. 25, 139.

Metzler, M. and Neumann, H.-G. (1971) Zur Bedeutung chemisch-biologischer Wechselwirkungen fur die toxische und krebserzeugende Wirkung aromatischer Amine. III Synthese und Analytik einiger Stoffwechselprodukte von trans-4-Dimethylaminostilben, cis-4-Dimethylaminostilben und 4-Dimethylaminobenzyl. Tetrahedron 27, 2225-2246.

Neumann, H.-G. (1983) Role of extent and persistence of DNA modifications in chemical carcinogenesis by aromatic amines. Rec. Res. Cancer Res. 84, 77-89.

Neumann, H.-G. (1986) The role of DNA damage in chemical carcinogenesis of aromatic amines. J. Cancer Res. Clin. Oncol. 112, 100-106.

Neumann, H.-G. Metzler, M., Brachmann, I., Thomas, C. (1970) Zur Bedeutung chemisch-biologischer Wechselwirkungen fur die toxische und krebserzeugende Wirkung aromatischer Amine. I. Krebserzeugende

Wirksamkeit einiger 4-Aminostilben- und 4-Aminobibenzyl- Verbindungen. Z. Krebsforsch. 74, 200.
Pemberton, R.E., Liberti, P., Baglioni, C. (1975) Isolation of messenger RNA from polysomes by chromatography on oligo (dT) cellulose. Anal. Biochem. 66, 18-28.
Peraino, C., Staffeldt, E.F., Ludeman, V.A. (1981) Early appearance of histochemically altered hepatocyte foci and liver tumors in female rats treated with carcinogens one day after birth. Carcinogenesis 2, 463-465.
Roberts, B.E. and Paterson, B.M. (1973) Efficient translation of tobacco mosaic virus RNA and globin 9S RNA in cell-free system from commercial wheat-germ. Proc. Nat. Acad. Sci. USA 70, 2330-2334.
Ruthsatz, M. and Neumann, H.-G. (1988) Synergistic effects on the initiation of rat liver tumors by trans-4-acetylamino-stilbene and 2-acetylaminofluorene, studied at the level of DNA adduct formation. Carcinogenesis 9, 265-269.
Sell, S., Hunt, J.M., Knoll, B.J., Dunsford, H.A. (1987) Cellular events during hepatocarcinogenesis in rats and the question for pre-malignancy. Adv. Cancer Res. 48, 37-111.
Sugioka, Y., Fujii-Kuriyama, Y., Kitagawa, T., Muramatsu, M. (1985) Changes in polypeptide pattern of rat liver cells during chemical hepatocarcinogenesis. Cancer Res. 45, 365-378.
Takami, H., Busch, F.N., Morris, H.P., Busch, H. (1979) Comparison of salt-extractable nuclear proteins of regenerating liver, fetal liver, and Morris hepatomas 9618 A and 3924 A. Cancer Res. 39, 2096-2105.
Wirth, P.J., Benjamin, T., Schwartz, D.M., Thorgeirsson, S.S. (1986) Sequential analysis of chemically induced hepatoma development by two-dimensional electrophoresis. Cancer Res. 46, 400-413.
Wirth, P.J., Sambasiva, R., Evarts, R.P. (1987) Coordinate polypeptide expression during hepatocarcinogenesis in male F-334 rats: comparison of Solt-Farber and Reddy models. Cancer Res. 47(9-12), 2839-2851.
Zeindl-Eberhart, E. and Rabes, H.M. (1992) Variant protein patterns in hepatomas and transformed liver cell lines as determined by high resolution two-dimensional gel electrophoresis (2DE). Carcinogenesis 13(7), 1177-83.
Zeindl-Eberhart, E., Jungblut, P.R., Otto, A., Rabes, H.M. (1993) Further analysis of protein variants in chemically induced rat hepatomas and transformed liver cell lines. J. Cancer Res. and Clin. Oncol. Suppl. 2, Vol. 119.

LIVER γ-GLUTAMYL TRANSPEPTIDASE ACTIVITY AFTER CYCLOSPORINE A AND AMLODIPINE TREATMENT

Jerzy G. Maj, Jeremiasz J. Tomaszewski and Agnieszka E. Haratym
Department of Clinical Biochemistry and Environmental Toxicology
School of Medicine
Jaczewskiego 8,
PL-20950 Lublin,
Poland

INTRODUCTION

γ-Glutamyl transpeptidase (GGT) (E.C.2.3.2.2) is a cell - surface enzyme that catalyzes transfer of the γ-glutamyl moiety of glutathione and other γ-glutamyl compounds to a variety of amino acids and peptides. It has also been suggested that the action of GGT may be involved in protein synthesis, amino acid transport, collagen formation and degradation of peptides (Meister, 1974; Orlowski and Meister, 1970). The GGT, a heterogenous sialoglycoprotein is localized in many mammalian tissues and cells (Kottgen et al., 1976; Tate and Meister, 1976). The pattern of GGT sugar chains can alter during development and in pathological states (Kottgen et al., 1976; Yamashita et al., 1993). The GGT activity is different in individual organs, tissues and cells and is dependent on physiological or pathological states, drugs and many others (Chung et al, 1990; Nishimura and Teschke, 1983; Paolicchi et al., 1993; Stastny et al., 1992; Weber et al., 1992). The GGT is involved in γ-glutamyl cycle and more recent studies have suggested that the enzymes of the cycle participate in a variety of essential cellular reactions and defense mechanisms (Meister, 1974). These include conjugation of toxicants and xenobiotics,

NATO ASI Series, Vol. H 88
Liver Carcinogenesis
Edited by G. G. Skouteris

leading to their detoxification or activation of toxicity, facilitation of nonenzymatic reduction of free radicals and enzymatic elimination of hydrogen peroxide, involvement in the synthesis of the prostaglandins, leukotriens and deoxyribonucleotides (Cohen, 1983; Deleve and Kaplowitz 1990; Koob and Dekant, 1991; Shi et al., 1993; Tate, 1980).

Cyclosporine A (CsA), is an immunosuppressive agent used to prevent allograft rejection and, more recently, to treat some autoimmune diseases. CsA blocks T cell activation essentially by inhibition of transcription of cytokine genes, including interleukin-2, interleukin-3, interleukin-4 and interferon-γ (Kahan, 1989; Liu, 1993; Waldmann, 1993). Recent advances in the mechanism of action of the CsA indicate that CsA inhibits calcium - calmodulin dependent protein, phosphatase calcineurin (Liu, 1993). CsA binds to a cytosolic binding protein called cyclophilin, which has peptidyl-prolyl cis-trans isomerase activity and which is inhibited upon binding to the drug (Fischer et al., 1989; Handschumacher et al., 1984; Takahashi et al., 1989). The CsA-cyclophilin complex then binds to calcineurin and inhibits its protein phosphatase activity (Liu, 1993). One theory proposes that calcium dependent intracellular processes represent the target of CsA action (Antoni et al., 1993; Colombani et al., 1985; Lin et al., 1993; Nicchitta et al., 1985). Calcium ions play an important role in the regulation and control of many cellular functions. The cellular processes affected by calcium ions are diverse and include muscle contraction, release of hormones, neurotransmitters, cytokines, activation or inactivation of enzymes (e.g. phospholipases, proteases, endonucleases), facilitation of transport processes and many facets of cell growth and mitosis (Lotersztajn et al., 1990; McCabe et al., 1992). Recent studies have demonstrated that increased intracellular calcium ion concentrations can play a critical role in cell death (McCabe et al., 1992). Three components participate in the control of the cytosolic calcium level: 1/Ca^{2+} entry into the cell through voltage-dependent or receptor operated channels; 2/Ca^{2+} storage and release by intracellular organelles (mitochondria and endoplasmic reticulum); 3/Ca^{2+} exit, which is driven by two specific systems, and ATP-dependent Ca^{2+} pump and a Na^{+}/Ca^{2+} exchanger (Carafoli, 1991; Lotersztasjn et al., 1990, Niggli and Lederer, 1991; Putney, 1990). Therefore, drugs or other agents can change cell metabolism by increasing or decreasing intracellular calcium ion concentration via one or more mechanisms.

Thus, we investigated the effect of calcium modulation by CsA and dihydropyridine calcium channel blocker amlodipine on the GGT activity in the liver.

MATERIALS AND METHODS

Animals

Female Swiss albino mice 8-12 weeks old, weighing 20-30 g were used. Animals were housed in plastic cages under 12h/12h light/dark cycle at 21 +/- 2°C temperature and 50% relative humidity with standard granulated food and water available ad libitum. Mice were allowed to acclimatize for 1 week.

Drugs and Chemicals

Cyclosporine A (CsA) (Sandoz Ltd., Basel, Switzerland) and amlodipine (Pfizer, Orsay, France) were used in the experiment. Amlodipine was dissolved in Tween 60 and diluted in 0.9% NaCl (Polfa, Poland). CsA was dissolved in DMSO and diluted in 0.9% NaCl. All other chemicals were of reagent grade quality and were purchased from: TRIS from Serva (Serva, Heidelberg, Germany) and all other from Sigma (Sigma Chemical Co., St. Louis, MO, USA).

Experimental Procedure

Mice received CsA intraperitoneally (i.p.) once daily for 5 days in doses: 10 mg/kg body weight (b.w.), 20 mg/kg b.w., 40 mg/kg b.w. The next groups of animals received i.p. once daily for 5 days CsA in above doses in combination with amlodipine in a dose of 0.07 mg/kg b.w. Control group received i.p. 0.9% NaCl (0.1 ml) per day for 5 days.

Preparation of Material

Mice were sacrificed 24 h after final administration of drugs. Livers were immediately removed from sacrificed animals and samples were homogenized at 4°C in a Potter-Elvehjem homogenizer equipped with Teflon pestle in 5 ml of 0.1M TRIS/HCl buffer, pH 8.0. Homogenates were centrifuged for 15 minutes at 400g.

Biochemical Assays

GGT activity was determined in liver homogenates using γGT 20 S kit (bioMerieux, Marcy-l'Etoile, France). The reaction temperature was 37°C and absorbance of p-nitroaniline was determined at 405 nm. One unit of enzyme activity was defined as 1 μmol of p-nitroaniline produced from L-γ-glutamyl-p-nitroanilide per minute. Protein concentrations were determined by the method of Lowry et al. (1951) with bovine serum albumin as a standard. GGT activity is expressed as mU/mg of protein.

Statistical Analysis

Statistical differences between groups were estimated using unpaired two-tailed Mann-Whitney U-test. $P<0.05$ was considered statistically significant. Values are expressed as mean +/- SEM.

RESULTS

After 5 days of treatment with CsA at doses 10 mg/kg b.w. and 20/kg b.w., the activity of liver GGT was increased to 4.03 +- 0/21 mU/mg of protein and 4/80 +/- 0.36 mU/mg of protein respectively, and were statistically significant compared with control group 2.94 +/- 0.18 mU/mg of protein ($p<0.05$) (Fig. 1). When animals were treated with CsA at a dose 40 mg/kg b.w., the activity of GGT was significantly decreased to 1.34 +/- 0.14 mU/kg of protein (Fig. 1). Effects of the treatment with CsA combined with amlodipine are shown in Fig. 2. Animals that received two drugs were compared with groups of animals that received only CsA in the same doses. Combined therapy increased the activity of GGT compared with groups treated with CsA and the control group (Table 1, Fig. 2). However, in groups receiving CsA and amlodipine, the GGT activity was CsA dose-dependent. In these groups the induction of GGT activity was higher at low dose of CsA.

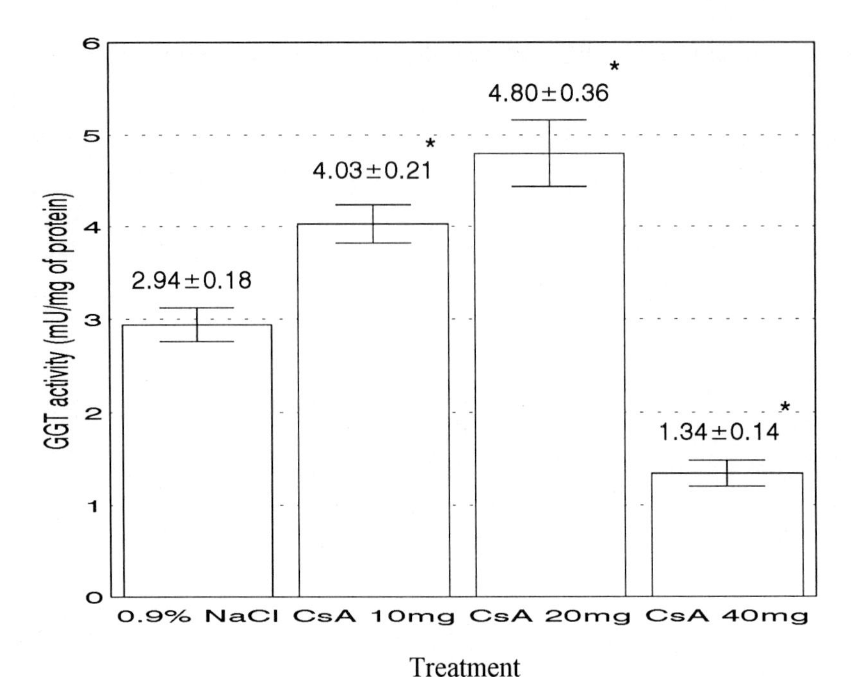

Fig. 1. The activity of GGT in liver after 5 days of CsA treatment. Values are mean +/- SEM. *$p<0.05$ compared with control.

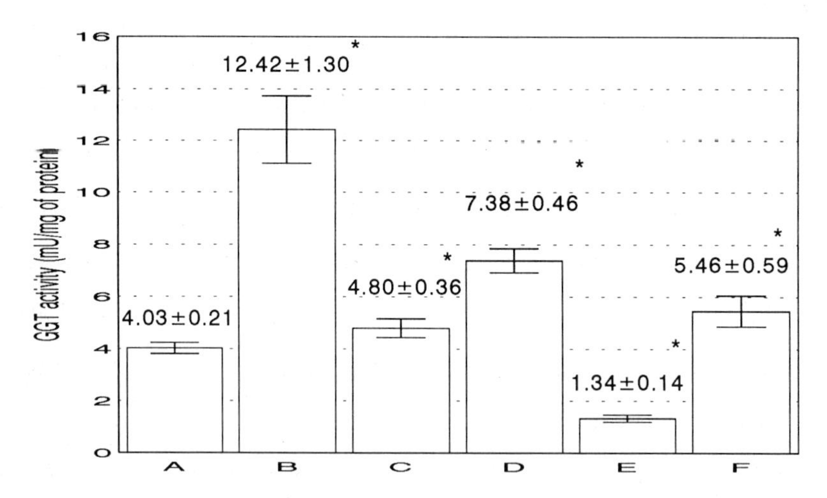

Fig. 2. The activity of GGT in liver after 5 days of CsA and amlodipine (Amlo) treatment. Groups of animals received: A - CsA 10 mg/kg, B - CsA mg/kg + Amlo 0.07 mg/kg, C - CsA 20 mg/kg, D - CsA 20 mg/kg + Amlo 0.07 mg/kg, E - CsA 40 mg/kg, F - 40 mg/kg + Amlo 0.07 mg/kg. Values are mean +/- SEM. *$p<0.05$ compared with groups receiving the same dose of CsA without amlodipine.

Table 1. The effect of CsA and amlodipine (Amlo) on the GGT activity in liver.

Doses of the drugs	GGT activity
0.9% NaCl (control)	2.94 +/- 0.18
CsA 10 mg/kg	4.03 +/- 0.21*
CsA 20 mg/kg	4.80 +/- 0.36*
CsA 40 mg/kg	1.34 +/- 0.14*
CsA 10mg/kg + Amlo 0.07mg/kg	12.42 +/- 1.30*
CsA 20mg/kg + Amlo 0.07mg/kg	7.38 +/- 0.46*
CsA 20mg/kg + Amlo 0.07mg/kg	5.46 +/- 0.59*

The activity of GGT is expressed in mU/mg protein. Results are expressed as mean +/- SEM. *$p < 0.05$ compared with control.

DISCUSSION

The data presented in this paper indicate that CsA at low doses increases the activity of GGT in the liver. This effect is augmented by calcium channel blockade by amlodipine. A high dose of CsA (40 mg/kg b.w.) decreases the GGT activity. The effect of CsA combined with amlodipine on the GGT activity is CsA dose - dependent. CsA is metabolized by isoenzymes of the liver cytochrome P-450 enzyme superfamily (Guengerich, 1992; Kronbach et al., 1988; Prueksaritanont et al., 1993). Coadministration of drugs that interact with the cytochrome P-450 system may affect CsA metabolism. Cytochrome P-450 inhibitors that decrease CsA metabolism include calcium channel blockers, ketoconazole, erythromycin, and androgens (Guengerich, 1992;

Kroemer et al., 1993; Kronbach et al., 1988; Lindholm and Henricssons, 1987). On the other hand, some agents that induce cytochrome P-450 enzymes increase CsA metabolism. These include phenobarbital, carbamazepine, valproate and dexamethasone (Kahan, 1989; Prueksaritanont et al., 1993). It was recently reported that cobaltic protoporphyrin IX, a cytochrome P-450 depletor, enhanced induction of the GGT activity in the liver by dexamethasone (Paolicchi et al., 1993). It may be possible that cytochrome P-450 participates in the regulation of GGT activity. However, CsA may act as an inhibitor of signal transduction by complexing with cyclophilin and subsequently inhibiting the calcineurin (Sering/threonine phosphatase 2B), in a Ca^{2+} - dependent reaction. An inhibition of calmodulin activation by CsA may limit the ability of the cell to extrude excess calcium and maintain normal calcium homeostasis. Recently, it was discovered that CsA inhibits the release of Ca^{2+} from mitochondria (Broekemeier et al., 1989). The above data may suggest that Ca^{2+} and Ca^{2+} - dependent biochemical pathways may play a role in the regulation of the activity of GGT in the liver. The action of CsA on the induction of the GGT activity is enhanced by amlodipine and indicates mechanisms depend on receptor - operated calcium channel. In our experiments, inhibition of Ca^{2+} influx into the cell by constant dose of amlodipine indicates that the CsA may affect the GGT activity in a dose - dependent manner. On the other hand, changes of the activity of GGT may modulate cellular level of glutathione. Thus, the effect of CsA and amlodipine on the cellular level of glutathione may be suggested.

In conclusion, the present study demonstrated that CsA may modulate the activity of GGT in the liver. Furthermore, the calcium channel antagonist amlodipine enhances the action of CsA, and mechanisms may be calcium dependent.

ACKNOWLEDGEMENTS

This work was supported by the School of Medicine Lublin, Poland, grant PW 4/93.

REFERENCES

Antoni, F.A., Shipston, M.F., Smith, S.M. (1993). Inhibitory role for calcineurin in stimulus - secretion coupling revealed by FK506 and cyclosporin A in pituitary corticotrope tumor cells. Biochem. Biophys. Res. Commun. 194:226-233.

Broekemeier, K.M., Dempsey, M.E., Pfeffer, D.R. (1989). Cyclosporin A is a potent inhibitor of the inner membrane permeability transition in liver mitochondria. J. Biol. Chem. 264: 7826-7830.

Carafoli, E. (1991). The calcium pumping ATPase of the plasma membrane. Ann. Rev. Physiol. 53: 531-547.

Chung, G.H., Lee, H.J., Yang, K.H. (1990). Regulation of the hydrolitic and transfer activities of gamma-glutamyl transpeptidase. Life Sci. 46: 1343-1348.

Cohen, G. (1983). The pathology of Parkinson;s disease: biochemical aspects of dopamine neuron senescence. J. Neural. Transm. [Suppl.] 19: 213-217.

Colombani, P.M., Robb, A., Hess, A.D. (1985). Cyclosporine A binding to calmodulin: a possible role site of action on T lymphocytes. Science 228: 337-339.

Deleve L.D., Kaplowitz, N. (1990). Importance and regulation of hepatic glutathione. Semin. Liver Dis. 10: 251-266.

Fischer, G., Wittman-Liebold, B., Lang, K., Keifhaber, T., Schmid, F.X. (1989). Cyclophilin and peptidyl-prolyl cis-trans isomerase are probably identical proteins. Nature 337: 476-478.

Guengerich, F.P. (1992). Human cytochrome P-450 enzymes. Life Sci. 50: 1471-1478.

Handschumacher, R.E., Harding, M.W., Rice, J., Drugge, R.J., Speicher, D.W. (1984). Cyclophilin: A specific cytosolic binding protein for cyclosporin A. Science 226: 544-547.

Kahan, B.D. (1989). Cyclosporine. N. Engl. J. Med. 321: 1725-1738.

Koob, M., Dekant, W. (1991). Bioactivation of xenobiotics by formation of toxic glutathione conjugation. Chem. Biol. Interact. 77: 107-136.

Kottgen, E., Reutter, W., Gerok, W. (1976). Two different gamma-glutamyltransferases during development of liver and small intestine: A fetal (sialo-) and an adult (asialo-) glycoprotein. Biochem. Biophys. Res. Commun. 72: 61-66.

Kroemer, H.K., Gautier, J-C., Beaune, P., Henderson, C., Wolf, C.R. (1993). Identification of P450 enzymes involved in metabolism of verapamil in humans. Naunyn-Schmied. Arch. Pharmacol. 348: 332-337.

Kronbach, T., Fischer, V., Meyer, U.A. (1988). Cyclosporine metabolism in human liver: identification of cytochrome P450 III gene family as the major cyclosporine - metabolizing enzyme explains interactions of cyclosporine with other drugs. Clin. Pharmacol. Ther. 43: 630-635.

Lin, C.S., Boltz, R.C., Siekierka, J.J., Sigal, N.H. (1991). FK-506 and cyclosporin A inhibit highly similar signal transduction pathways in human lymphocytes. Cell Immunol. 133: 269-284.

Lindholm, A., Henricssons, S. (1987). Verapamil inhibits cyclosporinc metabolism. Lancet 1: 1262-1263.

Liu, J. (1993). FK506 and cyclosporin, molecular probcs for studying intracellular signal transduction. Immunol. Today 14: 290-295.

Loterszatjn, S., Brechler, V., Pavoine, C., Dufour, M. (1990). The role of plasma membrane Ca^{2+} pumps as targets for hormonal action. In: Nahorsky S.R. (ed) Transmembrane signalling. Intracellular messengers and implications for drug development. John Wiley & Sons Ltd., New York. 141-156; 1990.

Lowry, O.H., Rosebrough, N.J., Farr, A.L., Randall, R.J. (1951). Protein measurement with the Folin phenol reagent. J. Biol. Chem. 193: 265-275.

McCabe, Jr. M.J., Nicotera, P., Orrenius, S. (1992). Calcium - dependent cell death. Role of the endonuclease, protein kinase C, and chromatin conformation. Ann. NY Acad. Sci. 663: 269-278.

Meister, A. (1974). Glutathione; Metabolism and function via the γ-glutamyl

Meister, A. (1974). Glutathione; Metabolism and function via the γ-glutamyl cycle. Life Sci. 663: 269-278.
Nicchitta, C.V., Kamoun, M., Williamson, J.R. (1985). Cyclosporine augments receptor - mediated cellular Ca^{2+} fluxes in isolated hepatocytes. J. Biol. Chem. 260: 13613-13618.
Niggli, E., Lederer, W.J. (1991). Molecular operations of the sodium - calcium exchanger revealed by conformation currents. Nature 349: 621-624.
Nishimura, M., Teschke, R. (1983). Alcohol and gamma-glutamyltransferase. Klin. Wochenschr. 61: 265-175.
Orlowski, M., Meister, A. (1970). The γ-glutamyl cycle: a possible transport system for amino acids. Proc. Natl. Acad. Sci. USA 67: 1248-1255.
Paolicchi, A., Chieli, E., Rugin, E.S., Tongiani, R. (1993). Inducibility of gamma-glutamyltransferase by dexamethasone in rat liver: relationship with the cytochrome P-450 content. Life Sci. 52: 631-637.
Prueksaritanont, T., Correia, M.A., Rettie, A., Swinney, D.C., Thomas, P.E., Benet, L.Z. (1993). Cyclosporine metabolism by rat liver microsomes. Evidence for involvement of enzyme(s) other than cytochromes P-450 3 A. Drug Metab. Dispos. 21: 730-737.
Putney, Jr. J.W. (1990). Receptor - regulated calcium entry. Pharmac. Ther. 48: 427-434.
Shi, M. Gozal, E., Choy, H.A., Forman, H.J. (1993). Extracellular glutathione and γ-glutamyl transpeptidase prevent H_2O_2-induced injury by 2,-dimethoxy-1,4-naphtoquinone. Free Radical Biol. Med. 15: 57-67.
Stastny, F., Pitha, J., Lisy, V., Hilgier, W., Kaucka, I., Albrecht, J. (1992). The effect of ammonia and pH on brain γ-glutamyl transpeptidase in young rats. FEBS Lett 300: 247-250.
Takahashi, N., Hayano, T., Suzuki, M. (1989). Peptidyl-prolyl cis-trans isomerase is the cyclosporin A - binding protein cyclophilin. Nature 337: 473-475.
Tate, S.S. (1980). Enzymes of merkapturic acid formation. In: Jakoby W.B. (ed) Enzymatic basis of detoxication. Vol. II. Academic Press, New York, 95-120.
Tate, S.S., Meister, A. (1976). Subunit structure and isozymic forms of γ-glutamyl transpeptidase. Proc. Natl. Acad. Sci. USA 73: 2599-2603.
Waldmann, T.A. (1993). The IL-2/IL-2 receptor system: a target for rational immune intervention. Immuno. Today 14: 264-269.

Weber, L.W.D., Lebowsky, M., Stahl, B.U., Kettrup, A., Rozman, K. (1992). Comparative toxicity of four chlorinated dibenzo-p-dioxins (CDDs) and their mixture. Arch. Toxicol. 66: 476-486.

Yamashita, K. Hitoi, A., Taniguchi, N., Yokosawa, N., Tsukuda, Y., Kobata, A.(1993). Comparative study of the sugar chains of γ-glutamyltranspeptidases purified from rat liver and rat AH-66 hepatoma cells. Cancer Res. 43: 5059-5063.

ENHANCED EXPRESSION OF THE 27 Da HEAT-SHOCK PROTEIN DURING DENA-INDUCED HEPATOCARCINOGENESIS IN RAT AND IN HUMAN NEOPLASTIC AND NONNEOPLASTIC LIVER TISSUES.

N.Mairesse*, M.Delhaye **/***, B.Gulbis*, P.Galand */***

* I.R.B.H.N., Universite' Libre de Bruxelles (ULB), ** Medicosurgical Gastroenterology, Erasme Hospital and *** Lab. of Cytology and Experimental Cancerology, School of Medicine, Bldg. C, 808 route de Lennik, B-1070 Brussels, Belgium.

INTRODUCTION

Mammalian cells respond to various kinds of stress by altering their normal patterns of protein synthesis and dramatically increasing amino acid incorporation into a series of specific polypeptides called "heat shock" or "stress" proteins (Hickey and Weber, 1982; Landry et al., 1982a; Welch, 1993) that seem to transiently protect the cell against further stress (Landry et al., 1982b).

Several lines of evidence suggest that they are involved in a number of other cell functions related to folding, unfolding and translocation of proteins (Gething and Sambrook, 1992; Georgopoulos, 1992). These stress proteins (Hsps) include a family of 27 Da proteins. Like others, these are involved in thermotolerance and very likely in specific functions related to cell growth and differentiation (see Ciocca et al., 1993 for review). As part of our effort to characterize the polypeptide constituents of neoplastic liver

NATO ASI Series, Vol. H 88
Liver Carcinogenesis
Edited by G. G. Skouteris

cells during diethylnitrosamine (DENA)-induced hepatocarcinogenesis in rats, we previously reported on enhanced expression of the 27 Hsp protein in neoplastic nodules and hepatocellular carcinomas (Mairesse et al., 1990). Here we extended the study on the expression of this protein in different neoplastic and nonneoplastic human liver tissues as a potential marker of premalignant and malignant liver lesions.

MATERIALS AND METHODS

Human tissues. Human liver tissues were obtained at partial or total hepatectomy. The clinical situations being investigated are summarized in Table 1. Tumors and surrounding liver tissues were excised, immediately washed in Krebs-Ringer bicarbonate (KRB) medium and then cut into small pieces. A histological analysis of a sample from each tissue specimen was performed.

Radiolabeling of cell proteins. In all experiments, liver and tumor fragments were incubated for 2h at 37°C in methionine-free or leucine-free medium in the presence of ^{35}S-methionine (125 µCi/ml, 1.500 Ci/mmol; Amersham Ltd., Buckinghamshire, UK) or [^{3}H]leucine (50µCi/ml, 120 Ci/mmol; Amersham Ltd.). To investigate the effects of heat shock, we incubated tissue fragments for 20 min at 44°C (37°C for controls) before adding [^{3}H]leucine.

In all cases, after having discarded the incubation medium, the incubated samples were immediately frozen or incubated in a small volume of O'Farrell's lysis buffer (O'Farrell, 1975).

To evaluate phosphorylation of the proteins, we preincubated the samples in medium made 10^{-5} mol/L in phosphate and then further incubated for 2h in the presence of 50 μCi/ml of carrier-free $[^{32}P]H_3PO_4$ (Amersham Ltd.) before performing protein extraction as described above.

To determine the radioactivity incorporated into protein, 5 μl samples were precipitated in cold 10% trichloroacetic acid; the precipitate was washed successively with 5% trichloroacetic acid, alcohol and acetone and counted in a liquid scintillation counter (Mark II Searle counter; Searle, Withoorn, The Netherlands). The rest of the original samples was immediately frozen at -70°C until electrophoresis was carried out.

Protein electrophoresis. Proteins were solubilized directly in SDS buffer for one-dimensional (1-D) SDS-PAGE (Laemmli, 1970) or in O'Farrell lysis buffer for 2-D SDS-PAGE (O'Farrell, 1975). In 1-D SDS-PAGE, samples were electrophoretically analyzed on an SDS-polyacrylamide linear gradient slab gel, 10-17%, in Tris-buffered saline solution (TBS; 800 mmol/L)(Laemmli, 1970). In 2-D SDS-PAGE, analysis consisted of isoelectric focusing on an ampholyte gel with a pH 5-7 gradient. The samples were loaded onto the basic end of the gel and subjected to electrophoresis at 510 V for 16h, then for 1h at 1.000 V. At the end of the run, the gel rods were equilibrated for 20 min in the SDS sample buffer and transferred to the top of a slab gel consisting of SDS-polyacrylamide linear gradient, 10-17%, in TBS (800mmol/L) according to a slightly modified O'Farrell method (O'Farrell, 1975) and described previously (Mairesse and Galand, 1982). Gels containing ^{35}S-labeled or ^{3}H-labeled proteins were impregnated with En^3Hance (NEN, Boston, MA) dried and autoradiographed at -80°C on Kodak XAR-5 films (Eastman Kodak Co., Rochester, NY). Gels containing ^{32}P-labeled samples were autoradiographed at -80°C with an intensifying screen but without any enhancing treatment.

Immunoblotting. After electrophoresis, the proteins were electrophoretically transferred to nitrocellulose sheets with the aid of a Trans-Blot apparatus (Bio-Rad Labs.,Richmond) at 30 V and 100 mA at 4°C overnight with a transfer buffer containing Tris-glycine and 30% (vol/vol) methanol as described by Towbin (Towbin et al., 1979). The blots were subsequently incubated for 2h in 10% (wt/vol) of fat-free dry milk in 500 mmol/L TBS, then incubated with a rabbit polyclonal antibody (IgG affinity-purified) raised against human 27hsp purified from HeLa cells (a generous gift from Dr.W.J.Welch, San Francisco, CA). The antibody was diluted 1:1000 in 125 mmol/L TBS. The blots were incubated in contact with this antibody overnight at room temperature with constant shaking. After three washes in 500 mmol/L TBS, the immune complexes on the blot were detected using a botinylated donkey anti-rabbit Ig (diluted 1:500; 1h in contact) used as binding bridge to biotin-streptavidin peroxidase performed complexes (Amersham Ltd.) (used at 1:500 dilution; 1h contact). Peroxidase activity was revealed by the mixed chromogenic substrate method (Young, 1989). For specificity controls, we used normal rabbit Ig fraction (Dako, Copenhagen, Denmark) or rabbit antihuman albumin (Behring AS, Marburg, Germany) in place of the primary human 27 hsp antibody. Measurement of protein abundance was performed with laser densitometry (SLR 1D-2D; Biomed Instrument Zeineh, Analis, Belgium) on immunoblots from the 1-D SDS-PAGE of all liver samples studied. Protein content was estimated on the basis of peak area.

Results and Discussion

As illustrated in Fig.1, the 27 KD phosphorylated protein detected in non-neoplastic and neoplastic lesions during DENA-induced hepatocarcinogenesis in rats, was identified by us as a member of the low molecular weight Hsp family (Mairesse et al., 1990). This assessment was based on the physico-chemical characteristics (MW, pl), the resolution into several isoforms after [^{3}H]-leucine labeling, the low methionine content, the phosphorylation of the acidic isoform and the inducibility upon cadmium treatment.

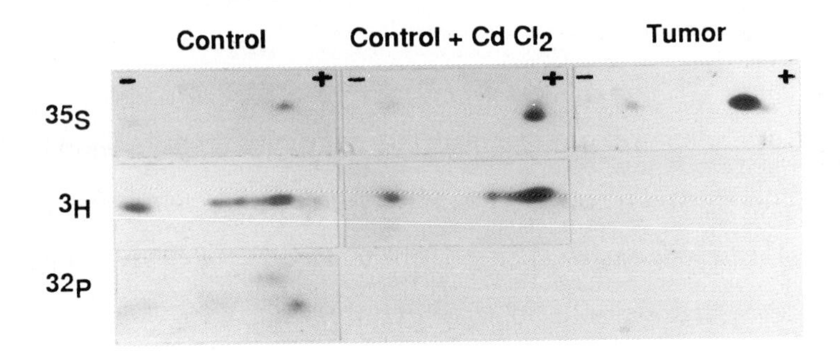

Figure 1.: 2 D-E fluorograms of [^{35}S], [^{32}P] labeled proteins from normal rat and from hepatocellular carcinoma. Normal liver fragments were incubated with or without cadmium chloride. Only the gel area corresponding to the region of 27 KD and pl 5.5-6 is shown.

The same approach (i.e. in vitro labeling of proteins and analysis of 2-DE patterns) was followed for investigating human liver samples in different pathological conditions (see table 1), with the advantage that it was possible to further identify the hsp 27 protein with a polyclonal antibody specific to the human 27 KD hsp (a generous gift from Dr.W.J.Welch,San Francisco, CA).

TABLE 1. Clinical and anapathological findings in study livers

Case no.	Age (yr)	Sex	Associated clinical condition	Anapathological findings	Tumor size (cm)
1	16	M	Serum liver enzymes normal; cerebral death	Normal liver	–
2	41	M	Alcoholism	Cirrhosis	–
3	29	F	Oral contraception	Liver cell adenoma (USL = normal)	2 × 2 × 1.5
4	0.5	M	Chemotherapy before surgery	Hepatoblastoma (USL = normal)	5.5 × 4 × 2
5	55	M	Viral hepatitis 13 yr before diagnosis	Poorly differentiated HCC (USL = normal)	23 × 22 × 5
6	33	F	Oral contraception	Trabecular, well-differentiated HCC in liver cell adenoma (USL = normal)	16 × 11 × 6
7	68	F	Gastric ulcus; gallbladder stones	Trabecular, moderately differentiated HCC (USL = cirrhosis)	5 × 5 × 4
8	58	M	Alcoholism	Trabecular, well-differentiated HCC (USL = cirrhosis)	3.5 × 2.5
9	52	F	Alcoholism	Acinar, well-differentiated HCC (USL = cirrhosis)	3.5 × 3 × 2.5
10	56	M	NANB hepatitis 4 yr before diagnosis	Multifocal, trabecular, well-differentiated HCC (USL = cirrhosis + dysplasia)	5 × 3 × 1
11	62	M	Diabetes	Trabecular, well-differentiated HCC (USL = hemochromatosis)	ND
12	47	M	HCV infection	Trabecular, well-differentiated HCC (USL = cirrhosis)	2 × 2

USL = uninvolved surrounding liver; NANB = non-A, non-B hepatitis; ND = not determined; HCV = hepatitis C virus.

The electrophoretic profiles of cellular proteins from hepatoma tissue (n=3), liver adenoma tissue (n=1) and the surrounding non neoplastic liver tissuc (n=4) were analyzed. The [^{3}H]-leucine labeling intensity of two 27 Da polypeptides with apparent isoelectric pH of 6.2 and 6 (p27a and p27b) was markedly increased in the tumor tissue relative to the normal liver (fig.2A).

In the lesions as well as in the surrounding normal tissue, heat shock resulted in a marked increase in the intensity of the two isoforms. A third polypeptide (p27c, pI5.8), barely detectable in heat-treated normal liver, appeared in the tumor after heat-shock (Fig.2). After $^{32}PO_4$ labeling, three labeled spots were revealed, corresponding to p27b, c and d, which therefore correspond to phosphorylated products.

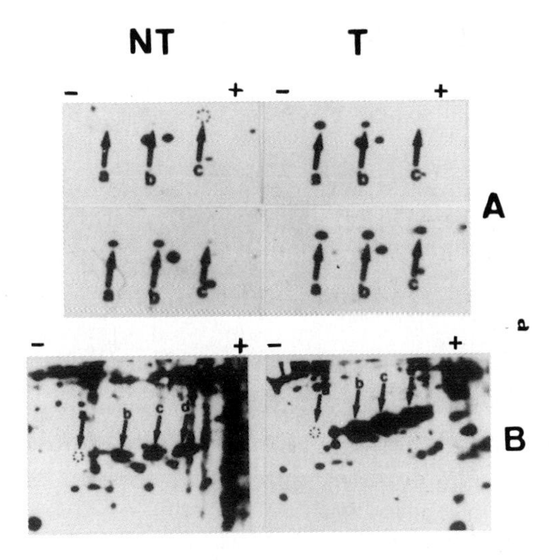

Figure 2:
(A). [^{3}H]-leucine labeling of liver cell adenoma (T) and non tumorous surrounding liver (NT) after incubation at 37°C or after 20 min exposure at 44°C before further labeling at 37°C (case 3 in table 1). The amount of radioactivity loaded on each gel was 100.000 cpm.

(B). [^{32}P]-labeling of proteins from a well-differentiated HCC and from the surrounding normal liver. The amount of radioactivity loaded on gels was around 500.000 cpm.

Western blot analysis of the electrophorograms with the specific anti-human hsp 27 antibody allowed the recognition of a series of spots corresponding in position to the above defined p27a, b, c and d and of one additional polypeptide p27a, not phosphorylated and probably of low turnover as it was not detected in [^{3}H]-patterns. The identification of individual isoforms was confirmed by autoradiography and superimposition of the film with the nitrocellulose sheet (Fig.3).

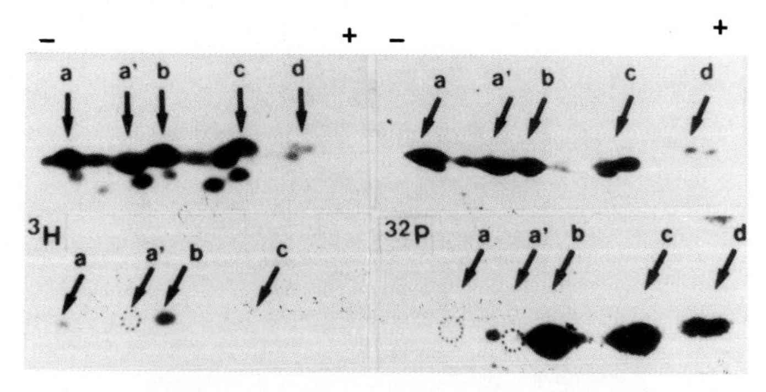

Figure 3: Western blot analysis using a polyclonal antibody specific to human Hsp 27 (top) followed by autoradiography of the nitrocellulose sheet (bottom) for identification of the different isoforms.

We propose that p27a' could be the product of another gene and/or of another messenger RNA resulting from an alternative splicing. Indeed, on Northern blot analysis, a double messenger RNA band at 95 kb was recognized by the complementary DNA probe specific to Hsp 27 gene (Hs 208-StressGen) (Mairesse N. et al., unpublished data, 1992).

The relative protein content in each isoform was evaluated by laser densitometry on immunoblots from 2-D SDS-PAGE in three different cases. Densitometry failed to reveal any significant change in the proportions of the individual isoforms of 27 hsp in the tumor compared with uninvolved tissue (data not shown). In accordance with that, the overall increase observed in the ^{32}P-labeled 27 Da Hsp in tumors related to that observed in the surrounding liver, also equally affected each of the isoforms. Measurement of protein abundance was performed on immunoblots from 1-D SDS-PAGE of 11 liver samples (fig.4, Table 1).

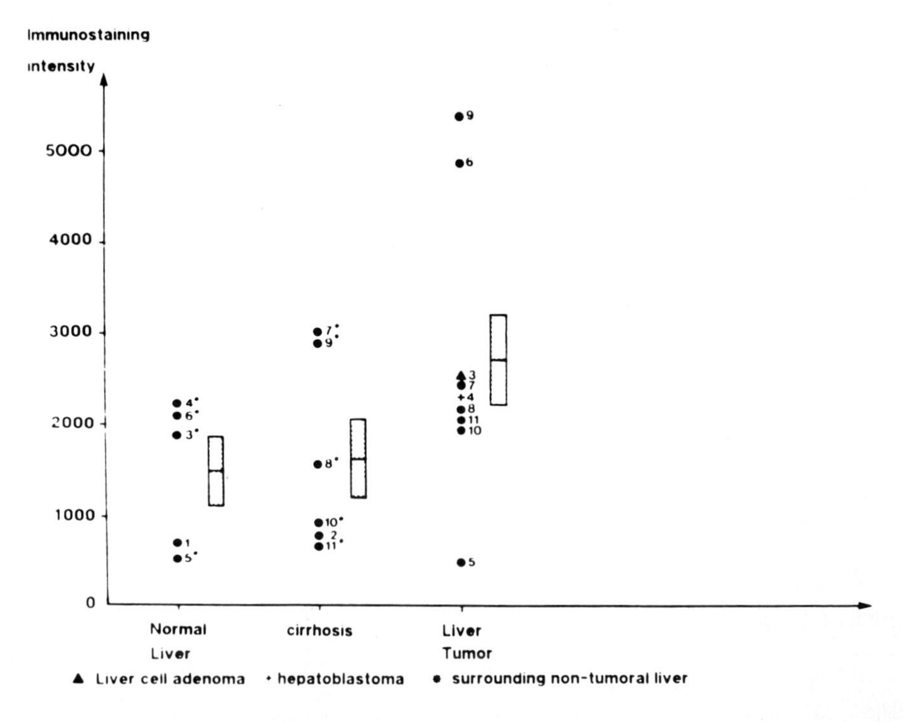

Figure 4: 27 hsp content in human liver tissues. 1-D electrophoregrams of all liver samples were immunoblotted with the polyclonal rabbit anti-human 27 hsp antibody and were scanned densitometrically to estimate 27 hsp on the basis of peak area. square=mean +/- S.E.M. The level of statistical significance was not reached for differences between normal liver or cirrhosis and liver tumor.

This showed a dramatic increase in the intensity of 27 hsp band in five of seven hepatomas compared with the nonneoplastic surrounding liver tissue. Interestingly, this was not found in the hepatoblastoma (case 4) or in the poorly or moderately differentiated HCCs (cases 5 and 7). This suggests a possible relationship with the degree of tumor differentiation.

Whereas increased constitutive amounts of an Hsp 70 related protein were observed in hepatoma cell lines (Schiaffonati et al., 1987; Cairo et al., 1989) and treatment with hepatocarcinogens was reported to increase Hsp 83 and Hsp 70 gene transcription in rat liver (Carr et al., 1986), no human well-differentiated hepatocellular carcinomas (HCCs) is observed in both the phosphorylated and the non-phosphorylated forms. In addition, our data validate our hypothesis about the identity of the Hsp 27-like protein detected in DENA-induced neoplastic nodules and HCCs in the rat (Mairesse et al., 1990).

The role of Hsp in cellular physiology (if any other that adaptation to stress) remains enigmatic. The high turnover rate of this protein (Edington and Hightower, 1990; Landry et al., 1982b) together with the rapid rate of phosphorylation-dephosphorylation in response to mitogenic signals, would fit with the regulatory role suggested in cellular growth and differentiation (Ciocca et al., 1993 for review; Spector et al., 1993; Knauff et al., 1992; Shakoori et al., 1992).

In differentiating systems, f.i. macrophage differentiation of HL-60 promyelocytic leukemic cells in response to phorbol ester (Spector et al., 1993), high level of phosphorylated hsp 27 coincides with the down-regulation of cell proliferation. It was recently hypothesized, that the phosphorylated form of hsp 27 may be related to growth arrest and the non-phosphorylated one should refer to cell proliferation (Faucher et al., 1993).

Finally, hsp 27 might intervene together with other hsps, in cellular resistance against the cytotoxicity of the carcinogen and various adverse conditions (such as oxidative injury resulting from ischemia) that prevail in the neoplastic tissues. A pleiotropic role of hsp 27 was clearly demonstrated in cells transfected with the human hsp 27 gene (Huot et al., 1991) and a line of evidence indicated that an hsp 27-like protein has in vitro antioxidant properties (Kim et al., 1989).

We propose that overexpression of hsp 27 might be related to the " resistant phenotype" acquired by some tumors during the malignant progression and might cooperate with other factors (such as glutathione and its associated enzymes - Pitot et al., 1989- or overexpression of multidrug resistance gene - Teeter et al., 1990-) and participates in the strategy of antioxidant defence for survival, thus conferring a selective advantage to the transformed cells.

REFERENCES

Cairo G, Schiaffonati L, Rappocciolo E, Tacchini L, Bernelli-Zazzera A (1989) Expression of different members of heat shock protein 70 gene family in liver and hepatomas. Hepatology 9:740-746

Carr BI, Huang TH, Buzin CH, Itakura K (1986) Induction of heat shock gene expression without heat shock by hepatocarcinogens and during hepatic regeneration in rat liver. Cancer Res 46:5106-5111

Ciocca DR, Oesterreich S, Chamness GC, McGuire WL, Fuqua SAW (1993) The small heat shock protein hsp27 is correlated with growth and drug resistance in human breast cancer cell lines. J Natl Canc Inst 85:1558-1570

Dougherty K, Spilmans D, Green C, Steward AR, Byard JL (1980) Primary cultures of adult mouse and rat hepatocytes for studying the metabolism of foreign chemicals. Biochem Pharmacol 29:2117-2124.

Edington BV, Hightower LE (1990) Induction of a chicken small heat shock (stress) protein: evidence of multilevel prostranscriptional regulation. Mol Cell Biol 10:4886-4898

Faucher C, Capdevielle J, Canal I, Ferrara P, Mazarguil H, McGuire WL, Darbon JM (1993) The 28-da protein whose phosphorylation is induced by protein kinase C activators in MCF-7 cells belongs to the family of low molecular mass heat shock proteins and is the estrogen-regulated 24-da protein. J Biol Chem 268:15168-15173

Gething HJ, Sambrook J (1992) Protein folding in the cell. Nature 355:33-45

Georgopoulos C (1992) The emergence of the chapnone machines. TIBS 17:295-299

Hickey ED, Weber LA (1982) Modulation of heat-shock polypeptide synthesis in HeLa cells during hyperthermia and recovery. Biochemistry 21:1513-1521

Huot J, Roy G, Lambert H, Chrσtien P, Landry J (1991) Increased survival after treatments with anticancer agents of Chinese hamster cells expressing the human Mr 27,000 heat shock protein. Cancer Res 51:5245-5252

Kim IH, Kim K, Thee SG (1989) Induction of an antioxydant protein of saccharomyces ceevisiae by O_2, Fe^{3+}, or 2-mercaptoethanol. Proc Natl Acad Sci USA 86:6018-6022

Knauff U, Bielka H, Gaestel M (1992) Over-expression of the small heat-shock protein, hsp 25, inhibits growth of ehrlich ascites tumor cells. FEBS Lett 309:297-302

Laemmli UK (1970) Cleavage of structural proteins during the assembly of the head of bacteriophage T4. Nature 227:680-685

Landry J, Chrσtien P, Bernier D, Nicole LM, Marceau N, Tanguay RM (1982a) Thermotolerance and heat shock proteins induced by hyperthermia in rat liver cells. Int J Radiat Oncol Biol Phys 8:59-62

Landry J, Bernier D, Chrσtien P, Nicole LM, Tanguay RM, Marceau N (1982b) Synthesis and degradation of heat shock proteins during development and decay of thermotolerance. Cancer Res 42:2457-2461

Mairesse N, Delhaye M, Galand P (1990) Enhanced expression of a 27 kD protein during diethylnitrosaminc-induccd hepatocarcinogenesis in rats. Biochem Biophys Res Commun 170:908-914

Mairesse N, Galand P (1982) Estrogen-induced proteins in luminal epithelium, endometrial stroma and myometrium of the rat uterus. Mol Cell Endocrinol 28:671-679

O'Farrell PH (1975) High resolution two-dimensional electrophoresis of proteins. J Biol Chem 250:4007-4021

Oesterreich S, Weng CN, Qiu M, Milsenbeck SG, Osborne CK, Fuqua SAW (1993) The small heat shock protein hsp 27 is correlated with growth and drug resistance in human breast cancer cell lines. Cancer Res 53:4443-4448.

Pitot HC, Goodspeed D, Dunn T, Hendrich S, Maronpot RR, Moran S (1989) Regulation of the expression of some genes for enzymes of glutathione metabolism in hepatotoxicity and hepatocarcinogenesis. Toxico

Pitot HC, Goodspeed D, Dunn T, Hendrich S, Maronpot RR, Moran S (1989) Regulation of the expression of some genes for enzymes of glutathione metabolism in hepatotoxicity and hepatocarcinogenesis. Toxico Appl Pharmacol 97:27-34

Reese JA, Bjard JL (1981) Isolation and culture of adult hepatocytes from liver biopsies. In Vitro 17(II):935-940

Scherer E, Emmelot P (1975) Foci of altered liver cells induced by a single dose of diethylnitrosamine and partial hepatectomy: their contribution to hepatocarcinogenesis in the rat. Eur J Cancer 11:145-154

Schiaffonati L, Bardella L, Cairo G, Rappocciolo E, Tacchini L, Bernelli-Zazzera A (1987) Constitutive and induced synthesis of heat shock proteins in transplantable hepatomas. Tumori 73:559-565

Shakoori AR, Oberdorf AM, Owen TA, Weber LA, Hickey E, Stein JL, Lian JB, Stein GS (1992) Expression of heat shock genes during differentiation of mammalian osteoblasts and promyelocytic leukemia cells. J Cell Biochem 48:277-287

Spector NL, Ryan C, Samson W, Levine H, Nadler LM, Arrigo AP (1993) Heat shock protein is a unique marker of growth arrest during macrophage differentiation of HL-60 cells. J Cell Physiol 156:619-625

Teeter LD, Becker FF, Chisari FV, Li D, Kuo MT (1990) Overexpression of the multidrug resistance gene mdr3 in spontaneous and chemically induced mouse hepatocellular carcinomas. Mol Cell Biol 10:5728-5735

Towbin H, Staehelin T, Gordon J (1979) Electrophoretic transfer of proteins from polyacrylamide gels to nitrocellulose sheets: procedure and some applications. Proc Natl Acad Sci USA 76:4350-4354

Welch WJ (1993) How cells respond to stress. Scientific American 5:34-41

Young PR (1989) Enhancement of immunoblot staining using a mixed chromogenic substrate. J Immunol Methods 121:295-296

HEPATIC RECOVERY OF DIMETHYLNITROSAMINE-CIRRHOTIC RATS AFTER INJECTION OF THE LIVER GROWTH FACTOR

Juan J. Diaz-Gil *, Carmen Rua #, Celia Machin #, M.Rosa Cereceda* , M.Carmen Guijarro * and Pedro Escartin **

* Laboratory of Experimental Hepatology
Department of Experimental Biochemistry
Clinica Puerta de Hierro
S.Martin de Porres, 4
28035 Madrid
Spain

Department of Cellular Biology
Faculty of Biological Sciences
Universidad Complutense
Madrid

** Department of Gastroenterology
Clinica Puerta de Hierro
Madrid

INTRODUCTION

Hepatocellular carcinoma (HCC) is a major worldwide health problem, with an estimated 250.000-1.000.000 new cases every year. The etiologic factors cover a wide spectrum, including chronic liver disorders caused by virus B,C and D, hepatotoxins such as aflatoxins and alcohol, certain genetic predisposition as in hemochromatosis, tyrosinosis and accumulation of collagen type I, and cirrhosis of variable etiology. Of all these factors, cirrhosis is the most frequent, and many authors have proposed it as the step immediately preceding the appearance of HCC (Johnson and Williams,

NATO ASI Series, Vol. H 88
Liver Carcinogenesis
Edited by G. G. Skouteris

1987). For this reason, any significant advance in the treatment of cirrhosis ought to have a clear impact in diminishing HCC incidence.

One of the models for production of liver cirrhosis in rats is injection of dimethylnitrosamine (DMN) at low doses over long periods of time.DMN is also a potent carcinogen that, although it spreads through different organs in the body, is metabolized principally in the liver (Magee, 1956). DMN is also considered an environmental carcinogen (Lijinsk et al., 1970). In the series of experiments our group is carrying out to study the therapeutic utility of the Liver Growth Factor, (LGF), we will show the improvement of several parameters in irreversibly cirrhotic rats after its injection.

LGF is a mitogen for liver cells, purified both from plasma of partially hepatectomised rats (Diaz-Gil et al., 1986a) and from plasma of patients with hepatobiliary disorders (Diaz-Gil et al., 1986b).It shows "*in vivo*" activity, after injection into rats or mice at low nanogram range, stimulating liver DNA and protein synthesis, mitotic index, PCNA-positive cells, its action being organ-specific. It also has "*in vitro*" activity, in rat liver cell cultures, at 30 pg LGF/ml, stimulating DNA synthesis, Na uptake, intracellular pH increase, etc.LGF has been characterized as an albumin-bilirubin complex (Diaz-Gil et al., 1987; Diaz-Gil et al., 1989).

MATERIALS AND METHODS

DMN (Sigma Co.) was injected i.p. (10 µl DMN/kg bw, using saline as vehicle) into male Wistar rats (170 g initial bw), according to a regimen of a single injection on three consecutive days (every Monday-Wednesday) for three weeks. DMN injection was discontinued for the following two weeks, and after which it was again injected for another three-four weeks. In a total of 8-9 weeks, rats had developed irreversible cirrhosis (Jenkins et al., 1985; Jezequel et al., 1987). The mortality of rats during the injection period was usually about 8-10%. At this point, we divided the whole batch of rats (36 per experiment) into three equal groups. Subgroup 1, serving as control cirrhotic rats, were uninjected, subgroup 2 was injected with 4 ng LGF/rat, two injections/week for three weeks, and subgroup 3 was injected with 3 μg LGF/rat, two injections/week for three weeks.

LGF was purified from serum of rats sacrificed 14 days after bile duct ligation.Serum concentrations of LGF and precursors were measured in these rats and in cirrhotic rats by a HPLC method (Singh and Bowers, 1986). The procedure for LGF purification was previously reported (Diaz-Gil et al., 1986a). LGF preparations, containing about 90% rat serum albumin, were checked for "*in vivo*" activity by injection into normal rats. 24 h later, ^{3}H-thymidine was injected into rats, to measure liver DNA synthesis rate by incorporation into DNA. LGF has bell-shaped dose-effect curves, so once we determined the optimal dose to stimulate liver DNA synthesis, it was used for injection into cirrhotic rats. LGF has two peaks of activity, at low ng or low μg doses, with apparently the same stimulatory capabilities. We used both doses to check their effects in cirrhotic rats.

Ether-anesthetized rats were exsanguinated by cardiac puncture. Serum enzymes (AST, ALT, γGT, ALP, total proteins and albumin) were monitored in a Technicon Autoanalyzer following standard procedures. After removal of the liver, some portions of hepatic lobules were fixed by immersion in 10% buffered formalin or Bouin's solution for light microscopy studies. Small pieces were dehydrated and embedded in paraffin, and 10 μm thick sections were stained with hematoxylin-eosin (H-E).Van Gieson's method was also performed.

The rest of the livers were immediately frozen in liquid nitrogen, and kept at -80°C until processed for liver collagen determination. For this purpose, the method of Rojkind and Gonzalez (1986) was followed.

Two-tailed student t-test was applied when indicated.

Results

Table 1 shows serum enzyme concentration of cirrhotic rats with and without injection of LGF. The major differences were observed between cirrhotic control rats (subgroup 1) and cirrhotic rats injected with μg doses of LGF (subgroup 3). Even in parameters such as γGT, total proteins and albumin, the concentrations in rats of subgroup 3 were closer to the healthy controls. Rats of subgroup 2 showed intermediate values.

TABLE 1

Serum enzyme concentration in DMN-cirrhotic rats three weeks after irreversible state

	AST (U/l)	ALT (U/l)	GGT(U/l)	ALP(U/l)	Tot. prot. (g/dl)	Alb (g/dl)
Cirrhotic controls						
Absolute value	235 ± 33	189 ± 33	8 ± 4	506 ± 73	5.4 ± 0.6	2.2 ± 0.4
Percentage[a]	154	219	200	231	86	64
Cirrh + ng LGF[b]						
Absolute value	258 ± 66	158 ± 37	6 ± 0.9	445 ± 53	5.3 ± 0.4	2.3 ± 0.1
Percentage[a]	169	183	150	203	84	67
Est. signif.[c]	N.S.	N.S.	N.S.	N.S.	N.S.	N.S.
Cirrh + μg LGF[b]						
Absolute value	184 ± 37	148 ± 25	6 ± 2.3	347 ± 20	5.8 ± 0.3	2.6 ± 0.4
Percentage[a]	121	172	150	158	92	76
Est. signif.[c]	$0.1<p<0.05$	$0.1<p<0.05$	N.S.	$0.02<p<0.01$	N.S.	N.S.

[a] Over healthy controls (n = 9 in all subgroups). [b] Doses and schedule are indicated in the Methods section. [c] Compared with cirrhotic control rats.

Morphological study of the three subgroups of rats revealed as main features: subgroup 1, generalized necrosis, with disappearance of hepatic parenchyma in some samples, inflammatory infiltrate, especially of neutrophils (figure 1, left), and fibrosis; subgroup 3, partial recovery of hepatic parenchyma coexisting with small areas of necrosis (indicated by star in Figure 1, right),clear decrease in inflammatory infiltrates and fibrosis. Findings in subgroup 2 were intermediate between those of the other two subgroups.

With respect to intraperitoneal acsites, we detected wide variations among different experiments. In series in which 10-20 ml of ascites or more in subgroup 1, rats in subgroup 3 lacked ascites. In series showing 30 ml of ascites or more in subgroup 1, rats in subgroup 3 had a remnant of 5-10 ml. Rats in subgroup 2 were in intermediate states.

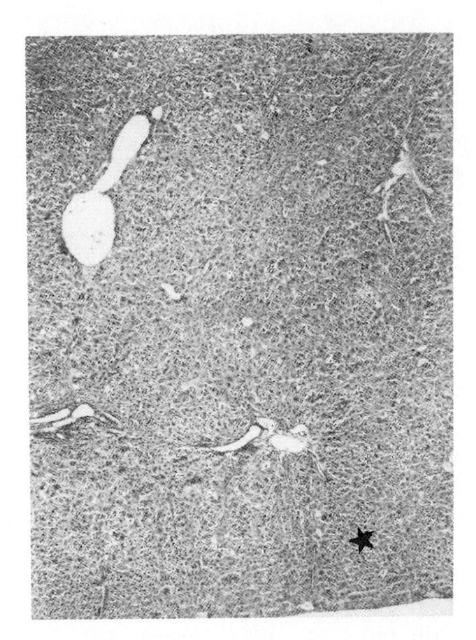

Figure 1 (left): General view of liver parenchyma of a cirrhotic control rat, t=7 weeks. Note generalized necrosis and numerous inflammatory infiltrates (arrows).(Van Gieson's stain, original magnification x48).
Figure 2 (right): Equivalent overview of the liver of a cirrhotic rat injected with µg LGF, t=7 weeks. Considerable hepatic recovery can be observed, with some remaining necrotic areas (star).(Van Gieson's stain, original magnification x37).

TABLE 2

Total collagen in liver of cirrhotic(DMN) rats

	Time	No. rats	mg Coll/g liv.	%
- Control (normal)		(n = 7)	0.88 = 0.13	100
- Cirrhotic	(t = 0)	(n = 5)	5,56 ± 2.64	631
- Cirrhotic	(t = 3w)	(n = 5)	5.43 ± 1.07	617
- " + μg LGF	(t = 3w)	(n = 5)	7.36 ± 1.31	836
- Cirrhotic	(t = 7w)	(n = 5)	5.87 ± 1.52	667
- " + μg LGF	(t = 7w)	(n = 6)	3.16 ± 0.68	359

Table 2 shows total liver collagen content of rats in the different subgroups studied. Cirrhotic rats remained constant in the interval studied, as in established cirrhosis. Seven weeks after cirrhosis onset, collagen content in rats injected with LGF was reduced by about 50%. At three weeks, we detected a transitory increase in μg LGF-injected rats. This point will be analyzed in the discussion section.

Table 3 discloses the so-called hepatic regeneration rate K, in the different subgroups of rats studied. This coefficient is equal to the ratio of LGF and its precursors, and indicates the rate at which LGF is produced and utilized. We analyzed this parameter in different situations, both in humans and in animal models, and it was closely related to the rate of liver regeneration. When coefficient K remained lower than in healthy controls for long periods of time, it was usually associated with poor prognosis. At t=3 weeks, K was clearly lower in cirrhotic rats with respect to LGF injected rats, and the same situation could be observed at seven weeks.

At this point, rats injected with μg LGF showed a hepatic regeneration rate close to healthy control rats. Coefficient K was lower in cirrhotic rats, both at 3 and 7 weeks, than in healthy control rats.

The improvement in cirrhotic rats injected with LGF was also evident in terms of survival. Figure 2 shows the survival profiles of the three subgroups from the point of irreversible cirrhosis. Seven weeks after this point, the survival of cirrhotic rats was 18%, while that corresponding to cirrhotic rats injected with ng or μg LGF was 50 and 61%, respectively. The profile of variation of these two subgroups was clearly different with respect to the corresponding curve in cirrhotic control rats.

TABLE 3

Hepatic regeneration rate in cirrhotic(DMN) rats

	Time	No. rats.	Coef. K (LGF/prec.)
- Control (normal)		(n = 4)	1.08
- Cirrhotic	(t = 0)	(n = 5)	(0.89-15.37)
- Cirrhotic	(t = 3w)	(n = 5)	0.87
- " + ng LGF	(t = 3w)	(n = 4)	1.31
- " + μg LGF	(t = 3w)	(n = 6)	1.56
- Cirrhotic	(t = 7w)	(n = 6)	0.65
- " + ng LGF	(t = 7w)	(n = 2)	0.69
- " + μg LGF	(t = 7w)	(n = 8)	0.92

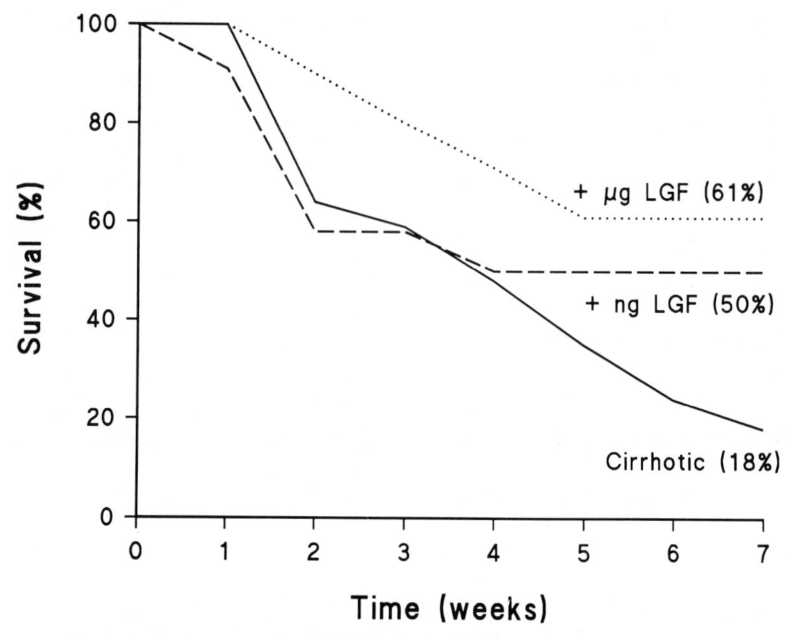

Figure 2: Survival curves of cirrhotic rats (continuous line), cirrhotic rats injected with ng LGF (dashed line), and cirrhotic rats injected with µg LGF (dotted line), n=35 in every subgroup at t=0. The survival percentage of each subgroup at t=7 weeks is indicated on each line.

DISCUSSION

In the model of DMN-cirrhosis used, two important aspects should be pointed out. First, the study was carried out in an established, irreversible state. This is essential, since the improvement in cirrhotic rats after LGF injection can not be ascribed to a spontaneous normalization of the parameters checked. This point is particularly clear in the stability of the liver collagen content of cirrhotic rats over the seven week period studied, and the high mortality of these rats, (over 85%) , in the same period of time. Second, the effects of LGF must be considered therapeutic because it was injected at the point of irreversibility. Most of the experiments referred to in the literature dealing with the study of the beneficial effects of a given chemical on a model of liver injury, involve injecting both compounds simultaneously or, in some cases the chemical prior to hepatotoxin injection. This strategy can mask hepatoprotective actions as therapeutic. This is not our case. Another important point is that the first injection of LGF was carried out five days after the last injection of DMN, so LGF effects can not be ascribed to any interference with DMN metabolism, since the hepatotoxin is metabolized within a period of 24 hours (Magee, 1956).

The reduction of total liver collagen in rats injected with µg of LGF at t=7 weeks (Table 2) is highly significant, both in quantitative and qualitative terms. This fact by itself seems to suggest a recovery of cell mass, which could justify the improvement of liver functionality shown by a certain normalization of serum enzymes. In any case, total liver collagen can be affected by variations in collagenase activity (degradation) and gene expression of the different types of collagen (synthesis). We have not studied whether LGF affects one or both processes simultaneously. Experiments are in progress to determine some of these points. On the other hand, in preliminary experiments in a model of established CCl_4-cirrhosis, LGF also produced a dramatic decrease in total liver collagen content, which seems to indicated a clear effect of LGF on different models of cirrhosis.

The reduction of ascites in cirrhotic rats after LGF injection was very clear in all six experiments carried out so far, but it was very difficult to quantitate due to the high variability among control cirrhotic rats in different experiments. This could be due to the marked activity of DMN, which makes the cirrhotic state very difficult to standardize among different experiments.

Finally, the increased survival of cirrhotic rats after injection of LGF, important by itself, is another consideration to be taken into account.At every point at which we compared parameters in the different subgroups, we were using the healthier, or less affected individuals, among the cirrhotic control rats; but in the LGF treated rats, we considered a higher percentage of a whole initial batch. This fact could disguise the real differences among subgroups, and consequently minimize the real importance of the changes produced by LGF injection.

Acknowledgement

We thank Ms.M.Messman for the preparation of the manuscript.This work was supported by Grant no.91/0556 from the Fondo de Investigaciones Sanitarias (FISS), Spain.

REFERENCES

Diaz-Gil JJ, Escartin P, Garcia-Canero R, et al. (1986a). Purification of a liver DNA-synthesis promoter from plasma of partially hepatectomized rats. Biochem J. 235: 49-55.

Diaz-Gill JJ, Sanchez G, Garcia-Canero R, Trilla C, Guerra MA, Escartin P (1986b). Characterization of a liver mitogenic factor from human plasma of subjects with hepatitis B. Hepatology 6:769.

Diaz-Gill JJ, Gavilanes JG, Sanchez G, et al. (1987).Identification of a liver growth factor as an albumin-bilirubin complex. Biochem J. 243:443-448.

Diaz-Gill JJ, Sanchez G, Trilla C, Escartin P (1988).Identification of biliprotein as a liver growth factor. Hepatology 8:444-448.

Jenkins SA, Grandison A, Baxter JN, Day DW, Taylor I, Shields R (1985).A dimethynitrosamine-induced model of cirrhosis and portal hypertension in the rat. J.Hepatology 1:489-499.

Jezequel AM, Mancini R, Rinaldesi ML, Macarri G, Venturini C, Orlandi F (1987). A morphological study of the early stages of hepatic fibrosis induced by low doses of dimethylnitrosamine in the rat. J.Hepatology 5:174-181.

Johnson PJ, Williams R (1987). Cirrhosis and the aetiology of hepatocellular carcinoma. J Hepatology 4:140-147.

Lijinsky W, Epstein SS (1970). Nitrosamines as environmental carcinogens. Nature 225:21-23.

Magee PN (1956) Toxic liver injury: The metabolism of dimethylnitrosamine. Biochem. J. 64:676-682.

Rojkind M, Gonzalez E (1974). An improved method for determining specific radioactivities of Proline-^{14}C and Hydroxyproline-^{14}C in collagen and in noncollagenous proteins. Anal. Biochem.57:1-7.

Singh J, Bowers LD (1986).Quantitative fractionation of serum bilirubin species by reversed-phase high performance liquid chromatography. J.Chromatogr. 380:321-330.

CELL CONTACT-MEDIATED REGULATION OF HEPATOCYTE DIFFERENTIATION / PROLIFERATION : ROLE OF LRP(s).

A. CORLU, P. LOYER , S. CARIOU, G.P. ILYIN, I. LAMY, M. CORRAL-DEBRINSKI and C. GUGUEN-GUILLOUZO.
INSERM U 49, Hôpital Pontchaillou, 35033 RENNES, FRANCE.

INTRODUCTION

Thc participation of proximal and/or contacting cells in the control of development and differentiation of various tissues, including liver, is now well-established (Edelman, 1987). Cell interactions can process through the production of soluble factors or the deposition of extracellular matrix components or directly, through cell-cell recognition signal(s) located on the plasma membranes.

Regarding liver differentiation, there is strong evidence that close contact between endodermal and mesenchymal cells is required. The specific recognition of cells, which occurs throughout liver organogenesis, is mediated, in part, by cell surface molecules. L-CAM, one of the primitive CAM's found in early embryonic epochs, is a major CAM expressed by embryonic liver cells (Gallin *et al.*, 1983). Also, at the adult stage, hepatocyte cords remain under the influence of mesodermal cells present in the sinusoids. Indeed, their activity was found to depend on both heterotypic hepatocytes-non parenchymal cells interactions and homotypic hepatocytes-hepatocytes interactions (Guguen-Guillouzo,1986). In the last few years, major progress has been made in acquiring knowledge about the molecules involved in such interactions, mainly those of the homotypic type. Besides L CAM, other cell surface molecules not involved in intercellular junctions like cell-CAM 105, have been found to mediate hepatocyte functional activity. Moreover, we have identified the Liver Regulating Protein (LRP) that plays a critical role in the functional activity of mature hepatocytes when cocultured with rat liver epithelial cells (RLEC) (Corlu *et al.*, 1991). LRP differs from the other cell adhesion molecules expressed in the liver as described above, since a specific monoclonal antibody directed against LRP does not influence either hepatocyte aggregation or hepatocyte adhesion to various extracellular matrix components. We have further characterized this molecule and defined its role on the proliferation/differentiation balance at both fetal and adult stages. Its expression in hepatocarcinoma has been analyzed.

NATO ASI Series, Vol. H 88
Liver Carcinogenesis
Edited by G. G. Skouteris

Concerning cell proliferation, our understanding of the molecular mechanisms that control the cell cycle in tissues has greatly increased in the last 10 years, owing to a better definition of the molecules directly involved in the cycle, protein kinases and cyclins, mainly those participating to the transitions G1/S and G2/M. The evidence is setting that the assembly of sub-units constituting the cdks/cyclins complexes as well as the events leading to their activation, and responsible for mitosis, represent an universal process (Draetta and Beach, 1988). In contrast, it is generally assumed that the G1 phase progression and the G1/S transition result from a series of competency properties and transduction signals, in response to extracellular mitogen factors that differ from one cell type to another (Sherr, 1993). In normal liver, hepatocytes are arrested in a quiescent state (G0) but are capable of dividing, following a highly controlled regenerative process. The mechanisms which contribute to this control remain, at least in part, unknown (Fausto, 1992). We have analyzed *in vivo* and *in vitro*, both expression and activation of the cell cycle-associated proteins, mainly $p34^{cdc2}$ and $p33^{cdk2}$ in proliferating rat hepatocytes during regeneration.
In addition, involvement of mitogenic factors in the G0/G1 transition of hepatocytes is still questionable. The G1 phase is generally divided into subphases and progression from a subphase to another one appears controlled by signals which define cell cycle blocks characteristic of each cell type. We have investigated the growth factor-dependence of the hepatocyte entry and progression through G1, using primary culture. A restriction point in mid-late G1 has been defined, using EGF as a growth signal. Furthermore, a cell interaction-dependent block located in mid G1 has been evidenced. Arguments for a possible involvement of LRP at this step, are provided.

Materials and methods

Animals.

Male Sprague-Dawley adult rats (approximately 180-200 g) were obtained from Charles River Laboratories, Cléon, France. 12 day-old fetuses were removed from bred females of the same origin. Partial (two-thirds) hepatectomy was performed according to the method of Higgins and Anderson (1931). As controls, sham operations (laparatomy and liver manipulation without tissue removal) were also performed. Liver carcinogenesis has been induced by giving 3-diethylnitrosamine (DENA) doses to females after a two-third hepatectomy as previously described (Pitot *et al.*, 1978). Rats were sacrified from 10 to 12 months after the carcinogen administration.

Cell isolation and culture.

Adult rat hepatocytes, isolated by collagenase perfusion of the liver (Guguen *et al.*, 1975), are suspended in a mixture of 75 % MEM and 25 % medium 199 supplemented with 10 % fetal calf serum (FCS) containing per ml : penicillin (100 IU), streptomycin sulphate (100 μg), bovine insulin (5 μg) and bovine serum albumin (BSA) (1 mg).

After 2 hours, medium was removed and replaced by a fresh medium. For hepatocyte proliferation assays a cocktail of EGF (50 ng/ml) and sodium pyruvate (20 mM) was used, as described by McGowan and Bucher (1983), and added to the cultures for different time periods.

The rat liver epithelial cell line (RLEC) was established from the liver of 10 days-old rats according to Morel-Chany *et al.* (1978).

For hepatocyte co-cultures, usually $8x10^5$ hepatocytes are plated in 2 ml of 10 % serum-supplemented medium in 3.5 cm Petri dishes. The medium containing unattached cells was removed 3 h later, and $1x10^6$ RLEC in fresh medium were seeded per dish in order to reach confluency with hepatocyte colonies within 24 h. The medium was thereafter supplemented with $3.5x10^{-6}$ M hydrocortisone hemisuccinate and changed every day. RLEC stop dividing when they reach confluency. It is also possible to seed hepatocytes into a confluent epithelial cell monolayer. For biochemical assays, hepatocytes were selectively detached from RLEC by incubation in a calcium-free HEPES-buffered collagenase solution (0.075 %) (pH 7.6) for 10 min. Livers from 12 days old fetuses were isolated, washed and digested in MEM supplemented with 10% FCS and enzymes (0.05% collagenase, 0.48% dispase I and 0.075% hyaluronidase) for 15 min under gentle stirring.

Antibodies.

The preparation and characterization of the Mab L8 used in this study have been described previously (Corlu *et al.*, 1991). Briefly, Balb/c mice were immunised against RLECs and their splenocytes were fused with Sp2-O/Ag 14 myeloma cells. Hybridoma secreting antibodies were selected. For indirect immunohistochemistry, mAb L8 ascitic fluid diluted to 1:500 was used. The anti $p34^{cdc2}$ used for immuno-blotting is a rabbit polyclonal antiserum obtained after injection of full-lenght $p34^{cdc2}$ purified from yeast (Draetta *et al.*, 1989). $p34^{cdc2}$ CT and $p33^{cdk2}$CT are two rabbit polyclonal antisera directed against the C-terminal part of human $p34^{cdc2}$ and $p33^{cdk2}$

respectively (Pagano *et al.*,1992). Cyclin A, B and E antisera are directed against the corresponding human proteins while cylin D1 antibody is raised against the murine analogue.

Indirect immunolocalisation and immunoprecipitation.
Cell cultures and liver fragments were fixed in a 4% paraformaldehyde solution buffered with sodium cacodylate 0.1M for 30 min at 4°C. LRP was localized using indirect immunoperoxidase (Guillouzo *et al.*, 1982). Molecular weight determination was performed on purified samples prepared from lysates of plasma membrane extracts following the procedure previously described (Corlu *et al.*, 1991).

Electrophoresis and western blotting of cell cycle proteins.
For each time point, biopsies were homogenized in homogenization buffer at a final concentration of 40 mg of biopsy tissue/ml corresponding to 5 mg of protein/ml. Cdc2-related kinases ($p34^{cdc2}$ and $p33^{cdk2}$) were purified by affinity on p9-beads. P9-beads purified proteins were recovered buffer, transferred to nitrocellulose paper in a transblot cell (Millipore) and used for incubation with antiserum. The p9-beads used in the studies were strikingly similar to those described by Azzi *et al.* (1992).

H1K activity assays.
P9-beads-purified cdc2/cdk2 kinases were used. Kinase assays were performed in a mixture containing 10 μl of histone H1 (5 mg/ml), 5 μl of [γ-32P] ATP and 20 μl of buffer C for 10 min, at 30°C, as described by Loyer *et al.* (1994).

Albumin production.
The amount of albumin released into the medium was measured by laser immunonephelometry, using a monospecific antiserum to rat albumin.

Extraction of RNA and Northern Blot Hybridization.
Total RNA was prepared using 5M guanidium thiocyanate/CsCL according to Chirgwin *et al.* (1979). Total RNA (10 or 20 μg) was resolved by electrophoresis, transferred onto a nitrocellulose filter and hybridized with ^{32}P nick-translated cDNA probes.

Results

Stability of hepatocytes in coculture with RLEC

Adult hepatocytes have a limited survival and lose the major part of their functional capacities within one week in culture. Although the changes reach a wide spectrum of functions, the more specific ones are preferentally modified. In various metabolic pathways, an early decrease of several specific enzyme activities occurs.

Different observations show that regardless the density and the composition of the medium used, cell viability and in vitro functional stability of hepatocyte monolayers do not exceed a few days and that neither homotypic interactions nor the presence of soluble factors are sufficient for maintaining liver specific gene transcription in postnatal and mature hepatocytes. Therefore, it has been questioned whether heterotypic cell interactions are playing a critical role in the regulation processes of hepatic differentiation in later stages of development and in adult life.

When cocultured with RLEC, hepatocytes from various species and from adult or fetus, survived for several weeks and retained high functional capacities. Among the functions preserved are the production of plasma proteins, cytochrome P-450 content, metabolism of drugs by phase I and phase II reactions, taurocholate uptake and metabolism of glycogen. In contrast to the functioning of hepatocytes maintained in a serum-free hormonally defined medium, hepatocytes cocultured with RLEC retained the capability of transcribing specific genes at a rate identical to that found in DMSO-treated cells and in cells seeded on Matrigel (Caron, 1990). It was found that the coculture system requires direct cell-cell contacts and that a spontaneous early production and deposition of extracellular matrix components was observed, reflecting the composition of the hepatic extracellular matrix.

In contrast, an early reappearance of fetal-like functions was detected, although DNA synthesis was almost absent throughout the coculture period. It included the fetal isoforms of enzymes like pyruvate kinase and glutathion-S-transferase. In addition, overexpression of oncogenes like c-myc and of some components of the extracellular matrix was clearly evidenced at starting the coculture and all along the culture period.

Involvement of LRP in hepatocyte functional activity.
To identify a cell surface protein involved in the interactions of hepatocytes with RLEC in coculture, mAbs were generated, using live RLEC as immunogen. From 24 hybridomas actively secreting antibodies directed against liver plasma membranes, only mAb L8 was able to markedly reduce the survival of hepatocytes in coculture. Most hepatocytes died after 5 or 6 days as in pure cultures, whereas RLECs were not affected. No effect on hepatocyte viability was detected with time in pure culture. These results indicated that mAb L8 did not induce toxic effects, but might alter cell-cell interactions between hepatocytes and RLEC. The capacity of the cells to secrete albumin was drastically reduced in cultures treated daily with mAb L8. The steady levels of albumin and aldolase B mRNAs were strongly increased from day 2 to day 5 in untreated cocultures, whereas these levels dramatically decreased in the presence of mAb L8 by day 5. This inhibitory effect was only evidenced when the antibody was present between days 1 and 2 after RLEC seeding, corresponding to the establishment of cell-cell contacts. Moreover, the effect of mAb L8 appeared fully reversible when the coculture was treated for 1 day.

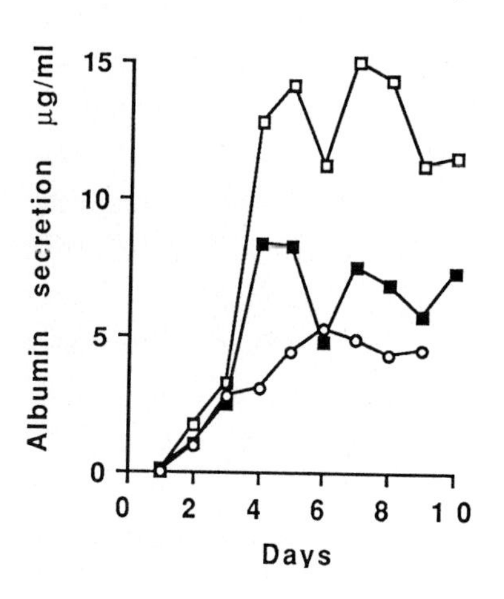

Fig. 1 : Comparison of albumin secretion by bipotential hepatoblasts maintained either in pure culture -o- , or in coculture -□- , or in coculture added with a mAb L8 monoclonal antibody directed against LRP -■- .

In addition, evidence was provided that LRP could be involved early in the regulating process of hepatocyte maturation by using bipotential hepatoblasts of 12 days old rat fetuses (Marceau, 1994). We observed that, when placed in coculture with RLEC, hepatoblasts were induced to differentiate in hepatocytes and never in biliary cells, while their proliferation activity was rapidly reduced to zero. Moreover, addition of anti LRP antibody strongly inhibited this differentiation process (fig. 1).

Characterization of the LRP molecule.

Immunopositive reaction with mAb L8 was located all along the plasma membrane of freshly isolated hepatocytes, of hepatocyte in pure cultures for only the 4 first days and of hepatocytes in cocultures for at least 2 weeks.

The antigen recognized by mAb L8 was also present in the normal liver. Antigenic sites were generally distributed in the hepatic lobule, but restricted to the sinusoids.

By analyzing the distribution of LRP in various tissues in the adult rat, it appeared that LRP was not liver-specific but it was also expressed in a limited number of other tissues, namely the exocrine pancreas, the gonads and the hemopoietic organs including spleen, thymus and bone marrow.

Interestingly, LRP remained present in nodules of hepatoma experimentally induced by a chemical carcinogen (DENA). However, a change in its distribution on plasma membranes of tumor cells was constantly observed, leading to postulate that modifications in its biological properties could occur in these cells (fig. 2).

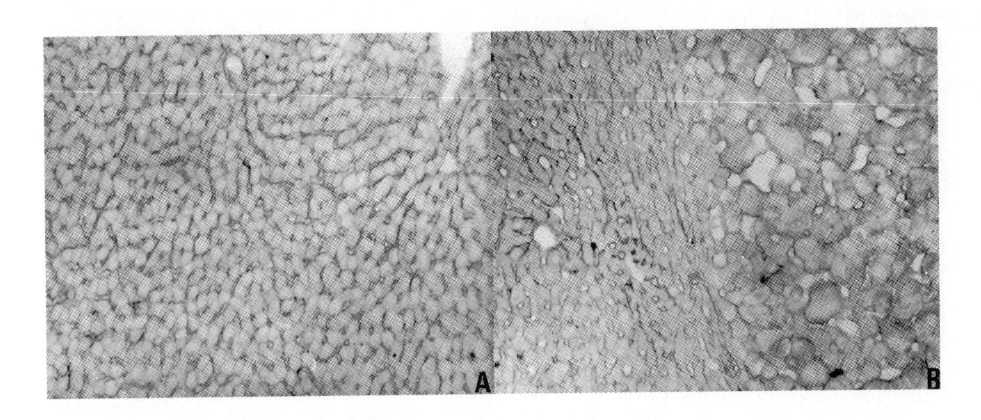

Fig. 2 : Light microscopy on serial cryostat sections of rat liver. Cell distribution of LRP in the lobule of normal liver (A) and in hepatic nodules 12 months after DENA administration to rats (B).

Immunoprecipitation was performed on detergent cell lysates after cell surface iodination of hepatocytes and RLEC. It revealed 2 peptide chains having apparent molecular masses of 85 and 73 kD from both hepatocytes and RLEC. The two polypeptides were preferentially recovered from the hydrophobic phase after Triton X-114 extraction, suggesting that both were integral plasma membrane proteins. The amino-acid sequence and cDNA cloning are under investigation. Comparison of our preliminary results to the international data bank led us to confirm that LRP is an original molecule. Furthermore, comparative analysis of the molecular weight of the mAb L8 immunoprecipitated protein in the different positive tissues indicated that cells such as Sertoli cells in the testis and thymocytes may express a molecule similar to that found in the liver, while in other cells the molecules appeared to be partly distinct with different molecular weights. This might reflect the existence of a family of LRP-like molecules.

EGF stimulation of hepatocyte proliferation.

Numerous studies have established that the *in vivo* proliferation temporal pattern in response to loss of tissue mass, consists of two waves, the first one corresponding to the synchroneous hepatocyte progression through the cell cycle. The time course of [^{3}H]thymidine incorporation obtained in this work confirmed the previous kinetics (Fabrikant, 1968) : the length of the G_1 and S periods were defined respectively from partial hepatectomy to 18 h and from 18 to 28 h . Analysis of the pattern of expression of the different G_1-related proteins indicated that $p34^{cdc2}$ was completely absent in resting hepatocytes and remained unexpressed until 18 h after partial hepatectomy, a time corresponding to G_1, and then accumulated in S, G2 and M phases. In contrast, with the $p33^{cdk2}$CT antibody we found that the $p33^{cdk2}$ protein was constantly expressed during the cell cycle and was enhanced during S, G2 and M phases. Cyclin A appeared as a faint band in non-dividing liver and during the G1 phase, drastically increased in S and G2/M, and decreased thereafter ; cyclin B was also transiently expressed between 22 and 30 h during S and M phases and was undetectable after this time. Interestingly, by investigating the H1K activity of p9-affinity-purified cdc2/cdk2 kinases we found that no H1 phosphorylation was detectable during the G1 phase while two peaks of p^{34cdc2} kinase activity was observed during the S and M phases, and only one peak of p^{33cdk2} kinase activity in the S phase (fig. 3). Surprisingly, cyclins D1 and E were present in resting liver and with modest variations throughout the cell cycle.

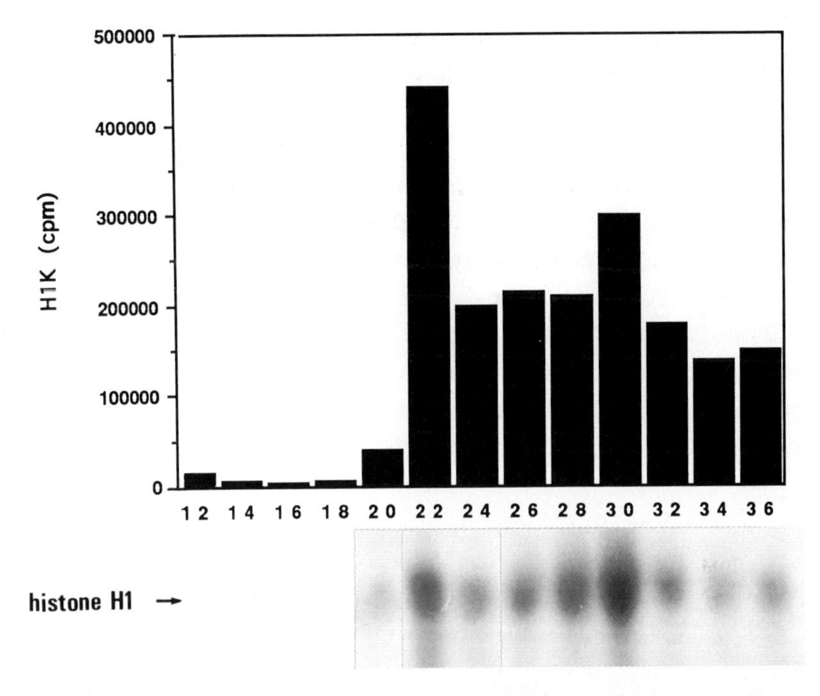

Fig. 3 : H1 kinase activity of cdc2 - related proteins purified on p9 beads and analyzed at different times after partial hepatectomy ; phosphorylation was quantified by measuring the amount of phosphorylated histone H1 and by direct visualization by autoradiography on SDS-PAGE gel.

In addition, we have investigated the growth factor-dependence of the hepatocyte entry and progression through G1, using primary culture. It was confirmed that the G0/G1 transition took place during collagenase perfusion independently of mitogenic factor, and was associated to the transient expression of early G1 protooncogenes, namely c-Fos, c-Jun and c-myc. The evidence was then provided that rat hepatocytes progressed through G1 regardless of growth factor stimulation, until a restriction point located in mid-late G1 phase. At this point, the cells were expressing c-myc and P53 while no p34cdc2 could be detected. In the absence of mitogen, the cells remained blocked at this point and did not replicate.
Furthermore, a series of experiments has been performed in order to define the effects on DNA replication, of a 6 h exposure to EGF in relation with the time of EGF addition. It appeared that the number of cells entering the S

phase was gradually increased until a maximum reached under the conditions where the cells were exposed to EGF at 18-24 h. Meanwhily, the entrance of the cells in S phase was constantly observed at about 60 h.

LRP and the proliferation/differentiation balance.

We have investigated whether LRP could play a role on the hepatocyte response to EGF. We have analyzed the hepatocytes maintained in coculture and exposed to EGF. We found that in coculture, adult rat hepatocytes were blocked in G1 and that they could not respond to EGF. In order to demonstrate that LRP was involved in this process cocultures were treated with mAb L8. Addition of this anti LRP monoclonal antibody restored the sensitivity to the mitogen and their capacity to enter in S phase and mitosis (fig. 4).

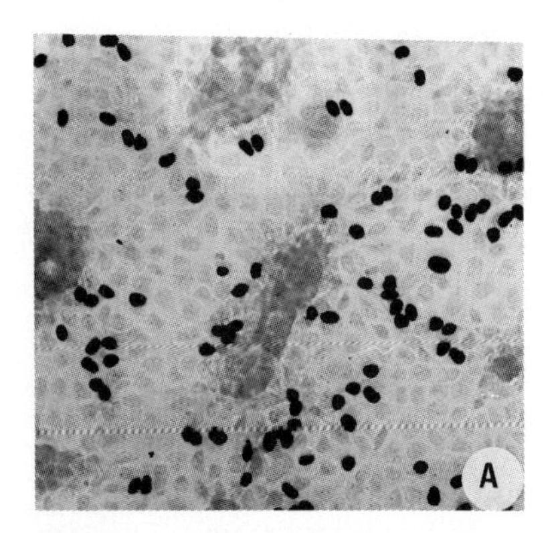

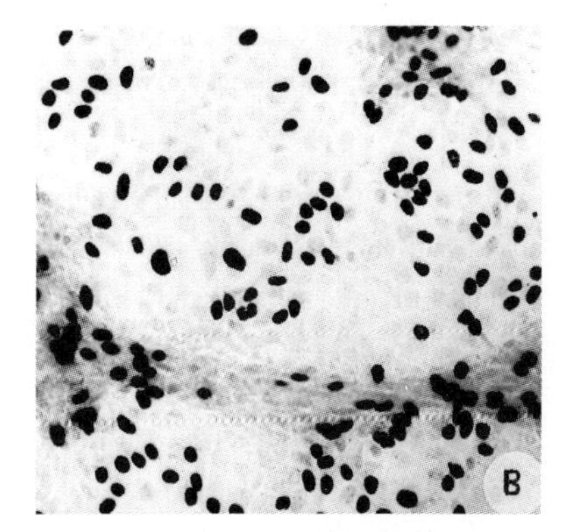

Fig. 4 : BrdU uptake in adult hepatocytes maintained in coculture and exposed either to EGF (50 ng/ml) (A), or to EGF + (50 μg/ml) mAb L8, a monoclonal antibody against LRP (B).

Evidence was provided that LRP could also be involved early in the balance of hepatocyte proliferation/differentiation. Hepatoblasts from 12 days old fetuses, were actively proliferating in pure culture while their proliferative activity became completely inhibited within two days of coculture.

Conclusion

These results described here demonstrate that very few, if any, tissue specific functions are able to be irreversibly expressed in differentiated adult hepatocytes and that cell interactions play a critical role in regulating the corresponding genes. As illustrated with the coculture system, the heterotypic cell interactions are concerned with this control and LRP is described as a new plasma molecule protein involved in the early step of recognition between the two cell types. Whether the transducing signal of LRP results in directly activating the transcription factors associated with liver specific genes is not established yet. Indeed, the possibility of a role of LRP throughout the deposition of some components of the extracellular matrix remains questionable.

An important observation is the role of the heterotypic cell interactions in the balance of hepatocyte proliferation/differentiation. A direct consequence of this would be the initiation of events that leading to DNA replication ; the first one is the G0/G1 transition which is related with metabolic changes like the induction of immediate-early oncogenes and of some other functions. It has been proposed that re-entry in G1 after partial hepatectomy could be the consequence of metabolic changes, regardless of growth factors (Fausto, 1992). Our results reinforce this hypothesis. Furthermore, they give strong arguments in favor of a major role of heterotypic cell interactions in the control of hepatocyte response to growth factors, and as a consequence, in the control of the transition to S phase. Thus, in coculture with RLEC, hepatocytes are blocked in mid G1 and do not respond to EGF.

LRP appears to be involved in the balance of proliferation/differentiation. From its distribution in few other tissues and changes in its molecular weight characteristics accordingly, it may be concluded that a family of LRP molecules does exist. It can be stressed that this molecule is distinct from cadherins and integrins in that it is involved in cell recognition but not in cell adhesion. Evidence that it favors the differentiation of bipotential hepatoblasts towards the hepatocytic type, at an early stage of fetal life, led us to postulate that this new glycoprotein would also play a pivotal role in the maturation process. Finally, its redistribution in hepatic nodules might suggest its involvement in hepatocarcinogenis.

Acknowledgements

We wish to thank D. Glaise for her technical assistance. This work was supported by INSERM, Ministère de la Recherche et de la Technologie, n° 90TO175 and Association pour la Recherche contre le Cancer.

REFERENCES

Azzi L, Meijer M, Reed S, Pidikiti R, Tung HY (1992) Interaction between the cell-cycle-control proteins p34cdc2 and p9CKShs2. Eur J Biochem 203:353-360

Caron JM (1990) Induction of albumin gene transcription in hepatocytes by extracellular matrix proteins. Mol Cell Biol 10:1239-1243

Chirgwin JM, Przybyla AE, MacDonald RJ, Rutter WJ (1979) Isolation of biologically active ribonucleic acid from sources enriched in ribonuclease. Biochemistry 18:5294-8299

Corlu A, Kneip B, Lhadi C, Leray G, Glaise D, Baffet G, Bourel D, Guguen-Guillouzo C (1991) A plasma membrane protein is involved in cell contact-mediated regulation of tissue-specific genes in adult hepatocytes. J Cell Biol 115:505-515

Draetta G, Beach D (1988) Activation of cdc2 protein kinase during mitosis in human cells : cell cycle-dependent phosphorylation and subunit rearrangement. Cell 54:17-26

Draetta G, Luca F, Westendorf J, Brizuela L, Ruderman J, Beach D (1989) Cdc2 protein kinase is complexed with both cyclin A and B : evidence for proteolytic inactivation of MPF. Cell 56:829-838

Edelman (1987) GM (1987) CAMs and Igs : cell adhesion and the evolutionary origins of immunity. Immunol Rev 100:11-45

Fabrikant J (1968) The kinetic of cellular proliferation in regenerating Liver. J Cell Biol 36:551-565

Fausto N (1992) Liver regeneration : models and mechanisms. Liver regeneration. D. Bernuau, G. Feldman (eds), John Libbey Eurotext, 1-16

Gallin WJ, Edelman GM, Gunningham BA (1983) Characterization of L-CAM, a major cell adhesion molecule from embryonic liver cells. Proc Natl Acad Sci 80:1038-1042

Guguen C, Gregori C, Schapira F (1975) Modification of pyruvate kinase isozymes in prolonged primary cultures of adult rat hepatocytes. Biochimie 9:1065-1071

Guguen-Guillouzo C (1986) Role of homotypic and heterotypic cell interactions in expression of specific functions by cultured hepatocytes. Isolated and cultured hepatocytes. John Libbey Eurotext Ltd/INSERM, 271-296

Guillouzo A, Beaumont C, Le Rumeur E, Rissel M, Latinier MF, Guguen-Guillouzo C, Bourel M (1982) New findings on immunolocalization of albumin in rat hepatocytes. Biol Cell 31:163-172

Higgins GM, Anderson RM (1931) Experimental pathology of liver ; restoration of the liver of the white rat following partial surgical removal. Arch Pathol 12:186-202

Loyer P, Glaise D, Cariou S, Baffet G, Meijer L, Guguen-Guillouzo C (1994) Expression and activation of CDKs (1 and 2) and cyclins in the cell cycle progression during liver regeneration. J Biol Chem 219:521-528

Marceau N (1994) Epithelial cell lineages in developing, restoring, and transforming liver : evidence for the existence of a "differentiation window". Gut 35:294-296

McGowan JA, Bucher NL (1983) Pyruvate promotion of DNA systhesis in serum-free primary cultures of adult rat hepatocytes. In Vitro 19:159-166

Morel-Chany E, Guillouzo C, Trincal G, Szajnert MF (1978) "Spontaneous" neoplastic transformation in vitro of epithelial cell strains of rat liver : cytology, growth and enzymatic activities. Eur J Cancer 14:1341-1352

Pagano M, Pepperkik R, Verde F, Ansorge W, Draetta G (1992) Cyclin A is required at two points in the human cell cycle. EMBO J 3:961-971

Pitot HC, Barsness L, Goldsworthy T., Kitagawa T (1978) Biochemical characterization of stages in hepatocarcinogenesis after a single dose of diethylnitrosamine. Nature 271:456-458

Sherr CJ (1993) Mammalian G1 cyclins. Cell 73:1059-1065

INTERACTIONS OF HUMAN HEPATOCYTES WITH HEPATITIS B VIRUS.

S. RUMIN, P.GRIPON, M. CORRAL-DEBRINSKI, J.CILLARD AND C.GUGUEN-GUILLOUZO.

INSERM U 49, Hôpital Pontchaillou, 35033 RENNES, France.

INTRODUCTION

The first interactions of the hepatitis B virus (HBV) with hepatocytes begin with the infection process. *In vivo*, HBV shows a very narrow host range since it can productively infect only some primates (Zuckerman, 1975). Moreover, a marked liver tropism has been observed which is nearly limited to normal hepatocytes. Furthermore, from several reports it appears that attempts at efficiently infecting hepatoma cells met with limited success, presumably because of changes in plasma membrane properties of these cells, leading to postulate that a specific binding of viral particles to a membrane receptor might exist. This receptor is not yet known.

The observation that viruses structurally related to HBV exist naturally in ducks, ground squirrels and woodchucks (Summers, 1981) has created new opportunities for the study of *in vitro* animal models. Furthermore, major progress was made when several laboratories achieved hepatitis B virus (HBV) replication by transfecting hepatoma cell lines with cloned viral DNA (Sureau *et al.*, 1986). However, these systems are not appropriate for studying the mechanism of initiation of infection in human cells.

Taking advantage of the high stability of hepatocytes, maintained in coculture or in the presence of DMSO (Isom *et al.*, 1985), we have defined conditions which allow *in vitro* infection of adult normal human hepatocytes. While the ability of viral particles to bind the cells was always high, a great variability in active replication was observed from one donor to another (Gripon *et al.*, 1988). This led us to improve the infection efficiency in culture by adding polyethylene glycol (PEG), a molecule known for its fusogenic properties (Aldwinkle *et al.*, 1982), and expected to induce virus-cell membrane fusion. We report an increased infection efficiency using these conditions. In addition, the ability of this model system to preserve the species and tissue specificities of natural infection, was analyzed.

Another important aspect of the interactions of HBV with human hepatocytes, besides infection, is related with the virus oncogenic properties. It has been well established that chronic hepatitis B virus (HBV) infection is strongly

NATO ASI Series, Vol. H 88
Liver Carcinogenesis
Edited by G. G. Skouteris

associated with a high incidence of hepatocellular carcinoma (Beasley *et al.*, 1981). However, mechanisms by which the virus leads to carcinogenesis, remain unclear. Among them, transactivation of cell genes by the virus (Kekulé *et al.*, 1993) and integration into the host genome, have been reported. These events were only described in few cases (Dejean *et al.*, 1986 ; Wang *et al.*, 1990) and *in vitro* studies have been mainly performed on immortalized or transformed cells (Höhne *et al.*,1990). An increase of reactive oxygen species (ROS) in infected cells could also be hypothesized. Here we investigate the possible functional alterations of human normal hepatocytes infected *in vitro* and maintained for a prolonged period in culture (2 months).

Material and methods

Cell isolation and culturing

Human hepatocytes were obtained from the livers of organ donors in agreement with French legislation. Cells were isolated by the procedure of Guguen-Guillouzo *et al.* (1982). A portion of the left lobe of the liver or biopsies were rapidly perfused with a calcium-free HEPES buffer and then with a HEPES-buffered solution containing 0,05 % collagenase and 5 mM $CaCl_2$. Adult rat hepatocytes were obtained from Sprague-Dawley male animals by also using the two-step collagenase perfusion, as previously described (Guguen-Guillouzo and Guillouzo, 1986).

Freshly isolated hepatocytes were seeded in the culture medium supplemented with 10 % fetal calf serum. The medium was composed of 75 % minimum essential medium and 25 % medium 199, added with 5 mg/l insulin, 300 mg/l penicillin, and 100 mg/l streptomycin. After 24 hr the medium was renewed by the same medium supplemented with 3.5×10^{-5} M hydrocortisone hemisuccinate. For human cell cultures, 10 % adult porcine serum and 2 % dimethyl sulfoxide (DMSO), were used until infection.

Long term cultures were also obtained by culturing the cells in the presence of DMSO. Because of its antioxidant properties, DMSO was not added to the cultures used for studies on lipid peroxidation.

Proliferation of hepatocytes was measured using the BrdU labelling method.

Human liver and skin fibroblast cell lines were obtained by cell overgrowth from immersed tissue fragments and were subcultured in the same medium as above, without corticoid. Fibroblasts were used for infection assays between passages 5 and 10. Rat liver epithelial cells (RLEC) were originally isolated by trypsinization of 10-day-old rat livers (Morel-Chany *et al.*, 1978).

Infection of cell cultures

Two sources of viral particles were used. One consisted of serum from one chronic HBV carrier with a high level of HBV DNA. The second source was a virion producing cell line derived from HepG2 cells (Sells *et al.*1987) : the medium from confluent cultures was collected every day. Precipitated Dane particles were suspended in phosphate buffer saline and used for infection. Infection was performed by overnight incubation of cell monolayers, at 37°C, in the culture medium containing the diluted infectious serum or the Dane particles. Cells were then washed three times with the culture medium.
PEG (MW 6000) was added to the culture medium at different concentrations and at the indicated times for all experiments.

Assays for HBV specific proteins and viral DNA forms

Hepatitis B surface (HBsAg) and e (HBeAg) antigens were identified by radioimmunoassay kits obtained from Abbott Laboratories.
For DNA extraction, cells were lysed in 0.5 % SDS in a buffer containing 200 µg of proteinase K and incubated overnight at 37°C. The DNA was precipitated using the Hirt fractionation procedure (Hirt, 1967). To demonstrate the presence of the covalently closed circular (CCC) form of HBV DNA, cells were lysed without treatment with proteinase K.
DNA was analyzed by electrophoresis in horizontal slab gels of 1.5 % agarose. Gels were neutralized and transferred to nitrocellulose filters. Hybridization was performed with a [^{32}P]HBV probe.
When indicated the purified DNA was digested with restriction endonucleases, using conditions recommended by the supplier.

Cytotoxicity evaluation

Peroxidation measurement was performed by quantification of free malondialdehyde (MDA) into the cells and the medium, according to the procedure described by Sergent *et al.*(1993), using HPLC. The cell leakage was estimated by measuring the LDH activity both in the cell lysate and in the culture supernatant, using a Roche kit (France).

RNA preparation and Northern blot hybridization

Total RNA was prepared using 5M guanidium thiocyanate/CsCl according to Chirgwin *et al.* (1979). Hybridization was performed with 32p-labelled complementary DNA probes : human c-myc(exon 3), mouse c-jun, HBV.

Results

High efficiency and specificity of HBV infection in the presence of PEG.

The first assays were performed by adding PEG (40 %) to the culture after 16 h exposure of cells to the infectious serum. Whatever the time of PEG treatment, no increase in viral infectivity was detected whereas a strong cytotoxicity was observed. The highest efficiency was obtained by reducing the concentration of PEG and increasing the time of treatment. The best protocol was defined as follows : 2 or 3 days old cells maintained with 2 % DMSO, were infected for 16 h with either infectious serum or Dane particles from a producing HepG2 cells supernatant, in the presence of 5 % PEG. Under these conditions, hepatits B virus penetration into the cells was greatly enhanced and infection was always possible, regardless of cell individual origin. The different markers of viral replication were examined. High levels of HBsAg were detected during the first 2 days following infection, it corresponded to the infectious source. Then, after a transient disappearance, high amounts of antigen were evidenced for the whole duration of culture. Both HBs and HBc/e antigens were clearly evidenced in all cases. Furthermore, the intracellular viral DNA, extracted by a procedure which selectively extracts DNA that is not covalently bound to proteins, consisted mainly of the CCC form. The CCC form is thought to represent the template for viral transcription. This form was absent from the infectious serum and thus was a reliable marker of viral infection. DNA was also extracted 7 days after infection and analyzed by Southern blot hybridization. High amounts of intracellular replicative intermediates were observed when cells were exposed to the virus 2 and 4 days after plating. These forms were present in much lower amounts in cells infected on the 6th and 8th days postplating. These results have led us to infect the cells preferentially on day 3 of culture.

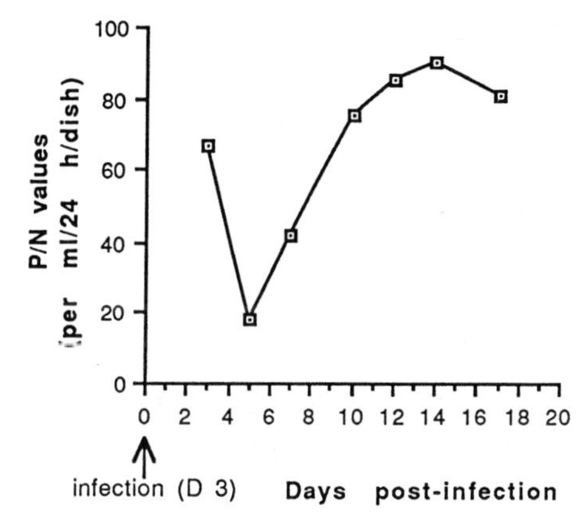

Fig. 1 : Kinetic of HBsAg production in the medium of cultured human hepatocytes experimentally infected with HBV and maintained in the presence of 3.5×10^{-5} M hydrocortisone.

In addition, we showed that cells infected according to the protocol above, could support active viral replication for one or two months if they were cultured in the presence of hydrocortisone or DMSO respectively. The cells produced high amounts of HBsAg (fig. 1), mature virions in the supernatant and contained abundant HBV DNA replicative intermediate forms in the nucleus and the cytoplasm.

It was important to examine whether the same species and tissue specificities were preserved *in vivo* and *in vitro* in the presence of PEG. Cells from various origins were incubated with HBV virions, without or with PEG. They included normal hepatocytes from rat, undifferentiated RLEC presumably derived from primitive biliary cells, and fibroblastic cells from human liver and skin. Different markers were used as followed : absence of detectable HBsAg in the culture supernatant suggested negative infection. Detection of the CCC DNA form in the cells by Southern blot hybridization and long term HBsAg production indicated virus internalization and viral gene expression.

Primary adult rat hepatocyte cultures were found to be resistant to HBV infection, either in the presence or absence of PEG, suggesting that the species specificity was preserved *in vitro* and unaltered by addition of PEG. Subcultures of human fibroblasts isolated from liver and skin and the undifferentiated rat liver epithelial cell line were also found to be resistant to infection, indicating that the tissular specificity was well preserved.

Long-term HBV infection and hepatocyte functional activity

We have used the long term culture conditions to examine whether the virus could alter the functional activity of normal human hepatocytes. We focused on two main aspects : proliferation and cytotoxicity.

One possibility would be that HBV could modify the proliferative activity of hepatocytes resulting in an increased number of cells entering the S phase. No difference could be detected between infected and non infected cultures, using BrdU labelling, while few cells spontaneously entered replication in the two situations.

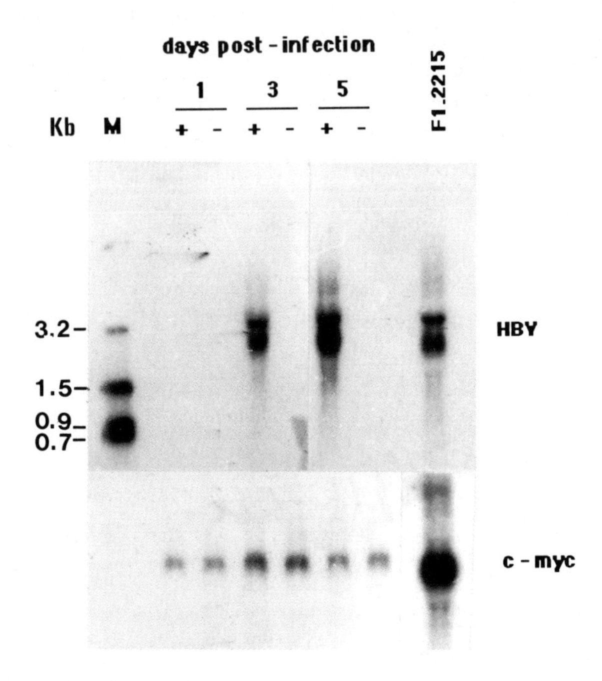

Fig. 2 : Pattern of HBVmRNAs and kinetic analysis of c-myc transcripts, present in infected (+) and non infected (-) hepatocytes maintained in the presence of DMSO and harvested 1, 3 and 5 days after infection.
The control is F1-2215 HepG2 hepatoma cells.
The size markers are 3.2, 1.5, 0.9 and 0.7 kb restriction fragments of cloned HBV DNA.

We, then, deviced experiments to detect changes in the expression level of proteins associated with the G1 phase, mainly the early G1 protooncogenes. We did not evidence any difference in the relative amounts of c-jun and c-myc mRNAs in infected and in control cells in short and long term cultures (fig. 2). This contrasted with previous reports showing transactivation of c-myc in immortalized cell lines supporting HBV replication (Höhne *et al.* 1990).

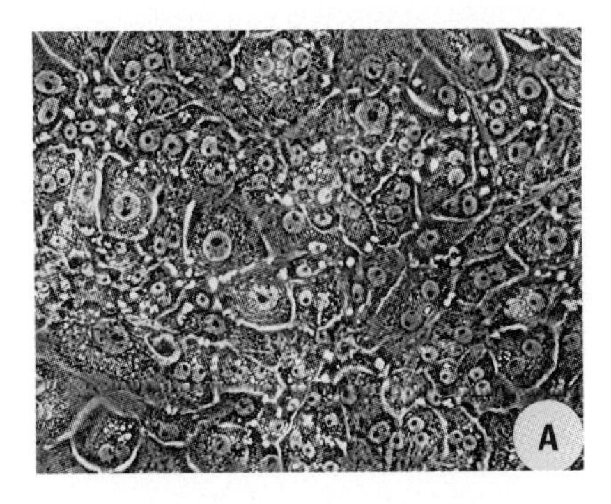

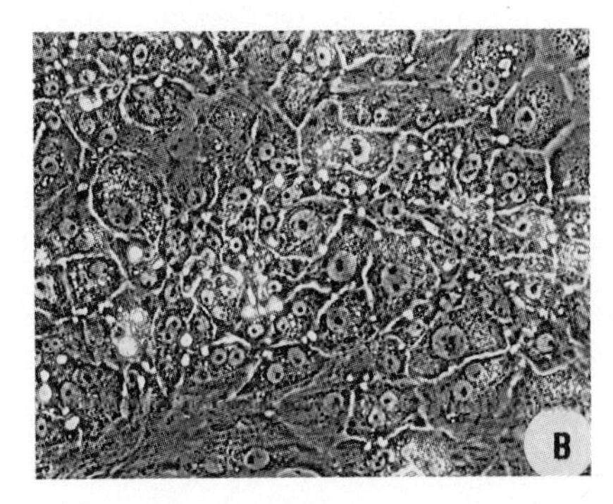

Fig. 3 : Phase-contrast micrograph of one month-old primary cultures of human hepatocytes maintained in the presence of $3.5x10^{-5}$ M hydrocortisone. (a) HBV- infected culture. (B) non infected culture.

Cytotoxicity was assessed using different criteria. Under the light microscopy cell morphology of infected and control cultures was similar all along the culture time (fig. 3). LDH activity measured either intracellularly or in the medium during the course of culture was strictly identical in both infected and non infected cells, leading to the conclusion that no major cell damage could be related to HBV replication. However, from the observations of Meyer *et al.*(1992) it might be hypothetized that gradual intracellular damage is induced by an increased oxidative stress in infected cells. To evidence lipid peroxidation we measured MDA production and release in the medium at different times of culture. Again, no detectable changes in MDA release levels were observed in the presence of the virus.

Process of viral DNA integration

Among the possible mechanisms by which HBV could be directly associated with oncogenesis, integration of the viral DNA in the host genome has been proposed ; till now, only two cases have been reported showing integration in the vicinity of genes involved in proliferation/differentiation (Dejean *et al.* 1986 ; Wang *et al.* 1990). We have looked for the presence of viral integrated forms in long term cultures. We have never succeeded in detecting these forms.

Conclusion

The high species and tissue specificity of natural infection by HBV leads to consider primary cultures of normal human hepatocytes as one of the most appropriate *in vitro* systems for analyzing the mechanism of initiation of infection and for investigating viral oncogenic properties. We show that it is possible to obtain a reproducible high level infection of human hepatocytes *in vitro* by using culture conditions which greatly favor the stability of the cell differentiation status, and by adding PEG during incubation of hepatocytes with the virions. Culturing the cells in the presence of 2% DMSO or with hydrocortisone hemisuccinate, allows to prolong their survival and to maintain an active viral replication up to 3 and 1 months respectively.
Whether internalization was the main limiting step of infection remains questionable. PEG could favor the virus fusion to the cells. Indeed PEG efficiency is correlated with a strong enhancement of viral particle binding. However, that PEG could directly favor a better interaction between virions and cell receptors or increased expression of cell receptors at the surface of hepatocytes cannot be excluded. Interestingly, PEG procedure preserves the specific features of HBV natural infection. Therefore, studies on the early steps of viral infection should be facilitated by the availability of human hepatocytes. Furthermore, this model is useful for analyzing the oncogenic effects of the virus in the context of normal adult human hepatocytes.
It is well known that persistent infection by hepatitis B virus is epidemiologically closely associated with the prevalence of hepatocarcinoma in man. However, the HBV-related mechanisms that convey the tumorigenic progression to hepatocytes are poorly understood. The first demonstration that HBV displays direct oncogenic potential was provided by showing that transformed foci appeared in immortalized parenchymal cells following transfection with HBV DNA (Höhne *et al.* 1990). However we failed to evidence any changes in functional activities in normal adult hepatocytes after long term HBV infection. Thus, no significant transactivation of protooncogenes could be detected. Either significant cytotoxicity or enhanced production of oxygen radicals were not observed. These results led to postulate that HBV alone could not lead mature hepatocytes to transformation. The development of transformation assays using carcinogenic chemical compounds in association with HBV infection is under progress.

The possibility that HBV might infect in vivo another liver cell population cannot be excluded.

Acknowledgements

We wish to thank M. Chevanne for her technical assistance. S. Rumin is recipient of the Ministère de la Recherche et de l'Enseignement. This work was supported by INSERM and the Association pour la Recherche contre le Cancer.

REFERENCES

Aldwinckle TJ, Ahkong QF, Bangham AD, Fisher D, Lucy JA (1982) Effects of poly(ethylene glycol) on liposomes and erythrocytes permeability changes and membrane fusion. Biochim Biophys Acta 689:548-560

Beasley RP, Lin CC, Hwang L, Chien C (1981) Hepatocellular carcinoma and hepatitis B virus : a prospective study of 22, 707 men in Taïwan. Lancet 2:1129-1133

Chirgwin JM, Przybyla AE, MacDonald RJ, Rutter WJ (1979) Isolation of biologically active ribonucleic acid from sources enriched in ribonuclease. Biochemistry 18:5294-8299

Dejean A, Bougueleret L, Grzeschick KH, Tiollais P (1986) Hepatitis B virus DNA integration in a sequence homologous to v-erb A and steroid receptor genes in a hepatocellular carcinoma. Nature 322:70-72

Gripon P, Diot C, Thézé N, Fourel I, Loréal O, Bréchot C, Guguen-Guillouzo C (1988) Hepatitis B virus infection of adult human hepatocytes cultured in the presence of dimethyl sulfoxide. J Vir 62:4136-4143

Guguen-Guillouzo C, Campion JP, Brissot P, Glaise D, Launois B, Bourel M, Guillouzo A (1982) High yield preparation of isolated human adult hepatocytes by enzymatic perfusion of the liver. Cell Biol Int Rep 6:625-628

Guguen-Guillouzo C, Guillouzo A (1986) Methods for preparation of adult and fetal hepatocytes. Isolated and cultured hepatocytes (A. Guillouzo and C. Guguen-Guillouzo, eds), Les Editions INSERM, Paris, Libbey, London :1-12

Hirt B (1967) Selective extraction of polyoma DNA from infected mouse cell cultures. J Mol Biol 26:365-369

LIVER-SPECIFIC ASPECTS OF HEPATITIS B VIRUS GENE EXPRESSION

Marshall J. Kosovsky, Hugh F. Maguire,
Bingfang Huan and Aleem Siddiqui*
Department of Microbiology and Immunology and
Program in Molecular Biology
University of Colorado Medical School
Denver, Colorado 80262, USA

INTRODUCTION

The human hepatitis B virus (HBV) is a member of the hepadnavirus family, which is compromised of DNA viruses that share structural and biological features. HBV infectivity is characterized by a marked hepatotropism, which is a prominent feature of hepadnaviruses. Although the majority of individuals infected with HBV resolve the primary infection and develop lasting immunity to subsequent infection, clinical studies have shown that 5-10% of individuals are chronically infected with the virus (Redeker, 1975). It has been estimated that there are over 250 million chronic HBV carriers worldwide (Rossner, 1992). A clinical consequence of chronic HBV infection is cirrhosis of the liver, which can lead to liver failure and death (Degroote et al., 1979). Furthermore, epidemiological studies have demonstrated that there is a strong association between hepatocellular carcinoma (HCC) and chronic HBV infection (Szmuness, 1978; Beasley et al., 1981). This association is supported by the following evidence: 1) there is a strong correlation in the geographical distribution of HCC and the occurence

*Corresponding author

NATO ASI Series, Vol. H 88
Liver Carcinogenesis
Edited by G. G. Skouteris

of chronic HBV infection, and 2) the hepatitis B surface antigen (HBsAg) is detected with a marked frequency in individuals with HCC. There is additional compelling evidence which correlates HBV with carcinogenesis. The HBV X protein (pX) has been shown to induce liver cancer in transgenic mice (Kim et al., 1991), and appears to function as a component of a tumor promoter signalling pathway (Kekule et al., 1993). Moreover, studies have demonstrated that HBV DNA is integrated in the cellular DNA of HCC (Brechot et al., 1980; Shafritz et al., 1981) and hepatoma cell lines (Edman et al., 1980; Marion et al., 1980). Finally, while the mechanism(s) responsible for HBV-induced oncogenesis is currently unknown, it is possible that the integration of the HBV enhancer triggers aberrant liver gene expression, which may then promote the events of liver neoplasia.

The HBV particle measures 42 nm in diameter, and contains a partially double-stranded, circular DNA genome that is 3,200 nucleotides in length (**Figure 1a**). The viral genome is enclosed within a 28 nm diameter nucleocapsid core, which is surrounded by a 7 nm hepatocyte membrane-derived envelope (Raney and McLachlan, 1991). There are four primary translational open reading frames {ORF}, and pX {S ORF} (Tiollais et al., 1985). As indicated in **Figure 1a**., the viral genome contains four promoter elements (P1, P2, Cp, and Sp) and two enhancer elements. Our laboratory (Jameel and Siddiqui, 1986; Bulla and Siddiqui, 1988; Hu and Siddiqui, 1991) and others (Antonucci and Rutter, 1989) have demonstrated that enhancer I facilitates transcriptional activation from the promoter elements of HBV genes. While enhancer II is capable of activating liver-specific transcription form heterologous promoters (Yee, 1989; Yuh and Ting, 1993), our studies have shown that enhancer II activity increases approximately 30 fold in the presence of enhancer I (Lopez-Cabrera et al., 1990; Lopez-Cabrera et al., 1991). Therefore, our findings suggest that HBV enhancer-mediated transcriptional activation may be differentially regulated by a cooperative interplay of the viral regulatory elements.

The HBV enhancer I element, which was first identified in 1985 (Shaul et al., 1985), spans nucleotides 966-1308 on the genome (HBV subtype adw). Enhancer I has been shown to exhibit liver-specific activity (Shaul et al., 1985; Jameel and Siddiqui, 1986; Bulla and Siddiqui, 1988; Antonucci and Rutter, 1989; Honigwachs et al., 1989; Patel et al., 1989; Trujillo et al., 1991; Hu and Siddiqui, 1991; Guo et al., 1991), and is capable of functioning in an orientation-independent manner (Shaul et al., 1985; Antonucci and Rutter, 1989). Enhancer I activity appears to be governed by trans-acting cellular factors, which interact with specific domains of the enhancer element [**Figure 1b.**, (Jameel and Siddiqui, 1986; Patel et al., 1989; Ben-Levi et al., 1989; Ostapchuk et al., 1989; Trujillo et al., 1991)]. Additionally, as demonstrated with other experimental systems, enhancer function may be regulated by complex mechanisms involving protein/protein interactions (Grueneberg et al., 1992; Roberts et al., 1993) and/or precise DNA conformational;bending requirements (Buratowski et al., 1991; van der Vliet and Verrijzer, 1993). Our laboratory has identified a sequence motif located at nucleotides 1125-1138 on enhancer I, designated as footprint V (FPV), which is the site of a liver-specific protein/DNA interaction (Patel et al., 1989; Trujillo et al., 1991). DNase I protection analyses have revealed the presence of a factor(s) in rat liver nuclear extracts and hepatoma cell lines, tentatively referred to as the hepatitis B enhancer-binding liver factor (BVLF), which interacts with the FPV motif. Nuclear extracts derived from non-liver cells, such as Hela and spleen, were devoid of such a binding activity. Additionally, we have

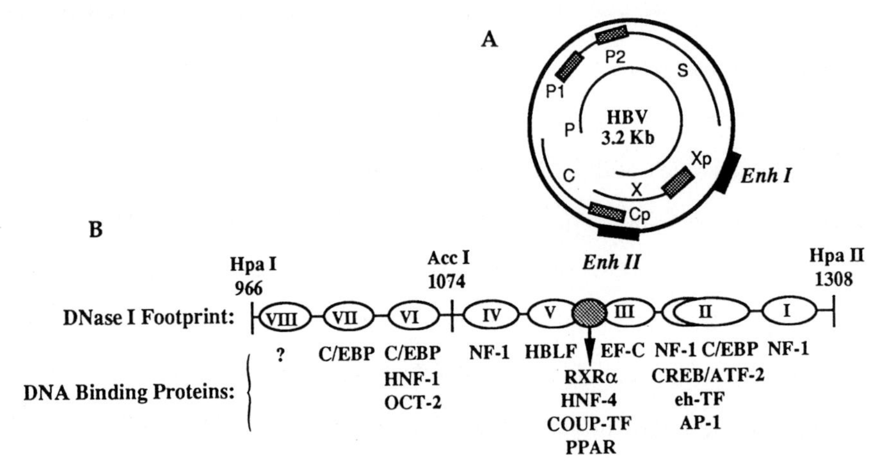

Figure 1. (A) Genomic organization of HBV. S, C, Pol, and X represent the genes for the hepatitis B surface antigen, core/e antigen, polymerase, and X proteins. P1, P2, Cp, and Xp represent promoter regions of the corresponding distal genes. The enhancer elements are designated as Enh I [spanning nucleotides 966 (Hpa I) to 1308 (Hpa II), HBV subtype adw] and Enh II [spanning nucleotides 1645 (Sty I) to 1885 (Sty I)]. Kb, kilobases. **(B)** Protein binding sites on enhancer I. The putative binding site of peroxisome proliferator activated receptor (PPAR) is shown (unpublished results). Restriction enzyme sites and corresponding nucleotide locations are indicated.

shown that the liver-enriched factors C/EBP (Trujillo et al, 1991) and retinoid X receptorα [RXRα, (Huan and Siddiqui, 1991)] transactivate enhancer I by interacting with specific sites that are distinct from FPV. As shown in **Figure 1b.**, the RXRα binding site overlaps the FPV and FPIII sequence motifs. The functional importance of the central region of the enhancer has been demonstrated by the following genetic studies: 1) transient transfection analyses of enhancer I deletion mutants indicated that the FPV-FPIII region comprises the enhancer core domain, which was shown to be critical for enhancer function (Trujillo et al., 1991); the enhancer core domain functioned at approximately 80% of the level exhibited by the full length enhancer I element; and 2) our recent studies have shown that point mutations in the HBLF or RXRα binding sites abrogate enhancer I activity (Huan and Siddiqui, 1991), which suggests that these sequence motifs are required for enhancer I-mediated transcriptional activation. Functional studies have further suggested that enhancer I activity is mediated by cooperative interactions between enhancer binding factors (Dikstein et al., 1990a; Dikstein et al., 1990b; Garcia et al., 1993). Therefore, the cellular factors may represent a functional repertoire of proteins that are capable of interacting with each other, in conjunction with enhancer binding, to differentially regulate viral gene expression.

Several lines of evidence indicate that the HBV X gene product, pX, is expressed in HBV infected cells (e.g., Moriarty et al., 1985; Vitvitski et al., 1988; Vitvitski-Trepo et al., 1990). Moreover, recent work has demonstrated that pX is required *in vivo* for viral replication by the mammalian hepadnavirus, woodchuck hepatitis virus (Zoulim et al., 1994). Since pX may play a critical role in the events which promote HCC in individuals chronically infected with HBV (Kim et al., 1991; Kekule et al., 1993), this viral protein has been the subject of intensive study. pX appears to exert its influence by functioning as a transcriptional transactivator. Transactivation by pX has been shown to occur through RNA polymerase II (class II) and III (class III) promoter elements (Aufiero and Schneider, 1990). Interestingly, truncated forms of pX, which may be synthesized *in vivo* based upon the presence of alternate translational start sites, are capable of differentially transactivating class II and III promoters (Kwee et al., 1992). **Table 1** lists representative examples of viral and cellular targets of pX transactivation (for review see Rossner, 1992); These findings were primarily based upon transient transfection analyses using a variety of hepatocyte and non-hepatocyte derived cell lines.

Table 1

Viral and Cellular Transcriptional Regulatory Elements and Cellular Genes Responsive to pX Transactivation

Responsive Element/Gene	Reference
VIRAL	
Hepatitis B Virus Enhancer I	Siddiqui et al.(1989) Faktor and Shaul (1990) Maguire et al. (1991)
Human Immunodeficiency Virus-1, Long Terminal Repeat (LTR)	Seto et al. (1988) Siddiqui et al. (1989) Twu et al. (1989)
Rous Sarcoma Virus LTR	Spandau and Lee (1988) Colgrove et al. (1989) Siddiqui et al. (1989)
Simian Virus 40 (SV40) Enhancer	Siddiqui et al. (1989) Luber et al. (1991)

CELLULAR	
AP-1/AP-2 binding sites	Seto et al. (1990)
c-fos promoter	Avantaggiati et al. (1993)
c-jun promoter	Twu et al. (1993)
Epidermal Growth Factor Receptor gene	Menzo et al. (1993)
Interleukin-8 gene (via NF-kB and C/EBP-like elements)	Mahe et al. (1991)
β-Interferon gene	Twu et al. (1987)
Intracellular Adhesion Molecule-1 (ICAM-1) gene	Hu et al. (1992)
Major Histocompatibility Complex (MHC) Class I gene, H-2K[b]	Hu et al. (1990)
MHC Class II gene, Human Leukocyte Antigen (HLA)-DR	Hu and Siddiqui,
Serum Amyloid Alpha (acute phase gene)	unpublished results.

Furthermore, pX has been shown to transactivate each of the HBV promoter and enhancer elements (Rossner, 1992; Nakatake et al.,1993). However, since pX does not appear to directly interact with DNA, the activation mechanism amy involve a direct protein-protein-protein interaction(s) between pX and trans-acting cellular factors. In support of this premise, studies from our laboratory (Maguire et al, 1991) and other (Natoli et al., 1994) have demonstrated that pX affects the DNA binding properties of the transcription factors CREB/ATF-2 and c-fos/c-jun (AP-1), respectively. The functional relevance of these effects upon DNA binding activity was established by transient transfection analyses, which showed that pX facilitates transcriptional activation through CREB/ATF-2 and AP-1 responsive elements.

There are other mechanisms that have been ascribed to pX to account for the transactivation function. A recent study demonstrated that pX transactivates transcription through AP-1 responsive elements in a manner that is dependent upon protein kinase C (PKC) activity (Kekule et al., 1993). Phorbol ester-induced depletion of PKC from human liver cells (CCL13) was shown to abrogate pX transactivation through AP-1 binding sites. This evidence, in conjunction with protein kinase inhibitor studies, supports the hypothesis that pX functions as a component of a PKC-activated signal transduction pathway. Furthermore, it has been demonstrated that pX transactivation involves serine/threonine kinases such as PKC and Raf-1 (Cross et al., 1993). While the work described above offers compelling evidence in support of a relationship between pX transactivation and signal transduction, recent studies indicate that pX functions in a PKC-independent manner (Murakami et al., 1994; Natoli et al., 1994). Therefore, given the controversial nature of these results, a widely accepted mechanism(s) delineating the molecular basis of pX activity remains elusive.

The work presented here explores the biochemical and functional properties of the HBV enhancer elements and pX. We have shown that enhancer I is required for the expression of HBV genes, and that enhancer I activates transcription in a liver-specific manner. Moreover, the regulation of transcription by members of the steroid/thyroid hormone nuclear receptor superfamily will be discussed in the context of HBV gene expression. This report also demonstrates that pX transactivates enhancer II through an ATF-2 responsive element. Finally, full length and truncated forms of pX have been analyzed for their ability to engage in protein-protein interactions with cellular proteins.

MATERIALS AND METHODS

Plasmids

The construction of plasmids of pHBV and pHBV-ΔE has been previously described (Hu and Siddiqui, 1991). pSluc2, pEISluc, and pEIISluc were prepared according to Lopez-Cabrera et al. (1991). ATF-2 cDNA was cloned in front of the SV40 early promoter/enhancer to make pSVATF-2. The preparation of pEtX was previously described (Jameel et al., 1990). pSBDR is a pX expression vector which contains the X gene (nucleotides 966-1997, Hpa I-Bgl II) under the control of the HBV enhancer/X promoter.

Full length and truncated forms of the X gene were cloned into the C-terminal end of the glutathione-S-transferase (GST) gene in the plasmid vector pGEX-2T (Smith and Johnson, 1988). The full length X gene, flanked by restriction sites Nco I and Ggl II (blunt-ended), was cloned into the Sma I site of pGEX-2T to produce the plasmid pGST-XFL (encoding 154 amino acids of pX). A subgenomic fragment of the X gene, spanning nucleotides 1376 to 1575 (Nco I-Rsr II), was cloned into the Sma I site of pGEX-2T to generate an N-terminal truncated form of the X gene (pGST-XN, encoding 66 amino acids of pX). A C-terminal portion of the X gene, spanning nucleotides 1687 to 1997 (Hinc II-Bgl II/Hinc II), was also cloned into pGEX-2T at the Sma I site to produce a C-terminal truncated form of the X gene (pGST-XC, encoding 51 amino acids of pX).

Protein Expression

Bacterial samples transformed with the respective expression vectors were grown in 50 ml cultures at 37°C for approximately 3 hours (O.D. = 0.6-0.8). Bacteria were induced with 0.4 mM IPTG for 3 hours and then harvested.

Following bacterial lysis, proteins were prepared as follows. pX expressed from pEtX (either unlabeled or labeled with [^{35}S]-methionine) was renatured from a 6 M urea extract of the insoluble lysate material. ATF-2 DNA binding domain (DBD) was present in the soluble supernatant of the bacterial lysate; Heat-stable ATF-2 DBD was isolated from the soluble supernatant by boiling for 10 minutes, pelleting the denatured material, and then collecting the ATF-2 DBD enriched soluble fraction. GST-fusion proteins were prepared from the soluble supernatant of the bacterial lysate by incubating for 30 minutes at 4°C in the presence of glutathione-Sepharose beads. Immobilized GST-fusion proteins were extensively washed prior to being used for the protein-protein interaction assay and pX multimerization analysis.

Nuclear Run-On Assay

Recircularized HBV genomes were prepared by excising the HBV DNA sequences from the plasmids pHBV and pHBV-ΔE (lacing enhancer I), followed by self-ligation in the presence of DNA ligase. HBV HindII DNA fragments, including the 5' upstream sequences of the HBcAg/HBeAg (c/e), pX (X), and HBsAg (pre-S/S) genes, were immobilized onto a nitrocellulose membrane. Following transfections of HepG2 cells with either the whole (pHBV) or mutated (pHBV-ΔE) recircularized HBV genome, newly synthesized radiolabeled RNA was prepared from isolated nuclei and was then used to probe the nitrocellulose membrane according to a modified method of Clayton et al. (1985).

Gel Retardation Assay

A fragment of the HBV enhancer II element (spanning nucleotides 1645 to 1805, Sty I-Mst I) was end-labeled with γ-[32p]-ATP in the presence of polynucleotide kinase. Protein samples were analyzed as previously described (Huan and Siddiqui, 1992).

Cell Transfections

Plasmid vectors were transfected into cells using the calcium phosphate precipitation method. Cells were then harvested 48 hours post-transfection and analyzed for luciferase expression according to deWet et al. (1987).

Protein-Protein Interaction Assay

Cellular proteins were labeled with [^{35}S]-methionine and prepared in the following manner. Subconfluent monolayers of Huh-7 cells were starved for 30 minutes in methionine-free media, and were then labeled with 100 μCi/ml [^{35}S]-methionine for 3 hours at 37°C. The cells were then washed, harvested, pelleted, and subsequently resuspended with 1 ml of NP-40 lysis buffer. The cells were briefly sonicated, the debris was pelleted, and the supernatant was then collected for analysis. Samples were initially pre-cleared by incubating in the presence of GST alone to remove proteins that were capable of non-specifically interacting with GST; The unbound proteins were then carried through the remainder of the assay. Pre-cleared protein samples (100 μg of [^{35}S]-methionine labeled cellular proteins or pX) were then incubated in the presence of immobilized GST-fusion proteins at 4°C for 1 hour. Following extensive washing, the bound proteins were eluted with SDS-polyacrylamide gel sample buffer, boiled for 5 minutes, and then analyzed on denaturing SDS-polyacrylamide gels.

RESULTS

The HBV enhancer I is a liver-specific regulatory element that is required for HBV gene expression

It has been established that the HBV enhancer I element is essential for the expression of HBV genes (Jameel and Siddiqui, 1986; Bulla and Siddiqui, 1988; Antonucci and Rutter, 1989; Hu and Siddiqui, 1991). This issue has been addressed by transfecting Hep G2 (hepatoblastoma) cells with either the whole HBV genome or a mutated genome lacking the enhancer I element, and subsequently analyzing HBV transcriptional activity using the nuclear run-on assay. **Figure 2a.** demonstrates that the transcriptional activity of HBV genes is markedly reduced in the absence of the enhancer I element. The levels of hybridization seen in lane 1 reflect the amount of HBcAg/HBeAg (c/e), pX (X), and HBsAg (pre-S/S) messenger RNA (mRNA) that was synthesized following the transfection of HepG2 cells with the whole HBV genome. When enhancer I was deleted from the genome, RNA synthesis from these genes was dramatically reduced (lane 2). Therefore, these results indicate that enhancer I activity is required for the expression of c/e, X, and pre-S/S mRNA.

To further examine the functional properties of the HBV enhancer I element, transient transfection analyses were carried out using hepatocyte and non-hepatocyte derived cell lines. Enhancer I was cloned into a plasmid vector upstream of the SV40 early promoter and the luciferase reporter gene.

Enhancer I was shown to activate luciferase expression by 40-50 fold in the hepatoma cell lines Huh-7 and HepG2, as compared to the normalized level of expression observed with the control vector **(Figure 2b.)**.

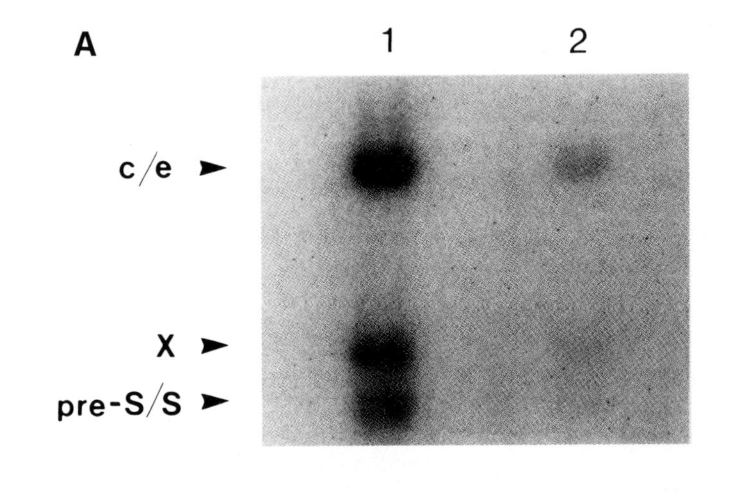

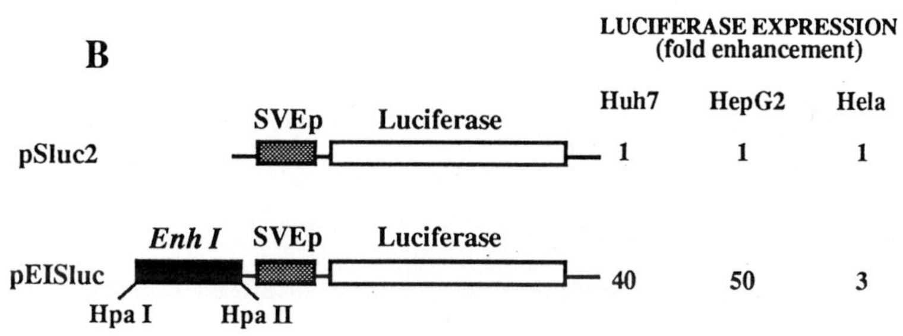

Figure 2. (A) Nuclear run-on analysis of HBV transcriptional activity. Following the transfection of HepG2 cells with the whole (lane 1) or mutated (lacking enhancer I, lane 2) recircularized HBV genome, labeled RNA was isolated and used to probe a nitrocellulose membrane. The membrane was prepared by immobilizing HBV DNA sequences representing the HBcAg/HBeAg (c/e, nt), pX (X), and HBsAg (pre-S/S) genes [Adapted from Hu and Siddiqui (1991)]. (B) Transient transfection analysis of the HBV enhancer I element. Expression of the firefly luciferase gene (DeWet et al., 1987) was under the control of the SV40 early promoter (SVEp) in the presence (pEISluc) or absence (pSluc2) of enhancer I. Luciferase expression was analyzed following transfections with Huh-7, HepG2, or Hela cells. Luciferase expression is expressed as fold enhancement, which reflects the level of expression above that observed with pSluc2. The luciferase expression by pSluc2 has been arbitrarily set at 1 [Adapted from Lopez-Cabrera et al., 1991].

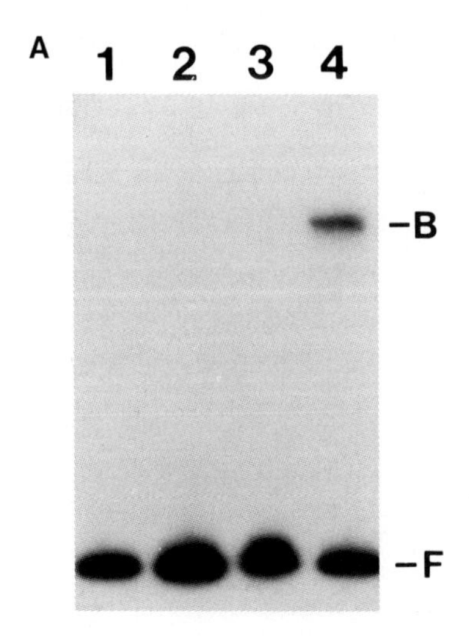

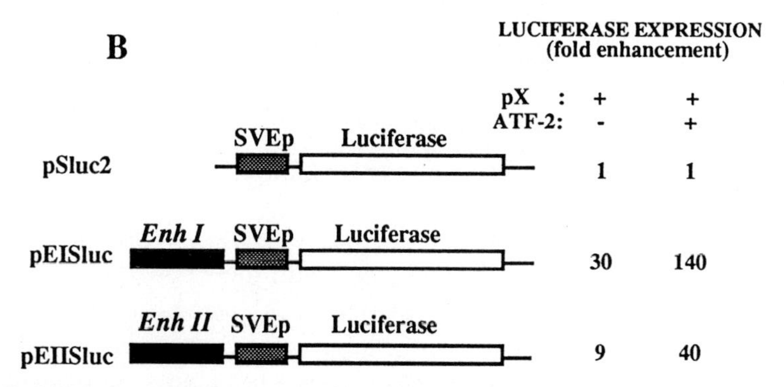

Figure 3. (A) pX facilitates an interaction between the ATF-2 DBD and the HBV enhancer II element. Gel retardation analysis was carried out using a 5'-[^{32}P]-labeled 160 base pair fragment of enhancer II. Proteins samples were incubated with the probe and were subsequently analyzed on native 5% polyacrylamide gels. Free probe (F) and protein/probe bound complexes (B) are shown. Lane 1, probe only; lanes 2-4 contain probe and the following protein samples: lane 2, 100 ng pX; lane 3, 100 ng ATF-2 DBD; lane 4, 100 ng pX + 100 ng ATF-2 DBD. **(B)** Transient transfection analysis of the HBV enhancer I and II elements. Expression of the firefly luciferase gene was under the control of the SV40 early promoter (SVEp) in the presence [pEISluc (enhancer I) or pEIISluc (enhancer II)] or absence (pSluc2) of HBV enhancer elements. Luciferase expression was analyzed following transfections with HepG2 cells (as described in the legend of Figure 2b.). Transfections were carried out with the respective reporter plasmids in the presence or absence of the indicated protein expression vectors [pX (pSBDR) and ATF-2 (pSVATF-2)].

Interestingly, these levels of induction were exclusively observed in hepatocyte-derived cells. The reporter plasmid containing enhancer I induced modest levels of luciferase expression (3 fold) following transfection into cells derived from a human carcinoma of the cervix (Hela). This work demonstrates that enhancer I is capable of activating transcription from a heterologous promoter element, and that this activity occurs in a liver-specific manner.

HBV pX transactivates enhancer I and II, and interacts with proteins present in HepG2 cells.

HBV pX has been shown to transactivate the enhancer I element. We show here for the first time that pX also transactivates the enhancer II element. Previous characterizations of enhancer II revealed that this regulatory element functions in an orientation and position-independent manner (Yee, 1989; Yuh and Ting, 1993). Since pX has been shown to transactivate enhancer I through an ATF-2 responsive element (Maguire et al., 1991), we considered the possibility that pX may transactivate enhancer II in a comparable manner. Gel retardation analysis has indicated that pX facilitates an interaction between the AFT-2 DNA binding domain (DBD) and enhancer II (**Figure 3a**). While pX (lane 2) or the DAFT-2 DBD (lane 3) alone did not bind to enhancer II, the ATF-2 DBD was capable of binding to the enhancer in the presence of pX (lane 4). This data suggests that pX directly interacts with ATF-2, which promotes an interaction between ATF-2, which promotes an interaction between ATF-2 and the HBV enhancer II element. TO examine the functional relevance of this putative interaction, transient transfection studies were carried out. **Figure 3b** demonstrates that pX transactivates enhancers I (pEISluc) and II (pEIISluc) by 30 and 9 fold, respectively, when compared to the control vector (pSluc2). When these transfections were repeated in the presence of an ATF-2 expression vector, the levels of transactivation through enhancers I and II were increased by 4.7 and 4.4 fold, respectively. Furthermore, transfections that were carried out in parallel with the ATF-2 expression vector and the respective reporter plasmids, in the absence of pX, did not show an appreciable increase in activation above that seen with the enhancer elements alone (data not shown). These results support the conclusion that pX transactivates the HBV enhancer II element through an ATF-2 responsive element.

Although there is some evidence to support the hypothesis that pX transactivates transcription by directly interacting with DNA binding proteins (Maguire et al., 1991; Natoli et al., 1994), there is only indirect evidence supporting the notion that pX interfaces with a signal transduction activation pathway (Kekule et al., 1993; Cross et al., 1993). While either or both of the above mechanisms may account for pX activity *in vivo*, the identification and characterization of cellular proteins which interact with pX may ultimately lead to the elucidation of the pX transactivation pathway. In order to address this issue, a protein-protein interaction assay was developed to identify pX binding proteins. An affinity resin was set up by immobilizing GST-pX fusion proteins (full length and truncated forms of pX) onto glutathione-Sepharose beads. Pre-cleared Huh-7 cell lysates, containing [^{35}S]-methionine labeled proteins, were prepared and processed through the affinity resin as described in the Materials and Methods. Bound and unbound proteins were then analyzed by SDS-polyacrylamide gel electrophoresis. Figure 4 that a distinct group of

Huh-7 proteins were selectively bound to the affinity resin (lanes 4-6), while the majority of the cellular proteins did not adhere to the resin (lanes 1-3). With the exception of approximately five proteins (labeled A, B, C, D, and E), the full length (lane 4), N-terminal truncated (lane 5), and C-terminal truncated (lane 6) been emphasized based upon the following distinctions. A particular protein was either bound to both the full length and C-terminal truncated forms of pX (D), the full length truncated form only (C), or the N-terminal truncated form only (A, B, and E; see small arrows). An argument is presented below (see Discussion) which suggests that pX specifically interacts with proteins C and D (see large arrows), whereas the other proteins that bound to the affinity resin appear to represent non-specific interactions.

The protein-protein interaction assay was further utilized to address the possibility that pX interacts with itself to form multimers. **Figure 5** shows that pX multimerizes with itself *in vitro*. Affinity resins were prepared using either GST alone or the GST-full length pX fusion protein. Radiolabeled pX was then prepared and subsequently incubated with each of the affinity resins. Following the elution of bound proteins, samples were analyzed on SDS-polyacrylamide gels. Labeled pX selectively bound to immobilized pX (lane 2), yet did not appreciably interact with immobilized GST (lane 1). Since the eluted samples were analyzed on denaturing protein gels, the extent of multimerization (dimer, trimer, etc.) is unknown. These results suggest that pX multimerizes with itself *in vivo*, which may represent a functionally active and/or inactive form(s) of pX.

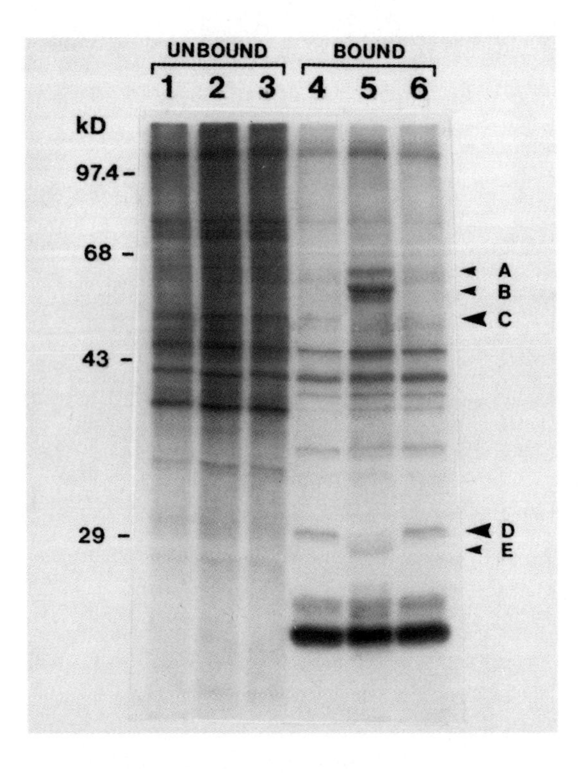

Figure 4. Hepatocyte-derived cellular proteins interact with pX. Pre-cleared [^{35}S]-labeled Huh7 cell lysates were incubated with pX affinity resins. Unbound (lanes 1-3) and bound (lanes 4-6) proteins were isolated and then analyzed on an SDS-polyacrylamide gel (13% acrylamide). Samples derived from incubations with the full length (lanes 1,4), the N-terminal truncated (lanes 2,5), and the C-terminal truncated (lanes 3,6) forms of pX are shown. Proteins of interest are designated A, B, C, D, and E. The following protein standards are indicated on the left: phosphorylase b (97.4 kilodaltons [kD]), bovine serum albumin (68 kD), ovalbumin (43 kD), and carbonic anhydrase (29 kD).

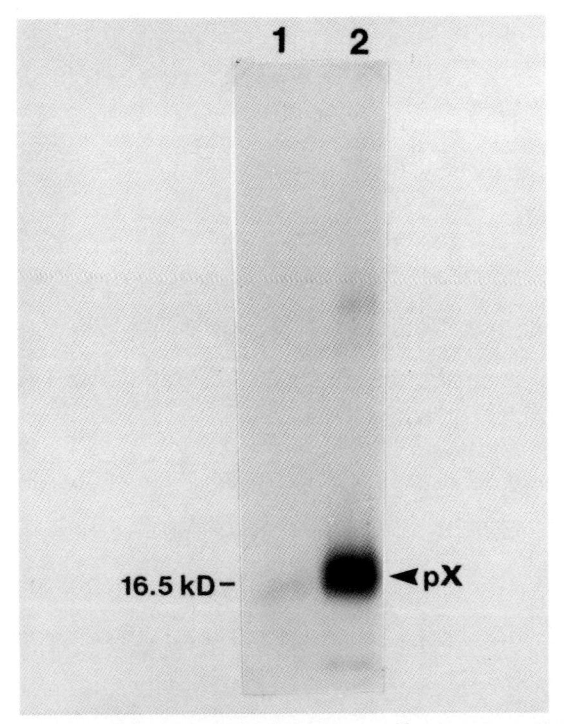

Figure 5. pX multimerizes with itself in vitro. [^{35}S]-labeled pX was incubated with affinity resins comprised of either immobilized GST or GST-full length pX fusion protein. Bound proteins were isolated and then analyzed on an SDS-polyacrylamide gel (12% acrylamide). Samples derived from incubations with immobilized GST (lane 1) and immobilized GST-full length pX (lane 2) are shown. The migration position of pX (16.5 kD) is indicated.

DISCUSSION

The hepatotropic nature of HBV is central to the biology of this pathogen. While a great deal of effort has focused upon gaining an understanding of the mechanisms which dictate this tissue-tropic behavior, progress towards this goal has been restrained by the lack of an *in vitro* propagation system for HBV. However, molecular analyses of the HBV genome have provided important information regarding the genetics of HBV gene expression. This chapter has discussed several aspects of HBV gene expression that are likely to contribute to the liver-specific manifestations of this virus.

The HBV enhancer I element functions in a liver-specific manner to facilitate the expression of HBV genes (**Figure 2**). Recircularized HBV genomes lacking enhancer I were shown to exhibit very low levels of mRNA synthesis following transfections into HepG2 cells. While the HBV genome is very small and compact, the removal of the enhancer I element does not disrupt the c/e, X, or pre-S/S genes (see **Figure 1a**). Therefore, the reduced expression of these genes was due to the absence of the enhancer element. The liver-specific activity of enhancer I appears to be mediated by liver-enriched factor C/EBP (Dikstein et al., 1990b; Trujillo et al., 1991) and the putative liver-specific factor HBLF (Trujillo et al., 1991), studies have revealed that liver-enriched members of the steroid/thyroid hormone receptor superfamily, including RXRα (Huan and Siddiqui, 1992; Garcia et al., 1993) and hepatocyte nuclear factor 4 (HNF-4, Garcia et al., 1991), transactivate the HBV enhancer I element by binding to the same site (see **Figure 1b**). Another member of the receptor superfamily, the chicken ovalbumin upstream promoter transcription factor (COUP-TF), appears to antagonize transcriptional activation by competing with RXRα and HNF-4 for binding to the same site on enhancer I (Garcia et al., 1993). Furthermore, preliminary work from our laboratory suggests that another liver-enriched member of the receptor superfamily, PPAR, also facilitates enhancer I-mediated transcriptional activation. These studies suggest that nuclear receptors play an important role in regulating HBV gene expression. Moreover, the proximity of the hormone receptor binding sites on enhancer I to the binding sites for cellular factors supports the hypothesis that a member(s) of the steroid-thyroid hormone receptor superfamily may facilitate enhancer I transactivation by engaging in a cooperative interaction with another cellular factor(S) that interacts with the enhancer. The implication of this hypothesis is that HBV gene expression may be modulated by a combinatorial mechanism involving a cellular factor(s) and a hormone-mediated regulatory pathway. The resultant activation of gene expression may subsequently promote the onset of the pathological manifestations of chronic HBV infection.

We have further examined the biochemical and functional properties of pX. Evidence is presented which supports the conclusion that pX transactivates HBV enhancer II through an ATF-2 responsive element (**Figure 3**). Previous work has suggested that there is a cooperative interplay between the HBV enhancer I and II elements (Lopez-Cabrera et al., 1990; Lopez-Cabrera et al., 1991; Su nd Yee, 1992). Therefore, since pX appears to transactivate both enhancers I (see **Table 1**) and II, it is possible that the putative interplay between these regulatory elements involves differential transactivations by pX. Furthermore, the cooperativity between these two enhancers may be subject to multiple levels of regulation, which may involve pX and a complex network of interactions between cellular factors and the enhancer elements.

As discussed in the introduction, the biochemical mechanism exhibited by pX is a subject that is replete with uncertainty. Further ambiguity extends to the question of the cellular localization of pX. It has been reported that pX is present in the cytoplasm (Vitvitski et al., 1988; Katayama et al., 1989; Hu et al., 1990), the nucleus (Hu et al., 1990), and is also associated with the nuclear periphery (Siddiqui et al., 1987; Levrero et al., 1990). This generalized localization is consistent with the hypothesis that pX functions differentially in a manner that is dependent upon its cellular location. The evidence supporting a relationship between pX activity and signal transduction implies that pX interacts directly with a component(s) of the cytoplasm. *In vitro* and *in vivo* studies support the conclusion that pX transactivates transcription by engaging in protein-protein interactions with DNA binding proteins in the nucleus. However, while the cellular localization studies suggest that pX functions in both the cytoplasm and the nucleus, there is no conclusive evidence to support this notion. In spite of the complexity of this issue, pX appears to play an important role in the regulation of viral and cellular gene expression (see **Table 1**).

We have initiated studies which focus on identifying cellular proteins that interact with pX. **Figure 4** shows that a distinct group of proteins present n Hun-7 cell lysates selectively bound to GST-pX affinity resins. We suggest that proteins C and D specifically interact with pX. This interpretation is based upon the following assumptions: 1) binding to the full length form of pX may represent a physiologically relevant interaction; and 2) a protein that interacts wit hone or both of the truncated forms of pX, yet is not considered to represent a physiologically relevant interaction. In this case, the C-and/or N-terminal truncated forms(s) of pX may have adopted non-physiological conformations, which could promote aberrant interactions., Based upon the above assumptions, since proteins A, B, and E did not bind to the full length form of pX, these interactions may not be physiological. Therefore, we suggest that under these experimental conditions, proteins that interact with N-terminal truncated form of pX may not be physiologically relevant. If one accepts this rationale, only proteins C and D appear to be physiologically relevant pX binding proteins. The purification and characterization of these proteins may ultimately reveal important aspects of the pX transactivation pathway within the cell.

ACKNOWLEDGEMENTS

This work has been supported by grants from the American Cancer Society, the National Institutes of Health, and the Lucille P. Markey Charitable Trust (to A.S.). M.J.K. is the recipient of a postdoctoral fellowship award from the National Institutes of Health; B.H. is the recipient of a postdoctoral grant from the Cancer League of Colorado.

REFERENCES

Antonucci, T.K. and Rutter, W.J. (1989) Hepatitis B virus (HBV) promoters are regulated by the HBV enhancer in a tissue-specific manner, *J. Virol.* **63**, 579-583.

Aufiero, B. and Schneider, R.J. (1990) The hepatitis B virus X-gene trans-activates both RNA polymerase II and III promoters, *EMBO J.* **9**, 497-504.

Avantaggiati, M.L., Natoli, G., Balsano, C., Chirillo, P., Artini, M., DeMarzio, E., Collepardo, D. and Levrero, M. (1993) The hepatitis B virus (HBV) pX transactivates the c-fos promoter through multiple cis-acting elements, *Oncogene* **8**, 1567-1574.

Beasley, R.P., Lin, C.-C., Hwang, L.-Y. and Chien, C.-S. (1981) Hepatocellular carcinoma and hepatitis B virus, *Lancet* **2**, 1129-1133.

Ben-Levy, R., Faktor, O., Berger, I. and Shaul, Y. (1989) Cellular factors that interact with the hepatitis B virus enhancer, *Mol. Cell. Biol.* **9**, 1804-1809.

Brechot, C., Pourcel, C., Louis, A., Rain, B. and Tiollais, P. (1980) Presence of integrated hepatitis B virus DNA sequences in cellular DNA in human hepatocellular carcinoma, *Nature* **286**, 533.

Bulla, G.A. and Siddiqui, A. (1988) The hepatitis B virus enhancer modulates transcription of the hepatitis B virus surface antigen gene from an internal location, *J. Virol.* **62**, 1437-1441.

Buratowski, S., Sopta, M., Greenblatt, J. and Sharp, P.A. (1991) RNA polymerase II-associated proteins are required for a DNA conformation change in the transcription initiation complex, *Proc. Natl. Acad. Sci. USA* **88**, 7509-7513.

Clayton, D.F., Harrelson, A.L. and Darnell, J.E. (1985) Dependence of liver-specific transcription on tissue organization, *Mol. Cell. Biol.* **5**, 2623-2632.

Colgrove, R., Simon, G. and Ganem, D. (1989) Transcriptional activation of homologous and heterologous genes by the hepatitis B virus X gene product in cells permissive for viral replication, *J. Virology* **63**, 4019-4026.

Cross, J.C., Wen, P. and Rutter, W.J. (1993) Transactivation by hepatitis B virus X protein is promiscuous and dependent on mitogen-activated cellular serine/threonine kinases, *Proc. Natl. Acad. Sci. USA* **90**, 8078-8082.

Degroote, J., Fevery, J. and Lepoutre, L. (1979) Long-term follow-up of chronic active hepatitis of moderate severity, *Gut* **19**, 510-513.

deWet, J.R., Wood, K.V., Deluca, M., Helinski, D.R. and Subramani, S. (1987) Firefly luciferase gene: Structure and expression in mammalian cells, *Mol.Cell. Biol.* **7**, 725-737.

Dikstein, R., Faktor, O., Ben-Levy, R. and Shaul, Y. (1990a) Functional organization of the hepatitis B virus enhancer, *Mol. Cell. Biol.* **10**, 3683 - 3689.

Dikstein, R., Faktor, O. and Shaul, Y. (1990b) Hierarchic and cooperative binding of the rat liver nuclear protein C/EBP at the hepatitis B virus enhancer, *Mol. Cell. Biol.* **10**, 4427-4430.

Edman, J.C., Gray, P., Valenzuela, P., Rall, L.B. and Rutter, W.J. (1980) Integration of hepatitis B virus sequences and their expression in a human hepatoma cell lin, *Nature* **286**, 535.

Faktor, O. and Shaul, Y. (1990) The identification of hepatitis B virus X gene responsive elements reveals functional similarity of X and HTLV-I tax, *Oncogene* **5**, 867-872.

Garcia, D.A., Natesan, S., Alexandre, C. and Gilman, M.A. (1992) Functional interaction of nuclear factors EF-C, HNF-4, and RXRa with hepatitis B virus

enhancer I, *J. Virol.* **67**, 3940-3950.
Grueneberg, D.A., Natesan, S., Alexandre, C. and Gilman, M.Z. (1992) Human and drosophila homeodomain proteins that enhance the DNA binding activity of serum response factor, *Science* **257**, 1089-1095.
Guo, W., Bell, K. and Ou, J.-H. (1991) Characterization of the hepatitis B virus enh I enhancer and X promoter complex, *J. Virol.* **65**, 6686-6692.
Honigwachs, J. Faktor, O., Dikstein, R., Shaul, Y. and Laub, O. (1989) Liver-specific expression of hepatitis B virus is determined by the combined action of the core gene promoter and the enhancer, *J. Virol.* **63**, 919-924.
Hu, K.-Q., Vierling, J.M. and Siddiqui, A. (1990) Trans-activation of HLA-DR gene by hepatitis B virus X gene product, *Proc. Natl. Acad. Sci. USA* **87**, 7140-7144.
Hu, K.-Q. and Siddiqui, A. (1991) Regulation of the hepatitis B virus gene expression by the enhancer element I, *Virology* **181**, 721-726.
Hu, K.-Q., Yu, C.-H. and Vierling, J.M. (1992) Up-regulation of intercellular adhesion molecule 1 transcription by hepatitis B virus X protein, *Proc. Natl. Acad. Sci. USA* **89**, 11441-11445.
Huan, B. and Siddiqui, A. (1992) Retinoid X receptor RXRα binds to and trans-activates the hepatitis B virus enhancer, *Proc. Natl. Acad. Sci. USA* **89**, 9059-9063.
Jameel, S. and Siddiqui, A. (1986) The human hepatitis B virus enhancer requires trans-acting cellular factors for activity, *Mol. Cell. Biol.* **6**, 710-715.
Jameel, S. Siddiqui, A., Maguire, H.F. and Rao, K.V.S. (1990) Hepatitis B virus X protein produced in Escherichia coli is biologically functional, *J. Virology* **64**, 3963-3966.
Katayama, K., Hayashi, N., Sasaki, Y., Kasahara, A., Ueda, K., Fusamoto, H., Sato, N., Chisaka, O., Matsubara, K. and Takenobu, K. (1989) Detection of hepatitis B virus X gene protein and antibody in type B chronic liver disease, *Gastroenterology* **97**, 990-998.
Kekule, A.S., Lauer, U., Weiss, L., Luber, G. and Hofschneider, P.H. (1993) Hepatitis B virus transactivator HBx uses a tumour promoter signalling pathway, *Nature* **361**, 742-745.
Kim, C.-M., Koike, K., Saito, I., Miyamura, T. and Jay, G. (1991) HBx gene of hepatitis B virus induces liver cancer in transgenic mice, *Nature* **351**, 317-320.
Kwee, L., Lucito, R., Aufiero, B. and Schneider, R.J. (1992) Alternate translation initiation on hepatitis B virus X mRNA produces multiple

polypeptides that differentially transactivate class II and III promoters, *J. Virology* **66**, 4382-4389.

Levrero, M., Jean-Jean, O., Balsano, C., Will, H. and Perricaudett, M. (1990) Hepatitis B virus (HBV) X gene expression in human cells and anti-HBx antibodies detection in chronic HBV infection, *Virology* **174**, 299-304.

Lopez-Cabrera, M., Letovsky, J., Hu, K.-Q. and Siddiqui, A. (1990) Multiple liver-specific factors bind to the hepatitis B virus core/pregenomic promoter: Transactivation and repression by CCAAT/enhancer binding protein, *Proc. Natl. Acad. Sci USA* **87**, 5069-5073.

Lopez-Cabrera, M., Letovsky, J., Hu, K.-Q. and Siddiqui, A. (1991) Transcriptional factor C/EBP binds to and transactivates the enhancer element II of the hepatitis B virus, *Virology* **183**, 825-829.

Luber, B., Burgelt, E., Fromental, C., Kanno, M., Kanno, M. and Koch, W. (1991) Multiple simian virus 40 enhancer elements mediate the trans-activating function of the X protein of hepatitis B virus, *Virology* **184**, 808-813.

Maguire, H.F., Hoeffler, J.P. and Siddiqui, A. (1991) HBV X protein alters the DNA binding specificity of CREB and ATF-2 by protein-protein interactions, *Science* **252**, 842-844.

Mahe, Y., Mukaida, N., Kuno, K., Akiyama, M., Ikeda, N., Matsushima, K. and Murakami, S. (1991) Hepatitis B virus X protein transactivates human interleukin-8 gene through acting on nuclear factor kB and CCAAT/ enhancer-binding protein-like cis-elements, *J. Biol. Chem.* **266**, 13759-13763.

Marion, P.L., Salazar, F.H., Alexander, J.J. and Robinson, W.S. (1980) Sate of hepatitis B viral DNA in a human hepatoma cell line, *J. Virol.* **33**, 795.

Menzo, S., Clementi, M., Alfani, E., Bagnarelli, P., Iacovacci, S., Manzin, A., Dandri, M., Natoli, G., Levrero, M. and Carloni, G. (1993) Trans-activation of epidermal growth factor receptor gene by the hepatitis B virus X-gene product, *Virology* **196**, 878-882.

Moriarty, A.M., Alexander, H. and Lerner, R.A. (1985) Antibodies to peptides detect new hepatitis B antigen: serological correlation with hepatocellular carcinoma, *Science* **227**, 429-433.

Murakami, S., Cheong, J., Ohno, S., Matsushima, K. and Kaneko, S. (1994) Transactivation of human hepatitis B virus X protein, HBx, operates through a mechanism distinct from protein kinase C and okadaic acid activation pathways, *Virology* **199**, 243-246.

Nakatake, H., Chisaka, O., Yamamoto, S., Matsubara, K. and Koshy, R. (1993) Effect of X protein on transactivation of hepatitis B virus promoters and on viral replication, *Virology* **195**, 305-314.

Natoli, G., Avantaggiati, M.L., Chirillo, P., Costanzo, A., Artini, M., Balsanoi, C. and Levrero, M. (1994) Induction of the DNA-binding activity of c-Jun/c-Fos heterodimers by the hepatitis B virus enhancer region, *Mol. Cell. Biol.* **14**, 989-998.

Ostapchuk, P., Scheirly, G. and Hearing, P. (1989) Binding of nuclear factor EF-C to a functional domain of the hepatitis B virus enhancer region, *Mol. Cell. Biol.* **9**, 2787-2797.

Patel, N., Jameel, S., Isom, H. and Siddiqui, A. (1989) Interactions between nuclear factors and the hepatitis B virus enhancer, *J. Virol.* **63**, 5293-5301.

Raney, A.K. and McLachlan, A. (1991) The biology of hepatitis B virus, in *Molecular Biology of the Hepatitis B Virus*, Mclachlan, A., E., CRC Press, Boca Raton, Florida, 1-38.

Redeker, A.G. (1975) Viral hepatitis: clinical aspects, *Am. J. Med. Sci.* **270**, 9-16.

Roberts, S.G.E., Ha, I., Maldonado, E., Reinberg, D. and Green, M.R. (1993) Interaction between an acidic activator and transcription factor TFIIB is required for transcriptional activation, *Nature* **363**, 741-744.

Rossner, M.T. (1992) Review: Hepatitis B virus X-gene product: A promiscuous transcriptional activator, *Journal of Medical Virology* **36**, 101-117.

Seto, E., Benedict Yen, T.S., Matija Peterlin, B. and Ou, J.-H. (1988) Trans-activation of the human immunodeficiency virus long terminal repeat by the hepatitis B virus X protein, *Proc. Natl. Acad. Sci. USE* **85**, 8286-8290.

Shafritz, D.A., Shouval, D., Sherman, H., Hadziyannis, S.J. and Kew, M.C. (1981) Integration of hepatitis B virus DNA into the genome of liver cells in chronic liver disease and hepatocellular carcinoma, *N. Engl. J. Med.* **305**, 1067.

Shaul, Y., Rutter, W.J. and Laub, O. (1985) A human hepatitis B enhancer element, *The EMBO J.* **4**, 427-430.

Siddiqui, A., Jameel, S. and Mapoles, J. (1987) Expression of the hepatitis B virus gene in mammalian cells, *Proc. Natl. Acad. Sci. USA* **84**, 2513-2517.

Siddiqui, A., Gaynor, R., Srinivasan, A., Mapoles, J. and Wesley Farr, R. (1989) Trans-activation of viral enhancers including long terminal repeat of the human immunodeficiency virus by the hepatitis B virus X protein,

Virology **169**, 479-484.

Smith, D.B. and Johnson, K.S. (1988) Single step purification of polypeptides expressed in Escherichia coli as fusions with glutathione-S-transferase, *Gene* **67**, 31-40.

Spandau, D.F. and Lee, C.-H. (1988) Trans-activation of viral enhancers by the hepatitis B virus X protein, *J. Virology* **62**, 427-434.

Su, H. and Yee, J.-K. (1992) Regulation of hepatitis B virus gene expression by its two enhancers, *Proc. Natl. Acad. Sci. USA* **89**, 2708-2712.

Szmuness, W. (1978) Hepatocellular carcinoma and the hepatitis B virus: Evidence for a causal association, *Prog. Med. Virol.* **24**, 40-69.

Tiollais, P., Pourcel C. and Dejean, A., (1985) The hepatitis B virus, *Nature* **317**, 489-495.

Trujillo, M.A., Letovsky, J., Maguire, H.F., Lopez-Cabrera, M. and Siddiqui, A.(1991) Functional analysis of a liver-specific enhancer of the hepatitis B virus, *Proc. Natl. Acad. Sci. USA* **88**, 3797-3801.

Twu, J.-S. and Schloemer, R.H. (1987) Transcriptional trans-activating function of hepatitis B virus, *J. Virology* **61**, 3448-3453.

Twu, J.-S., Chu, K. and Robinson, W.S. (1989) Hepatitis B virus X gene activates kB-like enhancer sequences in the long terminal repeat of human immunodeficiency virus 1, *Proc. Natl. Acad. Sci. USA* **86**, 5168-5172.

Van der Vliet, P.C. and Verrijzer, C.P. (1993) Bending of DNA by transcription factors, *BioEssays* **15**, 25-32.

Vitvitski, L., Meyers, M.L., Sninsky, J.J., Berthillon, P., Chevalier, P., Sells, M.A., Acs, G. and Trepo, C. (1988) Expression of the X gene product of hepatitis B virus and WHV in infected livers and transfected 3T3 cells. Evidence for cross-reactivity and correlation with core/e gene expression. In Zuckerman, A.J. (ed.): "Viral Hepatitis and Liver Disease." New York: Alan R. Liss Inc., pp. 341-344.

Vitvitski-Trepo, L., Kay, A., Pichoud, C., Chevallier, P., De-Dinechin, S., Shamoon, B.M., Mandart, E., Trepo, C. and Galibert, F. (1990) Early and frequent detection of HBxAg and/or anti-HBx in hepatitis B virus infection, *Hepatology* **12**, 1278-1283.

Yee, J.-K. (1989) A liver-specific enhancer in the core promoter region of human hepatitis B virus, *Science* **246**, 658-661.

Yuh, C.-H. and Ting, L.-P. (1993) Differentiated liver cell specificity of the second enhancer of hepatitis B virus, *J. Virol.* **67**, 142-149.

Zoulim, F., Saputelli, J. and Seeger, C. (1994) Woodchuck hepatitis virus X protein is required for viral infection in vivo, *J. Virology* **68**, 2026-2030.

TRANSCRIPTIONAL INVOLVEMENT OF THE HEPATITIS B VIRUS PROTEIN IN CELLULAR TRANSDUCTION SYSTEMS. PROTEIN-PROTEIN INTERACTIONS WITH bZip TRANSACTIVATORS

Ourania M. Andrisani
Dept. of Physiology and Pharmacology
School of Veterinary Medicine
Purdue University
W. Lafayette, IN 47907
USA

The Hepatitis B Virus Encoded X Protein.

The Hepatitis B virus genome is 3.2 Kb and contains four recognized open reading frames, three of which encode virion structural proteins, Tiollais *et al.* (1985). These include the S gene (surface antigen), the C-gene (core antigen) and the viral polymerase. The fourth open reading frame, which is conserved among all mammalian hepadnaviruses, encodes a 16.5 KDa protein, termed X antigen. The X gene product is expressed during viral infection, producing a 1Kb mRNA, Tiollais *et al.* (1985). Direct evidence for the ability of the X gene to encode a protein has been obtained by its expression in both prokaryotic, Moriarty *et al.* (1985) and eukaryotic systems, Kay *et al.* (1985). The importance of the X gene product in viral infection has been demonstrated by frameshift X mutants of ground squirrel hepatitis virus, which failed to grow in animal hosts, Siddiqui *et al.* (1987). Recent evidence obtained from transgenic animal experiments supports

NATO ASI Series, Vol. H 88
Liver Carcinogenesis
Edited by G. G. Skouteris

the role of the HBV X gene product in the development of hepatocellular carcinomas, Kim *et al.* (1991), although its mechanism is currently unknown.

A number of studies have demonstrated that the HBV X gene product is a transactivator protein of various cellular and viral enhancers and promoters, Twu and Schloemer (1987), Seto *et al.* (1988), Spandau and Lee, (1988), Colgrove *et al.* (1989), Siddiqui *et al.* (1989), Twu and Robinson (1989) Twu *et al.* (1989). The X-responsive, viral cis-acting elements are diverse in sequence and belong to the category of signal-responsive elements, such as AP-1, NF-κB and CRE, Ben-Levy *et al.* (1989), Siddiqui *et al.* (1989), and Seto *et al.* (1990). Interestingly, the studies carried out to date have not demonstrated direct DNA-binding by the HBV X protein. However, the studies by Seto *et al.* (1990) and Unger and Shaul (1990) have shown that pX can act as a direct transactivator when fused to a heterologous DNA-binding domain. These observations could provide an explanation for the mechanism of pX action. i.e., that the mechanism hinges upon protein-protein interactions between the HBV X protein and cellular transcription factors.

The first study to demonstrate direct protein-protein interactions between the viral X protein and a cellular transcription factor was the study by Maguire *et al.* (1991). It was reported in this study that the HBV X protein interacts directly with the leucine zipper transcription factors CREB/ATF. The CREB/ATF cellular transcription factors act by binding to the cAMP-response element, CRE. Thus, this observation supports that the action of the HBV X protein resembles the action of other viral, non DNA-binding transactivator proteins, such as E1A, Flint and Shenk (1989) and VP-16, Coding and O'Hare (1991). More importantly, the study by Maguire *et al.* (1991), demonstrated that the presence of the X protein alters the sequence specific recognition of CREB and allowed CREB to bind to variant CRE sites, otherwise not recognized by CREB alone.

How does the protein-protein interaction between CREB and X alter the sequence specific recognition properties of CREB? This question prompted my laboratory to further examine the mechanism of the protein-protein interactions occurring in the CREB/pX system. Work in my laboratory focuses primarily on the structure/function of the cellular transcription factor CREB. CREB is a 43 KDa phosphoprotein which mediates the transcriptional induction of cAMP-responsive genes. We have available in the laboratory a number of CREB reagents; these include CREB mutants, described in Andrisani and Dixon (1991), antibodies, and CREB-dependent assay systems, described in Andrisani *et al.* (1988), (1989), Santiago-Rivera *et al.* (1993) and Williams *et al.* (1993). The availability of these molecular tools enables us to further investigate the interactions occurring between CREB and the HBV X protein.

The cAMP Transduction Pathway in Liver.

One of the crucial functions of the liver is the maintenance of blood glucose levels, within a very narrow range. The liver achieves this homeostatic state by being both a consumer of glucose, via glycogen synthesis, and a producer of glucose, via glycogenolysis and gluconeogenesis. Hepatic gluconeogenesis is regulated by glucagon via the cAMP transduction pathway, reviewed by Pilkis and Granner (1992). cAMP regulates key regulatory, gluconeogenic enzymes such as phosphoenolpyruvate carboxykinase, (PEPCK), fructose 1,2-bisphosphatase, (Fru-1,6-P_2ase), and glucose-6-phosphatase, (Glu-6-Pase), by regulating both their enzymatic activity and the amount of enzyme available, Pilkis and Granner (1992). The latter case requires alterations in the rate of transcription of the genes encoding these regulatory enzymes. It is well established that glucagon, via the second messenger cAMP, exerts stimulatory effects on the transcription of both the PEPCK gene, Quinn *et al.* (1988) and the Fru-1,6-P_2ase gene, El-Maghrabi *et al.* (1991). Both genes contain within their promoter the cAMP-response-element, (CRE). Transcriptional induction in response to glucagon by way of cAMP is mediated via the transcription factor CREB. In addition to these genes, an important growth regulating, cAMP-regulated gene in the liver is the proto-oncogene c-fos, Sassone-Corsi *et al.* (1988).

The CREB Transcription Factor.

The transcription factor CREB is a 43 KDa phosphoprotein which is a member of the leucine zipper family of transactivators. The CREB protein was identified, purified, and cloned by studying the prototype cAMP-regulated promoter of the rat somatostatin gene, Montminy *et al.*, (1986), Andrisani *et al.*, (1987). Deletion studies of the 5' upstream region of the somatostatin gene by Montminy *et al.*, (1986) demonstrated that the cis-acting genetic element, TGACGTCA (CRE), confers cAMP-inducible transcription in PC12 cells. The search for factors recognizing the genetic CRE motif lead to the identification, Montminy and Bilezikjian (1987), Andrisani *et al.* (1988), purification, Montminy and Bilezikjian (1987), Zhu *et al.* (1989), and cloning, Hoeffler *et al.* (1988), Gonzales *et al.* (1989), of a 43 kDa protein. This protein interacts in a sequence specific manner with the SST-CRE in *in vitro* DNA binding assays. It is named CREB, for cAMP Responsive Element Binding protein.

CREB protein has also been identified by *in vitro* DNA-binding assays in extracts of HeLa cells, Andrisani *et al.* (1988), in PC12 cells, Montminy and Bilezikjian (1987), and in rat brain cells, Andrisani *et al.* (1988), Zhu *et al.* (1989). It has been purified to apparent homogeneity from PC12 cells, Montminy and Bilezikjian (1987), HeLa cells, Hurst and Jones (1987) and rat brain cells, Zhu *et al.* (1989). Following the cloning of the CREB cDNA, it was demonstrated by Berkowitz and Gilman (1990), that CREB mRNA is constitutively expressed. CREB protein purified from rat brain activates transcription from the somatostatin promoter *in vitro*, in a CRE-dependent manner, as shown by Andrisani *et al.* (1989). This study demonstrated that CREB is the somatostatin gene transactivator.

cAMP Transduction Pathway and CREB Phosphorylation

Extracellular signals are transduced to the nucleus by a complex series of biochemical events. One such signalling event occurs through an increase in the intracellular cAMP concentration, via the adenylate cyclase system which leads, in turn, to the activation of the cAMP-dependent protein kinase A (PKA). The catalytic subunit of PKA translocates to the nucleus, Nigg (1990), where it activates by phosphorylation the CRE-binding transcription factor CREB, Montminy and Bilezikjian (1987), Zhu *et al.* (1989).

Two isoforms of the CREB protein have been identified via cDNA cloning by Hoeffler *et al.* (1988) and Gonzales *et al.*(1989). These isoforms are $CREB_{327}$ and $CREB_{341}$ which differ by one alternatively spliced exon, Yamamoto *et al.* (1990). Both CREB isoforms are found in all cell lines tested, as shown by Berkowitz and Gilman (1990) and Yamamoto *et al.* (1990). Phosphorylation of the CREB protein by PKA has been demonstrated *in vitro* by Montminy and Bilezikjian (1987) and Zhu *et al.* (1989), as well as *in vivo*, following treatment of PC12 cells with forskolin, Montminy and Bilezikjian (1987). Phosphorylation of the CREB protein by PKA, within the consensus site RRPSY does not affect the binding of the protein to the CRE site, Gonzales and Montminy (1990). Functional CREB-dependent transfection experiments in F9 teratocarcinoma cells demonstrated that phosphorylation in $CREB_{341}$ of Ser-133 located within the PKA phosphorylation site, is required for transcriptional activity, Gonzales and Montminy (1990).

Architecture of CREB

Cloning of the CREB cDNA by Hoeffler *et al.* (1988) and Gonzales *et al.* (1989), has demonstrated that the CREB protein is a member of the leucine zipper family of transcription factors. The members of this family of proteins include the C/EBP, Landschulz *et al.* (1988), the proteins recognizing the AP-1 site (fos, c-jun, junB, GCN4), Vogt and Bos(1990), and the CREB/ATF proteins, Hai *et al.* (1989), all of which bind to the CRE motif. The amino acid sequence of this family of proteins includes the N-terminal domain required for transcriptional activation and containing the phospho-acceptor sites and the C-terminal domain involved in DNA binding. In the C-terminal domain two features stand out: a highly basic and conserved region and an equally conserved periodicity of leucines spaced by seven residues. This motif has been named the leucine zipper, Landschulz *et al.* (1988), and is involved in dimer formation. For example, fos and jun form heterodimers, Rauscher *et al.* (1988), whereas CREB forms homodimers through the leucine zipper motif, Dwarki *et al.* (1990). Dimerization serves to bring into proper juxtaposition the adjacent basic region of the protein and DNA binding occurs. The basic region of the protein is the DNA contact region, as demonstrated for GCN4, Hope and Struhl (1986), fos/Jun, Neuberg et al. (1989) and Turner and Tjian (1989), and CREB, Dwarki *et al.* (1990). This DNA contact region determines not only the affinity, but the sequence specific recognition of the DNA binding site. However, it is not understood how the basic domain of CREB brings about the specificity of binding at the CRE site.

We have constructed site-directed mutants of distinct amino acid residues within the DNA domain of CREB, Andrisani and Dixon (1991), and shown that amino acid residues 289 and 291 are important in CRE recognition. Specifically, we showed that a single lysine to glutamine substitution at position 289 or 291 of $CREB_{327}$ alters the methylation interference pattern of the CRE site when bound to these mutant CREB proteins. Additional mutants constructed at these positions showed that only identical basic residues at both positions 289 and 291 of $CREB_{327}$ restore the wild type methylation interference pattern of the CRE motif. This analysis points to the importance of distinct amino acid residues within the basic domain of CREB in CRE recognition.

Further studies in our laboratory examined the structural features of the DNA-binding domain of CREB upon binding to the cognate CRE site. We have employed biophysical, circular dichroism and 2D-NMR methods to decipher the solution structure of a recombinant peptide encoding the basic region-leucine zipper motif of CREB, Santiago-Rivera *et al.* (1993). Circular dichroism analyses demonstrated that the bZip peptide of CREB increases its α-helical content by 20% when binding to the CRE DNA. In contrast, only a 5% increase in its α-helical content is observed upon binding to the related AP-1 site. The increase in the α-helical content upon binding to the CRE DNA is attributed to specific DNA-protein interactions.

Our current understanding of the CREB DNA-binding domain is that the basic region undergoes a coil to helix transition upon binding to the cognate CRE site. The K_{289} and K_{291} amino acid residues of $CREB_{327}$ are key determinants involved in CRE recognition.

<u>Mechanism of CREB/X Interactions.</u>

Employing the CREB assay system and the DNA-binding mutants described above, we have begun to examine the mechanism of the protein-protein interactions occurring between the cellular CREB and viral X protein. We have expressed the HBV X protein employing the bacterial expression vector $T_{7\text{-}7}$, Studier and Moffatt (1986), in Bl21 Lys S (DE3), Tabor and Richardson (1985), following IPTG induction. The recombinant X protein was found in the insoluble pellet and solubilized in buffer containing 6 M urea and 5 mM DTT as reported by Jameel *et al.* (1990) and Wu *et al.* (1990). Following renaturation of pX, the sample was chromatographed on S-sepharose. Elution was in 10 mM phosphate buffer pH 7.2 containing 0.2 M KCl and 5 mM DTT. Throughout the DNA binding assays, 10 μl aliquots of renatured, S-Sepharose chromatographed X protein were used.

To understand the mechanism of the CREB/X interactions, we employed recombinant $CREB_{327}$ and X proteins in *in vitro* DNA binding assays and analyzed by the gel retardation assay, as initially reported by Maguire *et al.* (1991). Fig. 1 shows that a defined amount of CREB (lane 1) displays enhanced binding to the CRE site in the presence of a defined amount of X protein (lane 2). In contrast, heat inactivated X protein looses the ability to enhance the binding of CREB to the CRE site (lane 3).

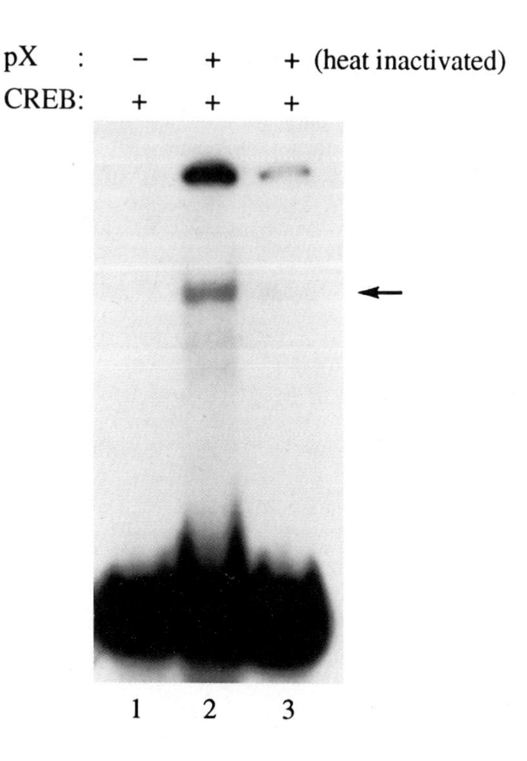

Figure 1: DNA binding assays employing the somatostatin CRE motif as the radiolabeled DNA probe. 10 ng of CREB protein was incubated with 15,000 cpm CRE DNA, in the presence of 1 g poly dIdC, 1 g BSA, in buffer containing 10 mM Hepes, pH 7.4, 1 mm DTT, 0.1 M KCl and 10% glycerol. 10 l aliquots of X protein were added where indicated. Heat inactivated pX was obtained by boiling the protein for 5 min. Analysis of the binding reactions was by electrophoresis on native 5% polyacrylamide gels.

Employing the bacterially expressed bZip domain of CREB in the above assay system with the HBV X protein, we observed a similar enhancement in CRE binding, suggesting that the target of the CREB/X interactions is the 70 amino acids residing at the carboxyl-end of the CREB protein (Williams *et al.*, unpublished results). K_d determination assays employing the CREB protein, Williams *et al.* (1993) and CREB/X recombinant proteins, quantitated a 10-fold enhancement in the affinity of CREB for the palindromic CRE site (Williams *et al.*, unpublished results). Crosslinking experiments of *in vitro* phosphorylated $CREB_{327}$ protein, in the presence of pX, failed to demonstrate a direct effect of pX on the rate of CREB dimerization. However, the methylation interference pattern of the CRE DNA bound to the CREB-pX complex differs from that bound to the CREB protein alone (Williams *et al.*, unpublished observations). These results support that pX targets the basic DNA-binding region of CREB and not the dimerization motif of the CREB protein.

Our working hypothesis is that pX promotes and maintains the correct folding pattern of the basic domain of CREB, potentially contributing to the enhanced binding affinity of CREB. We have demonstrated that the basic domain of CREB assumes a -helical configuration upon binding to the CRE motif, Santiago-Rivera *et al.* (1993). We propose that pX may act as a molecular chaperone by preventing the formation of incorrect folding intermediates or by stabilizing the active CREB conformation. Experiments to test these postulates are currently under way.

Transient transfection assays of mammalian expression vectors encoding the X gene, in conjunction with the CREB-dependent somatostatin promoter demonstrated a 10-fold enhancement in CREB-dependent transcription (Williams *et al.*, unpublished results). The *in vivo* transcriptional induction mediated by the viral X protein is of importance in order to understand the oncogenic potential of this molecule. The increased affinity of CREB due to interactions with pX will effect the activation of a network of cellular genes, including genes containing CRE-like elements which are not normally responsive to the physiological levels of CREB. Moreover, it is not yet known if other CRE-binding transcription factors, such as Jun/fos, Patel *et al.* (1990) and CREMs, Foulkes *et al.* (1991) which respond to different transduction pathways, Lamph *et al.* (1988), Ryder *et al.* (1988),Ryseck *et al.* (1988), interact with the viral X protein to enhance their binding to CRE elements. Thus it is intriguing to reflect on the magnitude of the transcriptional induction and deregulation that may result within the liver cell, by the action of the HBV X protein.

As described earlier, in the liver, a number of metabolically important genes are regulated at the transcriptional level via AP-1 and CRE genetic elements. Such genes include the key metabolic regulatory enzyme phosphoenolpyruvate carboxykinase, the proto-oncogene c-fos and others. A deregulated expression of key cellular regulators by pX has important implications in the pathogenesis of chronic liver disease and cancer.

The normal liver, a slowly regenerating tissue, is the result of a balance between the rate of hepatocyte loss and hepatocyte production. A small but persistent imbalance between hepatocyte cell death and hepatocyte cell growth can have serious consequences. Cirrhosis of the liver in chronic HBV patients and the subsequent development of hepatocellular carcinoma accompanies such a hepatocyte growth imbalance, due to the chronic expression of the viral gene products. The transcriptional involvement of the X gene product may prove to be a crucial parameter in the development of both chronic liver disease and hepatocellular carcinoma.

Acknowledgements:

This work was supported by grants to OMA from the National Science Foundation and the American Cancer Society #CN-69928.

REFERENCES

Andrisani, O.M and Dixon, J.E. (1991) Involvement of lysine residues 289 and 291 of the cAMP-responsive element-binding protein in the recognition of the cAMP-responsive element. J. Biol. Chem. 266: 21444-21450.

Andrisani, O.M., Hayes, T.E., Roos, B., and Dixon, J.E. (1987) Identification of the promoter sequences involved in the cell specific expression of the rat somatostatin gene. Nuc. Acids Res. 15: 5715-5728.

Andrisani, O.M., Pot, D.A., Zhu, Z. and Dixon, J.E. (1988) Three sequence-specific DNA-protein complexes are formed with the same promoter element essential for the expression of the rat somatostatin gene. Mol. Cell. Biol. 8:1947-1956.

Andrisani, O.M., Zhu, Z., Pot, D.A. and Dixon, J.E. (1989) In vitro transcription directed from the somatostatin promoter is dependent upon a purified 43 kDa DNA-binding protein. Proc. Natl. Acad. Sci. U.S.A. 86:2181-2185.

Ben-Levy, R., Faktor, O., Berger, I. and Shaul, Y. (1989) Cellular factors that interact with the hepatitis B virus enhancer. Mol. Cell. Biol. 9:1804-1809.

Berkowitz, L.A. and Gilman, M.Z. (1990) Two distinct forms of active transcription factor CREB (cAMP response element binding protein). Proc. Natl. Acad. Sci. USA 87:5258-5262.

Coding, C.R. and O'Hare, P. (1991) Herpes simplex virus VMW 65-octamer binding protein interaction. A paradigm for combinational control of transcription. Virology, 173:363-367.

Colgrove, R., Simon, G. and Ganem, D. (1989) Transcriptional activation of homologous and heterologous genes by hepatitis B virus X gene product in cells permissive for viral replication. J. Virol., 63: 4019-4026.

Dwarki, V.J., Montminy, M. and Verma, I.M. (1990) Both the basic region and the leucine zipper domain of CREB protein are essential for transcriptional activation. The EMBO J. 9:225-232.

El-Maghrabi, M.R., Lange, A., Kummel, L., Pilkis, S.J. (1991) The rat fructose-1,6-biphosphatase gene: structure and regulation of expression. J. Biol. Chem. 266:2115-2120.

Flint, J. and Shenk T. (1989) Adenovirus E1A protein paradigm viral transactivator. Ann. Rev. Genetics 23:141-161.

Foulkes, N.S., Borelli, E. and Sassone-Corsi, P. (1991) CREM gene: Use of alternative DNA-binding domains generates multiple antagonists of cAMP-induced transcription. Cell 64:739-749.

Gonzales, G.A. and Montminy, M.R. (1990) cAMP stimulates gene transcription by phosphorylation of CREB at serine 133. Cell 59:675-680.

Gonzales, G.A., Yamamoto, K.K., Fischer, W.H., Karr, D., Menzel, P. (1989) A cluster of phosphorylation sites on the cAMP-regulated nuclear factor CREB, predicted by its sequence. Nature 337:749-752.

Hai, T., Liu, F., Coukos, W.J. and Green, M.R. (1989) Transcription factor ATF cDNA clones: an extensive family of leucine zipper proteins able to selectively form DNA-binding heterodimers. Genes and Dev. 3:2083-2090.

Hoeffler, J.P. Meyer, T.E., Yun, Y., Jameson, L.J. and Habener, J.F.

(1988) Cyclic AMP-responsive DNA-binding protein: structure based on a cloned placental cDNA. Science 242: 1430-1433.

Hope, I.A. and Struhl, K. (1986) Functional dissection of a eukaryotic transcriptional activator protein, GCN4 of yeast. Cell 46:885-894.

Hurst, H. and Jones, N.C. (1987) Identification of factors that interact with the E1A-inducible adenovirus E3 promoter. Genes and Dev. 1: 1132-1146.

Jameel, S., Siddiqui, A., Maguire, H. and Rao, K. (1990) Hepatitis B virus X protein produced in *E. coli* is biologically functional. J. Virol. 64: 3963-3966.

Kay, A., Mandart, E., Trepo, C. and Galibert, F. (1985) The HBV HBX gene expressed in *E. coli* is recognized by sera from hepatitis patients. EMBO J. 4: 1287-1292.

Kim, C., Koike, K., Saito, Il, Miyarmura, T., Jay, G. (1991) HBVX gene of hepatitis B virus induces liver cancer in transgenic mice. Nature 351: 317-351.

Lamph, W.W., Wamsley, P., Sassone-Corsi, P. and Verma, I. (1988) Induction of proto-oncogene Jun/Ap-1 by serum and TPA. Nature 334: 629-631.

Landschulz, W.H., Johnson, P.F. and McKnight, S.L. (1988) The leucine zipper: a hypothetical structure common to a new class of DNA binding proteins. Science 240: 1759-1764.

Maguire, H.F., Hoeffler, J.P. and Siddiqui, A. (1991) HBV X protein alters the DNA binding specificity of CREB and ATF-2 by protein-protein interactions. Science 252: 842-844.

Montminy, M.R., and Bilezikjian, L.M. (1987) Binding of a nuclear protein to the cAMP response element of the somatostatin gene. Nature 328: 175-178.

Montminy, M.R., Sevarino, K.A., Wagner, J.A., Mandel, G., Goodman, R.H. (1986) Identification of a cyclic AMP responsive element within the rat somatostatin gene. Proc. Natl. Acad. Sci. USA 83: 6682-6686.

Moriarty, A.M., Alexander, H., Lerner, R.A., and Thornton, G.B. (1985) Antibodies to peptides detect new hepatitis B antigen: serological correlation with hepatocellular carcinoma. Science 227: 429-432.

Neuberg, M., Schuermann, M., Hunter, J.B. and Muller, R. (1989) A Fos protein containing the Jun leucine zipper forms a homodimer which binds to the AP-1 binding site. Nature 338: 589-590.

Nigg, E.A. (1990) Mechanisms of signal transduction to the cell nucleus. Adv. in Cancer Res. 55: 271-300.

Patel, L., Abate, C. and Curran, T. (1990) Altered protein conformation on DNA binding by fos and jun. Nature 347: 572-575.

Pilkis, S.J. and Granner, D.K. (1992) Molecular physiology of the regulation of hepatic gluconeogenesis and glycolsis. Ann. Rev. Phys. 54:885-909.

Quinn, P., Wong, T.W., Magnuson, M.A., Shabb, J.B. and Granner, D.K. (1988) Identification of the basal and cAMP regulation elements in the promoter of the PEPCK gene. Mol. Cell Biol. 8: 3467-3475.

Rauscher, F.J., III, Sambucette, L.C., Curran, T., Distel, R.J.and Spiegelman, B.M. (1988) Common DNA binding site for Fos-protein complexes and transcription factor AP-1. Cell 52: 471-480.

Ryder, K., Lau, L.F. and Nathans, D. (1988) A gene activated by growth factors is related to the oncogene V-jun. Proc. Natl. Acad. Sci. USA 85: 1487-1491.

Ryseck, R.P.,Hirai, S.I., Yaniv, M. and Bravo, R. (1988) Transcriptional activation of c-jun during the G_0/G_1 transition in mouse fibroblasts. Nature 334: 535-537.

Santiago-Rivera, Z.I., Williams, J.S., Gorenstein, D.G. and Andrisani, O.M. (1993) Bacterial expression and characterization of the CREB bZip module: Circular dichroism and 2D ^{1}H-NMR studies. Prot. Science 2: 1461-1471.

Sassone-Corsi, P., Visvader, J., Ferland, L., Mellon, P. and Verma, I.M. (1988) Induction of proto-oncogene fos transcription throught the adenylate cyclase pathway: Characterization of a cAMP-responsive element. Genes and Dev. 2:1529-1538.

Seto, E., Mitchell, P.J. and Yen, T.S.B. (1990) Transactivation by the hepatitis B virus X protein depends on AP-2 and other transcription factors. Nature 344: 72-74.

Seto, E., Yen, T., Peterlin, B. and Ou, J. (1988) Transactivation of the human immunodeficiency virus long terminal repeat by the hepatitis B virus X protein. Proc. Natl. Acad. Sci., USA, 85: 8286-8290.

Siddiqui, A., Gaynor, R., Srinivasan, A., Mapoles, J. and Farr, R.W. (1989) Transactivation of viral enhancers including long terminal repeat of the human immunodeficiency virus by the hepatitis B virus X protein. Virology, 169: 479-484.

Siddiqui, A., Jameel, S. and Mapoles, J. (1987) Expression of the hepatitis B virus X gene in mammalian cells. Proc. Natl. Acad. Sci.

USA, 84: 2513-2517.
Spandau, D. and Lee, C.H. (1988) Transactivation of viral enhancers by the hepatitis B virus X protein. J. Virol., 62: 427-434.
Studier, F.W. and Moffatt, B.A. (1986) Use of bacteriophage T_7 RNA polymerase to direct the expression of cloned genes. J. Mol. Biol. 189: 113-130.
Tabor, S. and Richardson, C.C. (1985) A bacteriophage T_7 RNA polymerase/promoter system for controlled exclusive expression of specific genes. Proc. Natl. Acad. Sci. USA 82: 1074-1078.
Tiollais, P., Pourcell, C. and Dejean, A. (1985) The Hepatitis B virus. Nature 317: 489-495.
Turner, R. and Tjian, R. (1989) Leucine repeats and adjacent DNA binding domain mediate the formation of functional c-fos/c-jun heterodimers. Science 243: 1689-1694.
Twu, J.S. and Schlomer, R.H. (1987) Transcriptional transactivating function of hepatitis virus. J. Virol., 61: 3448-3453.
Twu, J.S. and Robinson (1989) Hepatitis B virus X gene can transactivate heterologous viral sequences. Proc. Natl. Acad. Sci., USA, 86: 2046-2050.
Twu, J.S., Chu, K. and Robinson, W. (1989) Hepatitis B virus X gene activates NFB-like enhancer sequences in the long terminal repeat of human immunodeficiency virus 1. Proc. Natl. Acad. Sci., USA, 86: 5168-5172.
Unger, T. and Shaul, Y. (1990) The X protein of the hepatitis B virus acts as a transcription factor when targeted to its responsive element. EMBO J. 9:1889-1995.
Vogt P.K. and Bos T.J. (1990) Jun: oncogene and transcription factor. Adv. in Cancer Res. 55: 1-31.
Williams, J.S., Dixon, J.E. and Andrisani, O.M. (1993) Binding constant determination studies utilizing recombinant ΔCREB protein. DNA and Cell Biology 12: 183-190.
Wu, J.Y., Zhou, Z., Judd, A., Cartwright, C. and Robinson, W. (1990) The hepatitis B virus-encoded transcriptional trans-activator hbx appears to be a novel serine/threonine kinase. Cell 63, 687-695.
Yamamoto, K.K., Gonzalez, G.A., Menzel, P., Rivier, J. and Montminy, M.R. (1990) Characterization of a bipartite activator domain in transcription factor CREB. Cell 3, 611-617.

Zhu, Z., Andrisani, O.M., Pot, D.A. and Dixon, J.E. (1989) Purification and characterization of a 43 kDa transcription factor required for rat somatostatin gene expression. J. Biol. Chem. 264, 6550-6556.

ROLE OF DUCK HEPATITIS B VIRUS INFECTION, AFLATOXIN B_1 AND p53 MUTATION IN HEPATOCELLULAR CARCINOMAS OF DUCKS

Lucyna Cova
INSERM U271
151 Cours A. Thomas
69003 Lyon, France

INTRODUCTION

Risk factors such as chronic hepatitis B virus (HBV) infection and ingestion of a mycotoxin aflatoxin B1 (AFB1) are clearly involved in the etiology of hepatocellular carcinoma (HCC) in the high-incidence areas of Africa and Southeast Asia (Munoz and Bosch 1987). The role of dietary aflatoxins in the etiology of HCC has been highlighted by the recent detection of a high frequency of mutational hotspot in the third position of codon 249 of the p53 tumor suppressor gene in human HCCs from patients in Qidong, China (Hsu et al. 1991, Scorsone et al. 1992) and Mozambique (Bressac et al. 1991, Ozturk et al. 1991) where AFB1 exposure is high, but not in the HCCs from several other geographic locations in which AFB1 is not a risk factor (Kress et al. 1992, Challen et al. 1992).

HBV is a prototype member of the hepadnavirus family (Robinson 1980) which also includes the closely related mammalian viruses isolated from naturally infected woodchucks (WHV) (Summers et al. 1978) and ground squirrels (GSHV) (Marion et al. 1980) as well as avian viruses isolated from ducks (DHBV) (Mason et al. 1980) and herons (HHBV) (Sprengel et al. 1988). Although much is known about hepadnavirus structure and replication strategy, the mechanisms responsible for malignant transformation are not well understood (reviewed in Feitelson 1992). Chronic hepatitis following infection by HBV, WHV and GSHV frequently leads to HCC in the respective hosts. Among mammalian hepadnaviruses WHV

Agnes Duflot[1], Raj Mehrotra[2], Monica Hollstein[3], Christopher P. Wild[3], Shi-Fang Cao[4], Shun-Zhang Yu[4], Michel Prave[5], Ruggiero Montesano[3] Christian Trepo[1]
(1) INSERM U271, 151 Cours A. Thomas, 69003 Lyon, France, (2) KGMC, Lucknow, India, (3) IARC, 150 Cours A. Thomas 69008 Lyon, France, (4) Shanghai Medical University, Shanghai, PRC, (5) Ecole Nationale Veterinaire, Marcy L'Etoile, France.

NATO ASI Series, Vol. H 88
Liver Carcinogenesis
Edited by G. G. Skouteris

seems more oncogenic that HBV and GSHV since 100% of woodchucks experimentally infected with this virus develop HCC within 17-36 months after infection in the absence of other factors (Popper et al., 1987). The integration of viral DNA in tumor cells, often accompanied by rearrangements of both virus and cellular DNA, was found in almost all HCC examined and seems a prerequisite for the malignant transformation in the mammalian hosts.
The link between DHBV infection and HCC in ducks is less clear. Pekin ducks chronically infected with DHBV and followed in the USA, Europe and Australia for several years have not been found to develop liver tumors (Cullen et al. 1990, Cova et al. 1990, Lambert et al; 1991, Freiman & Cossart 1986). Interestingly, HCCs have been found in domestic ducks in only a single area of the world, Qidong, China (Omata et al. 1983, Marion et al. 1984). Although several factors such as DHBV carrier rate, breed and age of ducks and subtype of DHBV are suspected to contribute to this stiking difference in geographical repartition of liver cancer, the exposure to dietary aflatoxin was sugested to play a major role in Chinese duck liver carcinogenesis (reviewed in Cova et al. 1993).
Qidong is an area of high human HCC incidence in which both HBV and AFB_1 are risk factors and a p53 mutational hotspot is so prevalent (Hsu et al. 1991, Scorsone et al. 1992). Among animal models which are usefull for the understanding of the specificity of this mutation, the duck is of a particular interest since it is highly susceptible to the carcinogenic effect of AFB1 (Uchida et al. 1988, Cullen et al. 1990, Cova et al. 1990) and can be infected by DHBV, which is known to be the relevant model for the study of HBV infection and replication (reviewed in Schödel et al. 1991). However data on liver cancer in Chinese ducks are missing since only few such HCCs have been analysed so far (Omata et al. 1983, Marion et al. 1984) and only one case of integrated DHBV DNA has been reported (Yokosuka et al. 1985).
To better define the association between DHBV infection, AFB_1, HCC and p53 gene mutations we have analysed a large panel of liver samples obtained from domestic ducks in Qidong.

MATERIALS AND METHODS

Liver samples:
Two series of liver samples obtained from domestic ducks in Qidong were analysed: 1) - out of 1321 ducks, sacrificed during 1973 Qidong mid-fall festival, 59 livers showing a gross abnormality were formalin fixed,

2) - 16 frozen liver samples were collected from local farms in Qidong (1988-1989). The histopathological diagnosis and PCR analysis have been performed on both the formalin-fixed and frozen liver samples, while the DHBV DNA integration and AFB_1-DNA adducts were examined only in the frozen material.

Histopathological study of duck liver

Liver tissue was paraffin embedded and sections were stained with standard histopathological techniques. HCC were classified, according to Nakashima and Kojiro (1987), as trabecular, schirrous, pseudoglandular and undifferentiated types. The portal tract changes eg: cellular inflammation and bile duct proliferation were recorded.

DNA extraction from paraffin-embedded tissue

The tumor cell areas from 10-µm tissue sections of each paraffin-embedded block were selectively removed based on corresponding hematoxylin-eosin stained sections. The tissues were minced to a fine powder and deparaffinized with sequential washes in xylene and ethanol. The pellet was incubated with 1%SDS and proteinase K (400µg/ml) for 6 hours at 37°C, thereafter nucleic acids were extracted with phenol-chloroform, ethanol-precipitated and the DNA resuspended in 20 µl of water (Goelz et al. 1985).

Detection of DHBV by Polymerase Chain Reaction.

DHBV-specific primers MD03 and MD33, located in the highly conserved region of the polymerase gene (Mack and Sninsky 1988) were used (Table 1).

Table 1 PCR primers and probes used in this study

primer	sequence
MD03[a]	5'-ctcaagcttATCATCCATATA-3'
MD33	5'-cttggatCCAATGGGCGTCGGTCT-3'
MD10	5'-CAGCCCTTTTCTCCTCCATCTCTTCACTACTGCCCTCGGA-3'
R1[b]	5'-ctcggatccTACAACTTCATGTGC(T)AACAGT-3'
L1	5'-ctcgtcgaCTCCAGTGTAAGGATGGT-3'
S7	5'-GGGATGAACCGCCGCCCCATC-3'

a MD03 and MD33 are the primers and MD10 is the probe for DHBV DNA detection. b R1 and L1 are primers and S7 is the probe for exon7 of duck p53 amplification. The bases of primer sequences in the lowercase letters represent 5' extensions containing a restriction site for eventual cloning (Mack and Sninsky 1987).

Each reaction was performed as previously described (Baginski et al. 1991) in a total volume of 50µl using 35 cycles; 94°C (30s), 55°C (30s), 72°C (1min). For the analysis of amplified DNA, 10 µl of the PCR product were resolved in a 3% NiuSieve (FMC Corporation)-1% agarose (Sigma, U.S.A.) gel. The gel was transferred overnight on to nylon membrane (Hybond N^+, Amersham) and hybridized at 42°C, with the internal oligonucleotide probe (table 1) labeled by terminal transferase as previously described (Baginski et al. 1991). The sensitivity of this PCR-Southern blot (PCR-SB) assay was estimated as 0.8 fg, using serial dilutions of DNA from a DHBV-positive liver (data not shown).

PCR amplification and sequencing of p53

Genomic DNA was subjected to PCR amplification using primers R1 and L1 which derived from exon 7 of the chicken p53 gene sequence (Table 1). Amplification was performed using 40 cycles: 30s at 95°C, 1min at 55°C and 1min at 72°C, and analysed by PCR-SB with the S7 internal probe, as described above. The PCR procedure is described in detail elsewhere (Duflot et al. 1994).
Asymmetrical PCR (40 cycles) was performed using 1/5 of the purified amplification product for 40 cycles, in 100 µl with an uneven ratio of the two primers (1 pmol: 50 pmol) (Gyllensten and Erlich 1988). The asymmetrical PCR products, purified through a Sephadex G50 column, were sequenced directly by the Sanger dideoxy chain termination method using Sequenase version 2 kit reagents (U.S. Biochemical). The sequence data were further confirmed by analysis of PCR products cloned in to the pUC18 vector according to Boehringer Mannheim protocols. At least four individual clones were sequenced.

AFB1-DNA adduct analysis

Isolated DNA was acid hydrolysed to release AFB_1-imidazole ring-opened guanine (AFB_1-Fapy) adducts, the adducts were purified on Sep-pak C18 Cartridges (Waters, Montigny, France) followed by reverse phase HPLC and quantified by ELISA as described in detail elsewhere (Cova et al. 1994).

RESULTS

Liver histopathology

Out of 75 Chinese duck livers a total of 44 HCCs (36/59 formalin fixed and 8/16 frozen samples) were identified. The predominant type of HCC was characterized by trabecular arrangement of malignant cells resembling hepatocytes with eosinophilic cytoplasm, prominent nuclei and were often separated by fibrous bands. This trabecular type HCC was frequently associated with different morphological types of HCC: scirrhous, pseudoglandular, and poorly differentiated (data not shown). In addition 4 ducks had cholangiocarcinoma and one had bile duct adenoma. Cirrhosis was present in 10 ducks, 6 of which had an HCC. In 22 ducks varying degrees of portal inflammation, and portal fibrosis were observed in the absence of advanced liver disease. Bile duct proliferation was present in the majority of liver samples (56/75).

Detection of DHBV DNA in duck livers.

The formalin-fixed and frozen liver samples were analysed for DHBV DNA by PCR-SB using DHBV-specific primers located in the highly conserved region of the polymerase gene. As illustrated in Fig. 1 the predicted 124-bp band was observed in the DNA from three out of ten formalin fixed HCCs and in normal liver (lanes 3, 6, 8 and 1 respectively) as well as in the control DNA from a frozen Qidong duck HCC (lane 15).

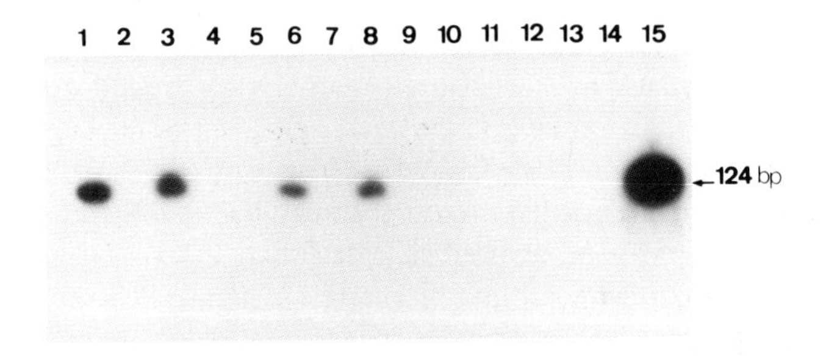

Fig. 1. PCR-SB analysis of DHBV DNA in formalin-fixed and frozen Qidong duck livers. Lanes 1-2 are DNA from normal livers. Lanes 3-12 are DNA from 10 HCCs. Lane 11 is DNA from a cirrhosis and lane 12 a DNA from cholangiocarcinoma. Lane 13 is PCR negative control (all reaction components except DNA), lane 14 no sample. Lane 15 is control DNA from the frozen liver of a Qidong duck, known to be DHBV-infected.

Using this PCR-SB test we have demonstrated the presence of DHBV DNA in 11 out of 42 HCCs analysed (Table 2). As a control, we have probed for p53 DNA by PCR-SB amplification and demonstrated the presence of the 92-bp p53-amplification product (data not shown) in all HCCs, suggesting that the absence of detectable DHBV DNA in 31 of these archival duck livers was not related to the DNA degradation. Altogether, these results suggest that liver cancer in Qidong ducks may occur in the absence of viral DNA detectable by PCR-SB (detection limit: 0.8 fg of DHBV DNA per μg of liver DNA).

AFB1-DNA adducts

DNA was extracted from eight HCCs which were stored as frozen tissue, although for all but one of these ducks only the tumorous tissue was available. The presence of AFB_1-Fapy was detected in one liver DNA sample with an adduct level of 6.38 pmoles of AFB_1-Fapy adducts per mg DNA. This positive result was found in the only non-neoplastic tissue available for analysis and was in a duck (no 31) which was negative for DHBV. These results have been described in detail elsewhere (Cova et al 1994).

Search for mutation in codon 249 of p53 gene

Taking advantage of p53 protein conservation in various species (Soussi et al. 1990), primers R1 and L1 (Table 1) located within exon 7 of the known chicken p53 sequence (Soussi et al. 1988), an avian species evolutionarily close to duck, were selected for duck p53 sequence determination around codon 249. For simplicity, in this report we refer to the duck equivalent of the human codon 249 (arg ^{249}) simply as "codon 249". The direct sequencing of a 92-bp PCR-amplified fragment from normal duck liver, encompassing codon 249 and located within exon 7 of the duck p53 gene, allowed us to determine this yet unknown sequence. The sequence data, further confirmed by analysis of cloned PCR-products, showed 72% DNA homology between duck and human p53 in this region. The predicted amino acid sequence of this fragment of duck p53 indicates that arginine in position 249 was conserved in duck p53; however due to the degeneracy of the genetic code the respective codon sequence was CGC while it is AGG in human (Duflot et al. 1994).

Ten formalin-fixed (5 DHBV-positive and 5-DHBV negative) HCCs of Qidong ducks were analysed for the presence of codon 249 mutations in the p53 gene by direct sequencing of the 92-bp PCR-product. None of the sequences we analysed differed from the wild type p53 sequence (Table 2).

Table 2 DHBV status and p53 mutation at codon 249 in Chinese duck HCCs

Histopathological diagnosis	number of ducks	DHBV-DNA PCR status	p53 mutation at codon 249[a]
HCC frozen samples[b]	8	4[c] +	NT
		4 -	0/1 DHBV[d] -
HCC formalin-fixed[b] samples	34	7 +	0/5 DHBV +
		27 -	0/5 DHBV -

[a] direct sequencing analysis of the PCR-amplified fragment was performed; [b] included trabecular, scirrhous, pseudoglandular, and undifferentiated HCC types; [c] integration of DHBV DNA sequences into cellular DNA was demonstrated in 1 of these 4 HCCs; [d] the presence of AFB_1-DNA adducts was detected in this frozen HCC sample (no 31) by HPLC

In addition, the p53 amplification products from two formalin-fixed HCCs of Qidong ducks were also cloned and 5 and 7 clones, respectively were sequenced. None showed a mutation in codon 249 of p53. Of particular interest was the frozen HCC of duck no 31 from Qidong (Table 2) for which the exposure to aflatoxin has been clearly demonstrated by detection of AFB_1-DNA adducts by HPLC in this liver DNA sample. The direct sequencing of amplification product from this HCC, further confirmed by cloning and sequencing of 9 individual clones, revealed an absence of mutations.

DISCUSSION

We report here observations on DHBV infection, AFB_1 exposure, p53 mutation and liver cancer in domestic Chinese ducks from Qidong county where HBV infection and AFB1 are risk factors and HCC is particularly frequent in humans. There is limited data on the liver pathology occuring in Qidong ducks since to date only four HCCs have been described and all of them were well differentiated HCC of trabecular type (Omata et al. 1983, Marion et al. 1984). In the present study, on a large panel of 44 HCCs, we have observed a range of different morphological types e.g. schirrous, pseudoglandular and even undifferentiated HCCs which were not previously reported. Another interesting pathological feature of our study was the biliary proliferation both in ducks with and without HCC from the Qidong area. This biliary proliferation was never reported as being associated with DHBV infection but has been observed by us and others in ducks experimentally exposed to AFB1 (Cullen et al.1990, Cova et al.1990).

We have analysed a large collection of Chinese duck HCC by highly specific and sensitive PCR-SB, using primers located within the viral polymerase gene, and demonstrated that although the DHBV carrier rate in Qidong is high, liver cancer occurs in Chinese ducks in the absence of DHBV infection. The absence of DHBV DNA in some duck HCCs was not related to the degradation of DNA, in the formalin-fixed paraffin-embedded samples collected 20 years ago, since we have detected by PCR the presence of a p53 sequence in all HCCs tested. In addition, to verify that the primer target sites were not altered or missing, virus-negatives samples have been tested for the DHBV DNA presence using another primer set located within the DHBV core gene (Duflot et al. manuscript in preparation). The lack of association between DHBV infection and HCC suggests that factors other than DHBV may be important in duck liver carcinogenesis in Qidong.

The previous reports on Chinese duck HCC revealed only a single case of DHBV DNA integration (Yokosuka et al.1985), and we have recently reported a second case (Cova et al. 1994). Unlike HCC of humans and woodchucks the integration of DHBV DNA does not seem to be a prerequisite for HCC development since it was observed in only one out of four DHBV-positive HCCs analysed (Cova et al. 1994). Recently, Gu et al (1992) have suggested the presence of human HBV sequences in some HCCs from Qidong ducks. Using the same approach, i.e. Southern blot analysis of total duck liver DNA followed by hybridization with HBV-specific probe, we have tested for but not detected human HBV sequences in any of the 8 frozen Qidong duck HCCs analysed (Cova et al. unpublished observation).

Given the high aflatoxin food contamination in Qidong (Zhu and Huang 1986), the AFB1-exposure of domestic ducks was suspected. Moroever it is an accepted practice in Qidong to nourish ducks with leftover or spent rice and corn originally destined for human consumption which are a major source of human dietary aflatoxin (Howard et al., 1973, Yu 1985). We have demonstrated that one out of 8 frozen HCCs analysed was positive for the presence of AFB_1-DNA adducts in liver. This was the one sample where non neoplastic tissue was available and we cannot rule out the possibility that other ducks were exposed to aflatoxin but that lower adduct levels were present in the tumor compared to that of normal tissue, this is discussed in detail elsewhere (Cova et al. 1994). Alternatively or in addition, differences in AFB_1 exposure between these ducks could have occured. The relationship between the initial AFB1 exposure, the amount of AFB_1-DNA adducts and the induction of liver cancer in the duck should be further investigated in the carcinogenicity experiments.

In contrast to the human data, we have not found a mutation at codon 249 in any of the eleven HCCs from Qidong ducks (including the sample for which the presence of AFB_1-DNA adducts was demonstrated). Furthermore we have not found mutations in four DHBV positive and negative HCCs from AFB_1-treated ducks (Duflot et al. 1994). However, in humans the preferential binding of AFB_1 to codon 249 (AGG*) occurs at the third nucleotide G* (Hsu et al. 1991, Puisieux et al. 1991, Aguilar et al. 1993), while in the duck, codon 249 is CGC and lacks this G* residue (Duflot et al.1994). Even though the amino acids corresponding to residues 248, 249 and 250 are identical in human and duck, due to the degeneracy of the genetic code, 4 of the 9 base pairs in the DNA sequence are different. Thus the absence of hotspot mutation is not surprising since the human hotspot is absent in duck from the viewpoint of base sequence context, this is discussed in detail elsewhere (Duflot et al. 1994)

Moreover species differences in metabolism and DNA repair might be responsible for infrequent codon 249 mutation in ducks as has been suggested for other AFB_1-exposed animal models such as rats (Hulla et al. 1993) and nonhuman primates (Fujimoto et al. 1992) for which this mutation has not been observed.

Although we have not found codon 249 mutation in duck HCCs relevant mutations may be found elsewhere in the duck p53 gene. The ongoing search for p53 mutations in exons 5-8 in HCCs from domestic Qidong ducks as well as ducks experimentally exposed to AFB_1 will be informative in this respect.

Acknowledgments

This work was supported in part by grant 6735 from ARC and by grant 104-1 from IFCPAR. A. Duflot is the recipient of a fellowship from Ligue Nationale contre le Cancer.

REFERENCES

Aguilar F, Hussain S.P and Cerutti P (1993) Aflatoxin B_1 induces the transversion of G to T in codon 249 of the p53 tumor suppressor gene in human hepatocytes. Proc. Natl. Acad. Sci. USA, 90, 8586-8590.

Baginski I, Chemin I, Bouffard P , Hantz O and Trepo C (1991) Detection of polyadenylated Rna in hepatitis B virus-infected peripheral blood mononuclear cells by polymerase chain reaction. J. Inf. Dis., 163, 996-1000.

Bressac B Kew M Wands J and Ozturk M (1991) Selective G to T mutation of p53 gene in hepatocellular carcinoma from southern Africa. Nature, 350, 429-431.

Challen C., Lunec J, Warren W, Collier J and Bassendine M F (1992) Analysis of the p53 tumor-suppressor gene in hepatocellular carcinomas from Britain. Hepatology, 16, 1362-1366.

Cova L, Wild CP, Mehrotra R, Turusov V, Shirai T, Lambert V, Jacquet C, Tomatis L, Trépo C and Montesano R (1990) Contribution of aflatoxin B_1 and hepatitis B virus infection in the induction of liver tumors in ducks. Cancer Res. 50:2156-2163.

Cova L, Duflot A., Prave M., Trépo C. (1993) Duck hepatitis B virus infection, aflatoxin B_1 and liver cancer in ducks. Arch. Virol. Suppl. 8, 81-87

Cova L, Mehrotra R, Wild CP, Chutimataewin S, Cao SF, Duflot A, Prave M, Yu SZ, Montesano R and Trepo C (1994) Duck hepatitis B virus infection, aflatoxin B_1 and liver cancer in domestic Chinese ducks. Brit. J. Canc., 69, 104-109.

Cullen JM, Marion PL, Sherman GJ, Hong X, Newbold JE (1990) Hepatic neoplasms in aflatoxin B_1-treated, congenital duck hepatitis B virus-infected and virus-free Pekin ducks. Cancer Res. 50:4072-4080.

Duflot A, Hollstein M, Mehrotra R, Trepo,C. and Cova L (1994) Absence of p53 mutation at codon 249 in duck hepatocellular carcinomas from the high incidence area of Qidong (China). Carcinogenesis in press

Feitelson M (1992) Hepatitis B virus infection and hepatocellular carcinoma . Clinical Microbiology Reviews, 5, 275-301

Freiman J.S and Cossart YE (1986) Natural duck hepatitis B virus infection in Austria. Aust. J. Exp. Biol. Med. Sci., 64, 477-484.

Fujimoto Y, Hampton LL, Luo L, Wirth PJ, Thorgeirsson SS (1992) Low frequency of p53 gene mutation in tumors induced by aflatoxin B1 in nonhuman primates. Cancer Res. 52:1044-1046.

Goelz SE, Hamilton SR and Vogelstein B (1985) Purification of DNA from formaldehyde fixed and paraffin embedded human tissue. Biochemical and Biophysical Research Communications, 130, 118-126.

Gu JR (1992) Viral hepatitis B and primary hepatic cancer. in Wen YN, Xu ZY and Melnick JL eds. Viral hepatitis in China; problems and control strategies, Basel, 19, 56-72

Gyllensten U.B and Erlich H A (1988) Generation of single-stranded DNA by the polymerase chain reaction and its application to direct sequencing of the HLA-DQA locus. Proc. Natl. Acad. Sci. USA, 85, 7652-7656.

Howard H, Weller MW, Humphrey PS and Clark GA. (1973) Domestic mallards. In The Waterfowl of the World. Delacour, J. (ed.). 4, pp 154-166.

Hsu IC, Metcalf RA, Sun T, Welsh JA, Wang NJ, Harris CC (1991) Mutational hot spot in the p53 gene in human hepatocellular carcinomas. Nature 350:427-428.

Hulla JE, Chen ZY, Eaton DL (1993) Aflatoxin B1-induced rat hepatic hyperplastic nodules do not exhibit a site-specific mutation within the p53 gene. Cancer Res. 53:9-11.

Kress S., Jahn UR., Buchmann A., Bannasch P. and Schwarz M. (1992) p53 mutations in human hepatocellular carcinomas from Germany. Cancer Res., 52, 3220-3223.

Lambert V., Cova L., Chevallier P, Mehrotra R and Trepo C (1991) Natural and experimental infection of wild mallard ducks with duck hepatitis B virus. J. Gen. Virol., 72, 417-420.

Mack DH, Sninsky JJ (1988) A sensitive method for the identification of uncharacterized viruses related to known virus groups: Hepadnavirus model system. Proc. Natl. Acad. Sci. USA 85:6977-6981.

Marion PL, Oshiro LS, Regnery DC, Scullard G H and Robinson W S (1980) A virus in Beecheyi ground squirrels that is related to hepatitisB virus in humans. Proc. Natl. Acad. Sci. USA, 77, 2941-2945.

Marion PL, Knight SS, Ho BK, Guo YY, Robinson WS, Popper H (1984) Liver disease associated with duck hepatitis B virus infection of domestic ducks. Proc. Natl. Acad. Sci. USA 81:898-902.

Mason W, Seal G and Summers J (1980) Virus of Pekin ducks with structural and biological relatedness to human hepatitis B virus. J. Virol., 36, 829-836.

Munoz N.M. and Bosch F.X. (1987) Epidemiology of hepatocellular carcinoma. In: Neoplasm of the Liver (K. Okuda and K. G. Ishak Eds.), Springer-Verlag, Tokyo, 3-19

Nakashima T, Kojiro M (1987) Histopathological appearences of hepatocellular carcinoma. In: Hepatocellular Carcinoma , Nakashima T, Kojiro M. (eds) Springer Verlag, pp 41-47.

Omata M, Uchiumi K, Ito Y, Yokosuka O, Mori J, Terao K, Wei-Fa Y, O'Connel AP, London WT, Okuda K (1983) Duck hepatitis B virus and liver diseases. Gastroenterology 85:260-267.

Ozturk M. and collaborators. (1991) p53 mutation in hepatocellular carcinoma after aflatoxin exposure. Lancet, 338, 1356-1359.

Popper H, Roth L, Purcell RH, Tennant BC Gerin JL. (1987). Hepatocarcinogenicity of the woodchuck hepatitis virus. Proc. Natl. Acad. Sci. USA, 84, 866-870.

Puisieux A, Lim S, Groopman J, Ozturk M (1991) Selective targeting of p53 gene mutational hotspots in human cancers by etiologically defined carcinogens. Cancer Res. 51:6185-6189.

Robinson,W.S. (1980) Genetic variation among hepatitis B and related viruses. Ann. NY Acad. Sci., 354, 371-378.

Schödel F, Weimer T, Fernholz D, Sprengel R, Wildner G and Will H. (1991) The biology of avian hepatitis B viruses. In McLachlan,A. (ed.) Molecular Biology of the Hepatitis B Virus, CRC Press, Boston London, pp. 53-80.

Scorsone K A, Zou Y, Butel JS and Slage B L (1992) p53 mutations cluster at codon 249 in hepatitis B virus-positive hepatocellular carcinomas from China. Cancer Res., 52, 1635-1638.

Soussi T, Begue A, Kress M, Stehelin D, May P (1988) Nucleotide sequence of a cDNA encoding the chicken p53 nuclear oncoprotein. Nucleic Acids Res. 16:11383.

Soussi T, Caron de Fromentel C and May P (1990) Structural aspects of p53 protein in relation to gene evolution. Oncogene., 5, 945-952.

Sprengel R Kaleta EF and Will H (1988) Isolation of a hepatitis B virus endemic in herons. J. Virol., 62, 3832-3839.

Summers J, Smoles JM and Snyder R (1978) A virus similar to human hepatitis B virus associated with hepatitis and hepatoma in woodchucks. Proc. Natl. Acad. Sci. USA, 75, 4533-4537.

Uchida T, Suzuki K, Esumi M, Arii M and Shikata T (1988) Influence of aflatoxin B_1 intoxication on duck livers with DHBV infection. Cancer Res. 48, 1559-1565.

Yokosuka O, Omata M, Zhou, YZ, Imazeki F, Okuda K (1985) Duck hepatitis B virus DNA in liver and serum of chinese ducks: integration of viral DNA in a hepatocellular carcinoma. Proc. Natl. Acad. Sci. USA 82:5180-5184.

Yu S.Z. (1985) Epidemiology of primary liver cancer. In Subclinical Hepatocellular Carcinoma, Tang, Z-Y. (ed) pp. 189-211. Springel-Verlag, Berlin, Heidelberg, New York, Tokyo.

Zhu, Y. R. & Huang, X., Y.(1986). Hepatocellular carcinoma in Qidong County. In: Tang Z.Y., Wu M.C. and Xia S.S. (ed), Primary liver cancer. pp 204-222, China Academic Publ. , Beying.

THE PREPARATION AND CULTURING OF RAT HEPATOCYTES

Paul Skett
Department of Pharmacology
West Medical Building
University of Glasgow
GLASGOW G12 8QQ
Scotland, U.K.

Isolated, cultured hepatocytes would seem at first sight to be an ideal model system for the study of hepatocarcinogenesis and the testing of chemicals that may be carcinogens as the liver is the major site of activation of many potential carcinogens and, thus can act as generator of the toxic metabolite as well as the target.The great potential of this model is, however, diluted by the diversity of approaches taken by various workers in the field to the preparation and culturing of hepatocytes.There is no accepted or recommended method for preparing the cells from intact liver (although many methods have been published) and little work on the optimisation of the medium and medium additions needed for culturing of hepatocytes.

One major problem with the use of isolated and cultured hepatocytes is that the very enzymes in the liver that can activate carcinogens are unstable in cultured cells and rapidly disappear from the cells (Paine, 1990).Many attempts have been made to slow or abolish this decline in enzyme activities such as co-culture, use of inhibitors and extracellular

NATO ASI Series, Vol. H 88
Liver Carcinogenesis
Edited by G. G. Skouteris

matrices but none have met with complete success and few can do more that slow down the apparently inevitable disappearance of phase I, oxidative metabolism of foreign chemicals (for review of these methods see Rogiers (1993), Guillouzo (1993), and Skett (1994)).

In an attempt to overcome some of these problems and formulate a recommended protocol for the preparation of rat hepatocytes, a meeting was held of scientists with expertise in this field (organized by the European Center for the Validation of Alternative Methods (ECVAM)).The full recommended protocol is given below:

Proposal for Hepatocyte Preparation from 250 g Rat

1. Deeply anaesthetise animal (sodium pentobarbital has most often been used but a better anaesthetic would be preferred due to known interaction with liver).
2. The use of heparin before cannulation is recommended to inhibit blood coagulation.
3. Use 2-step collagenase perfusion method.
   a.400 ml (in 10 min) Ca^{2+}-free Hank's BSS containing 0.1mM EGTA without recirculation.
4. Liver should blanch very quickly during this procedure (few seconds)- if not STOP.
5. *In situ* or *ex vivo* perfusion equally good.
6. Special attention to quality of collagenase-test batches first.
7. Special attention to temperature during perfusion- **liver** at 37^0 C not buffer.
8. Care for pH of buffer -7.4 to be maintained throughout.pH may change during perfusion- check!

9. Care for osmolality of buffer - 300mOsm throughout.
10. Low pulsatility of pump required.
11. It is not clear whether oxygenation is required or desirable during perfusion.To be checked.Hank's BSS either buffered with HEPES (no oxygenation) or HCO_3 (gassed with carbogen) routinely used.
12. End of perfusion judged by appearance of liver.
13. Cells dispersed by removing capsule and gentle combing or shaking in culture medium containing 10% FCS or 1% BSA (to stop action of proteases).Do not temperature shock the cells (either work at room temperature or on ice until culturing).
14. Filter through nylon mesh (60μm) and sediment at 50g for 1-2 min.Wash 2-3 times, final time with culture medium.
15. Cells should be viable (>80% by trypan blue) and numerous ($4x10^7$ cells/gm).
16. If culturing, cells should attach well within 4hr (75%).

This standard protocol is to be lodged with the INVITOX and Hepatocyte Users Group (HUG) databases and given out freely to all enquirers.The protocol is not rigid and will be subject to modification in light to further experience.It would be appreciated if any group trying this method could contact INVITOX or HUG and give their experiences with the above method so that a bank of information could be assembled.

Once viable rat hepatocytes have been prepared, it is equally if not more vital to harmonise the method of culture so that comparable results can be obtained.Otherwise each group using the cells will obtain results which are only valid for their own laboratory and cannot be looked at in a wider context.The Commission of the European Communities (CEC) has also been active in this field, funding preliminary studies on the optimum, serum-free medium for maintaining liver specific functions in rat hepatocytes in culture.The results obtained so far are given below:

Assessment of Culture Media

Isolated hepatocytes were prepared by the standard double perfusion method of Seglen (1976) and plated onto 9cm plastic dishes at 1×10^7 cells/plate.Medium, containing 2.5% foetal calf serum and 15% horse serum was used for the first 4h to allow cell attachment after which serum-free medium (containing only 0.1% bovine serum albumin) was used.The degree of maintenance of phase 1 xenobiotic metabolizing enzymes was assessed using androst-4-ene-3,17-dione as substrate as this is metabolized by a number of cytochrome P-450-dependent and flavin-dependent enzymes and, thus, a good range of enzymes can be assessed with the same substrate (Gustafsson & Stenberg, 1973).Media were chosen that had been used before to culture hepatocytes.

Figure 1 shows the lack of maintcnancc of cnzyme activities in Ham's F-10 medium over 4 days.All cytochrome P-450-dependent enzyme activities (7α-, 6β- and 16α-hydroxylases) fall markedly within 24 h and continue to fall.The flavin-dependent enzymes (17-OHSD and 5α-reductase) are maintained (or even rise a little)at 24h but subsequently fall markedly to reach levels somewhat above that of the cytochrome P-450-dependent enzymes at 48 and 72h.

Figure 2 shows the results with Leibovitz L-15 medium.A better maintenance of activities is seen but the 7α- and 6β-hydroxylase activities fall significantly within 24h and stay low.The 16α-hydroxylase activity, however, remains at control levels throughout the culture period.The flavin-dependent enzymes fall to about 50% of zero time values at 24h but rebound to control levels for the rest of the incubation period.This may be

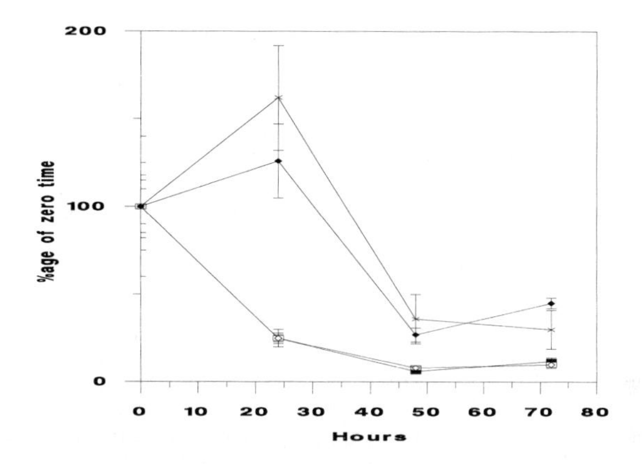

fig.1. The effect of culturing in Ham's F-10 nutrient medium supplemented with 0.1% bovine serum albumin on the metabolism of androst-4-ene-3,17-dione in hepatocytes isolated from an adult male rat. ■ = 7α-hydroxylase; □ = 6β-hydroxylase; ◇ = 16α-hydroxylase; ♦ = 17-hydroxysteroid dehydrogenase; x = 5α-reductase.

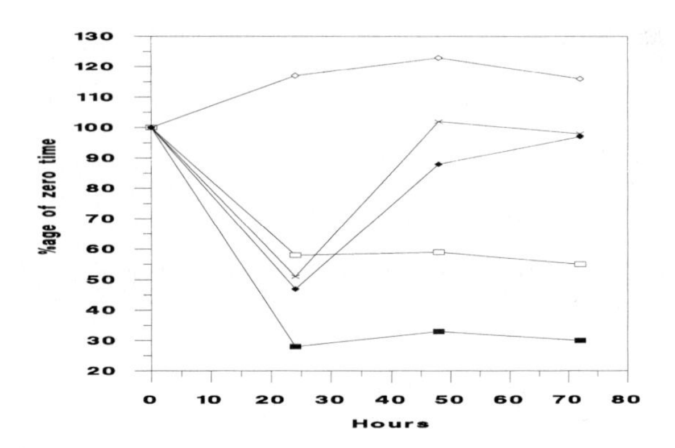

fig.2. The effect of culturing in Leibovitz L-15 medium supplemented with 0.1% bovine serum albumin on the metabolism of androst-4-ene-3,17-dione in hepatocytes isolated from an adult male rat. Symbols as in fig.1.

a function of the 4h period of culture in the presence of serum and an experiment would need to be performed with a shorter plating time (say 1h) to investigate this.This was not done due to the success of the other media and the fact that in the other media (see below) no dip in activity is seen at 24h even when serum is included for the first 4h.

Figure 3 shows the effect of culture in William's E medium.It is seen that activities are stable or even increase to a certain extent in this medium.The flavin-dependent enzymes (17-hydroxysteroid dehydrogenase and 5α-reductase) rise above 0 time control levels at 24h but fall back to control by 72h.The 7α- and 16α-hydroxylase also rise above zero time values by 24h but stay slightly elevated whereas the 6β-hydroxylase remains at control levels throughout.Two further experiments with William's E medium are shown in figures 4 and 5 and illustrate the consistency of the effects of this medium.In these latter tow experiments, the plating time was reduced to 1h and seemed to make little difference and the period of culture was extended to 144h with no loss of enzyme activities.

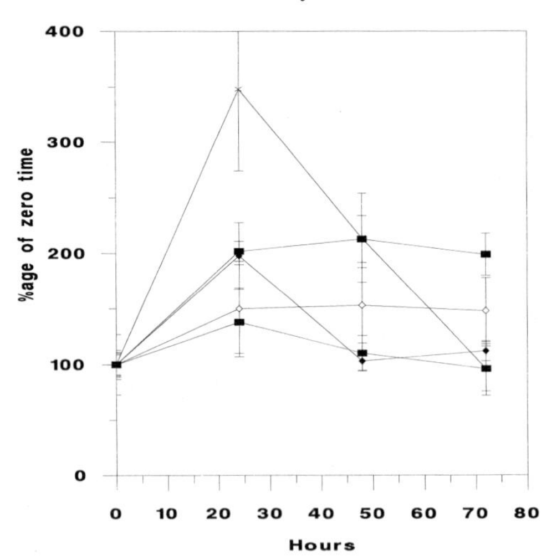

fig.3. The effect of culturing in Williams' E medium supplemented with 0.1% bovine serum albumin on the metabolism of androst-4-ene-3,17-dione in hepatocytes isolated from an adult male rat. Symbols as in fig.1.

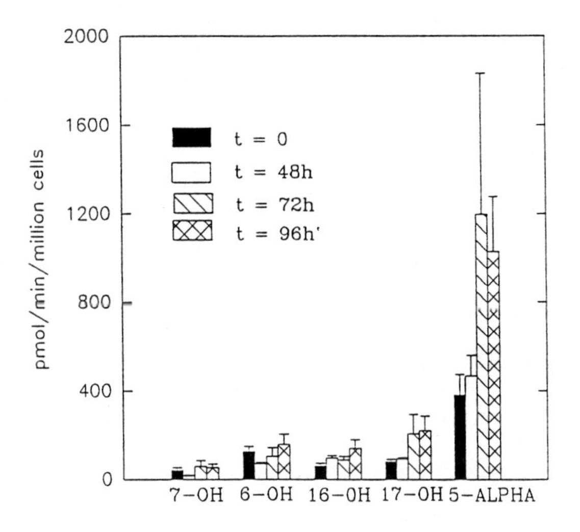

fig.4. The effect of culturing in Williams' E medium supplemented with 0.1% bovine serum albumin on the metabolism of androst-4-ene-3,17-dione in hepatocytes isolated from an adult male rat. OH = hydroxylase, 5-ALPHA = 5α-reductase.

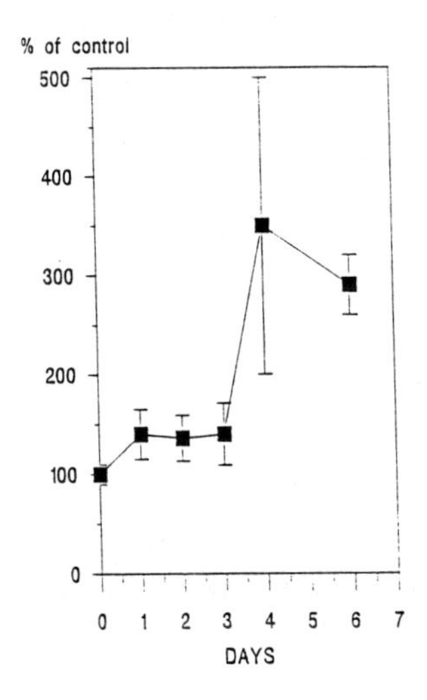

fig.5. The effect of culturing in Williams' E medium supplemented with 0.1% bovine serum albumin on the 16α-hydroxylase active on androst-4-ene-3,17-dione in hepatocytes isolated from an adult male rat.

The results obtained with Chee's medium are shown in Figure 6. Again all activities are maintained for at least 48h at virtually zero time levels except for the 5α-reductase which moved significantly above zero time values (250%) at 72h. Further experiments with this medium, using a plating time of 1h and extending the culture period to 144h showed that activity was maintained at 24h and was little changed at 144h. In one experiment, the flavin-dependent enzyme activities showed a fall in activity at 96h which was maintained at 144h.The 5α-reductase is known to be a particularly volatile enzyme with wide variations in activity seen in any liver preparation and, thus, little can be read into changes in this activity.

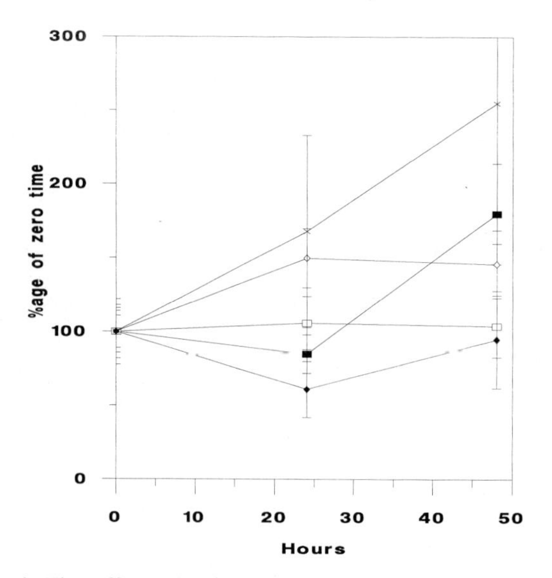

fig.6. The effect of culturing in Chee's medium supplemented with 0.1% bovine serum albumin on the metabolism of androst-4-ene-3,17-dione in hepatocytes isolated from an adult male rat. Symbols as in fig.1.

In summary, the preliminary results suggest that William's E or Chee's medium may be the most suitable for long-term culture of viable, functioning hepatocytes.

Hormone Additions

The use of insulin and glucocorticoid additions to help maintain hepatocyte functions is widespread. To assess the usefulness of this approach, we looked at the effects of insulin (at physiological to supraphysiological doses), hydrocortisone and dexamethasone (a synthetic glucocorticoid) on maintenance of androst-4-ene-3,17-dione metabolism in Ham's F-10 medium (the medium which did not maintain activities very well without hormone additions).

As can be seen from figure 7, insulin at all concentrations had little effect on maintenance of any enzyme activity.

The effects of glucocorticoids were, however, more interesting. The effect of the natural glucocorticoid, hydrocortisone, was dose-dependent with 10^{-6}M concentrations giving a positive effect on the cytochrome P-450-dependent activities and on the 5α-reductase but not on the 17-hydroxysteroid dehydrogenase (figure 8). Lower concentrations of hydrocortisone had less of an effect and, indeed,hastened the fall in enzyme activities in some cases (e.g. 7α-hydroxylase - figure 8). Dexamethasone was by far the best at maintaining 7α-hydroxylase activity but had a marked suppressive effect on 17-hydroxysteroid dehydrogenase activity (figure 8). Dexamethasone is known to induce cytochrome P-450 3A1, the enzyme performing the 7α-hydroxylation so it is not surprising that this activity is maintained in the presence of this drug.

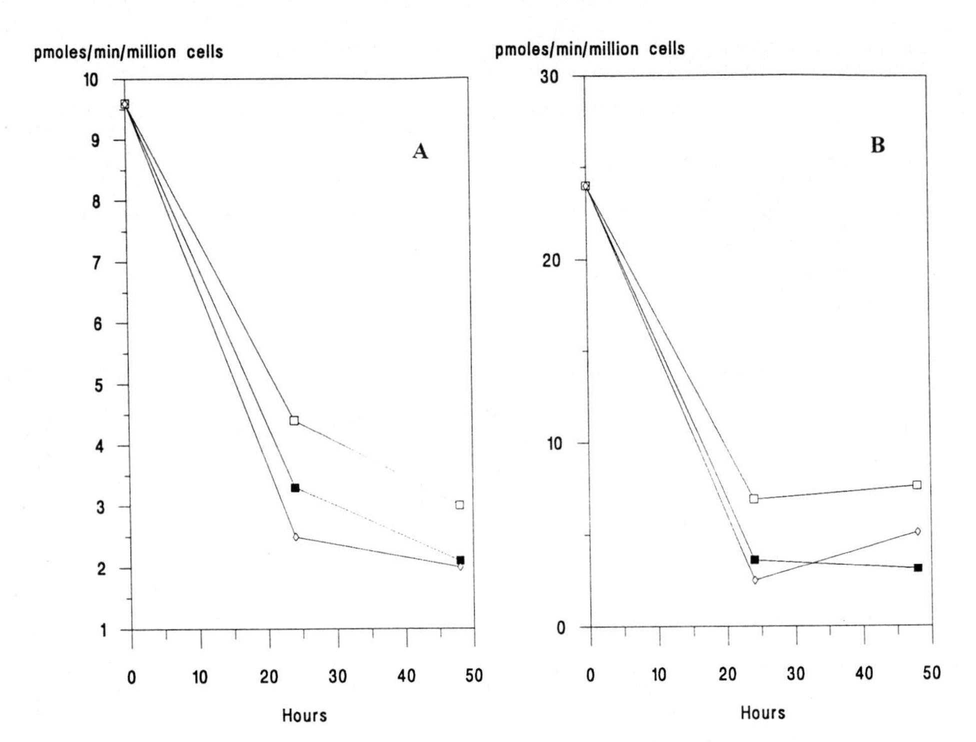
pmoles/min/million cells
10
9
8
7
6
5
4
3
2
1
A
0
10
20
30
40
50
Hours
pmoles/min/million cells
30
20
10
0
B
0
10
20
30
40
50
Hours

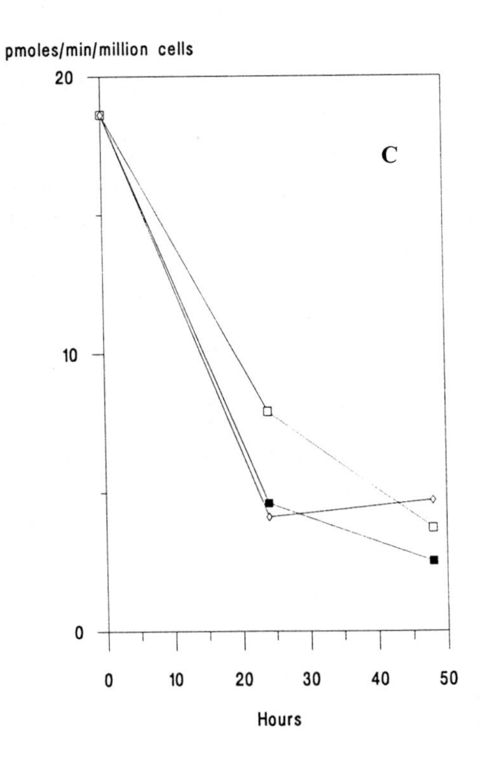
pmoles/min/million cells
20
10
0
C
0
10
20
30
40
50
Hours

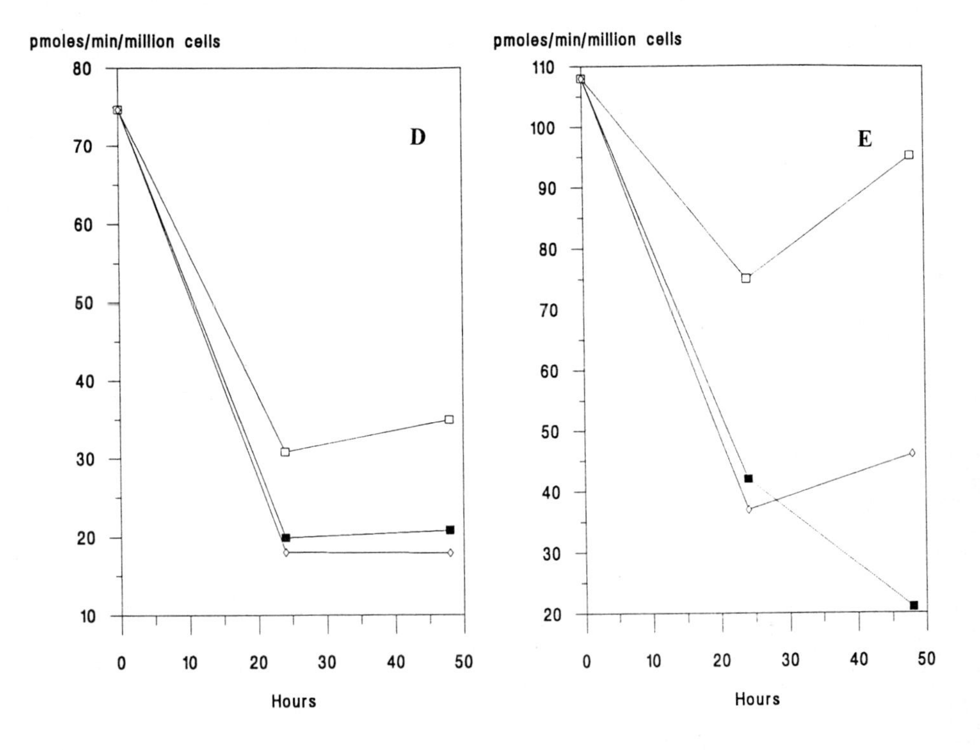

fig.7. The effect of insulin (10-9M) (□) or (10^{-6}M) (■) on the metabolism of androst-4-ene-3,17-dione in hepatocytes isolated from an adult male rat. A- 7α-hydroxylase; B- 6β-hydroxylase; C- 16α-hydroxylase; D- 17-hydroxysteroid dehydrogenase; E- 5α-reductase.

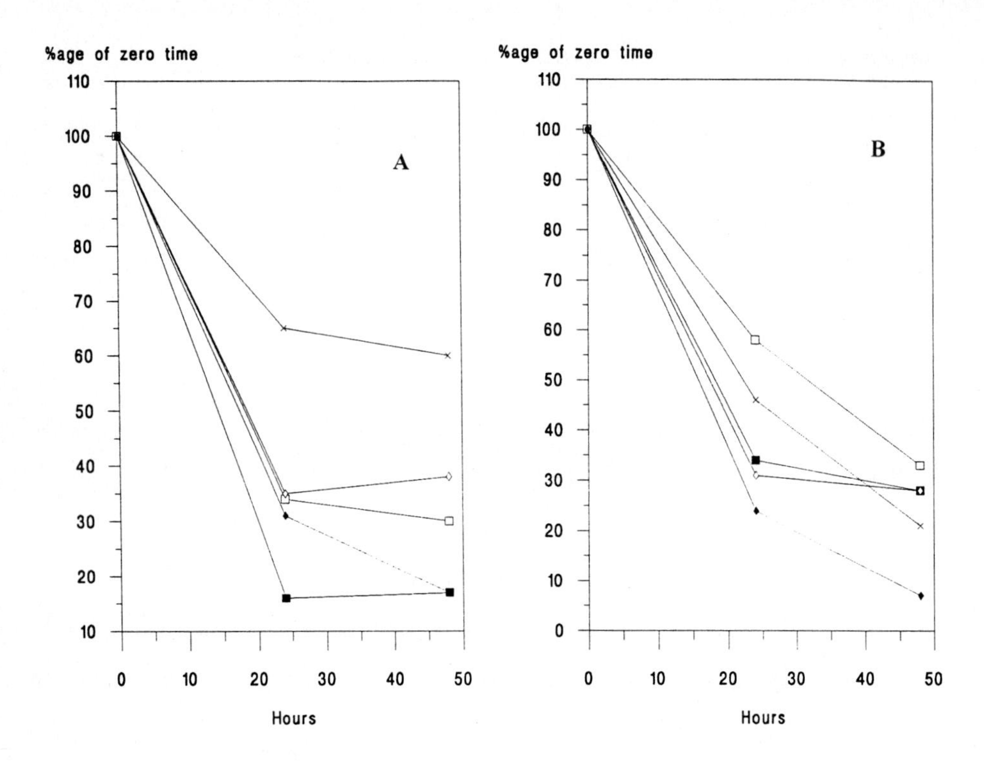
%age of zero time
A
Hours
%age of zero time
B
Hours

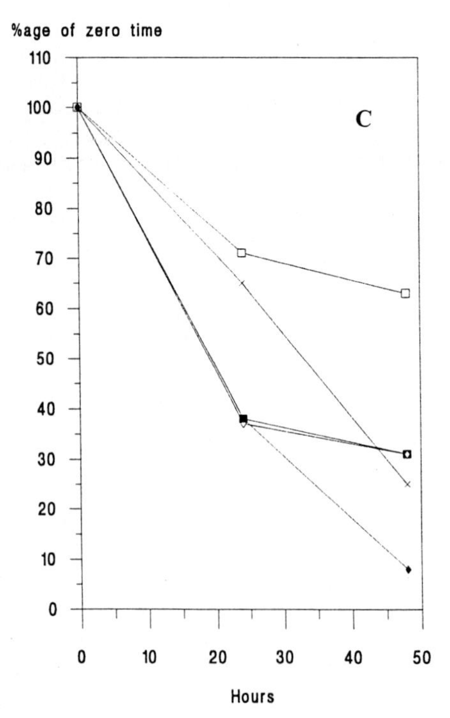
%age of zero time
C
Hours

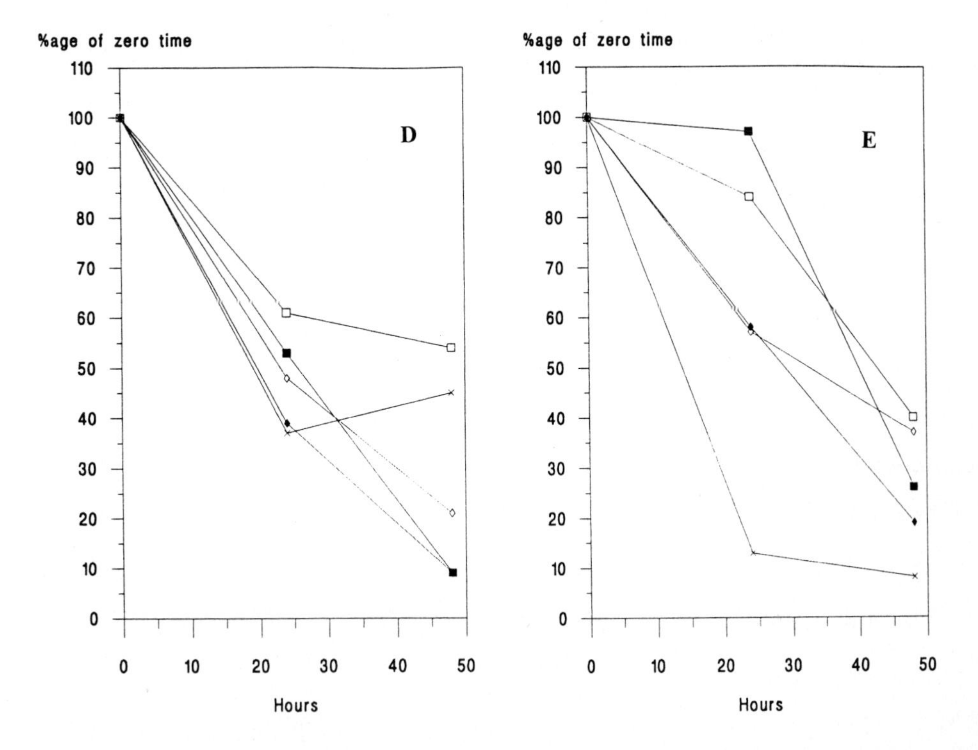

fig.8. The effect of dexamethasone (10-9M) (■) or hydrocortisone (10-8M) (♦), (10-7M) (◇) or (10-6M) (□) on the metabolism of androst-4-ene-3,17-dione in hepatocytes isolated from an adult male rat. A- 7α-hydroxylase; B- 6β-hydroxylase; C- 16α-hydroxylase; D- 17-hydroxysteroid dehydrogenase; E- 5α-reductase.

In summary, additions of insulin and/or glucocorticoids do not seem to assist in the general maintenance of xenobiotic metabolism in cultured rat hepatocytes but may be useful in maintained specific enzyme activities for specific purposes.

Use of Isolated Hepatocytes to Assess Hepatotoxicity

Hepatocytes were cultured originally in Ham's F-10 but have since been moved to William's E on the basis of results received from the assessment of media suitability. The anticancer drugs are added directly to the medium and left for various lengths of time. Viability of the cells is assessed by measurement of lactate dehydrogenase (LDH) in the culture medium, indicating damaged cells.

Owing to confidentiality agreements concerning the drugs, I am not able to discuss the results at length but the hepatotoxicity data obtained from these cells matched very closely that described from in vivo experiments in concentration and comparative terms.

The simple,reproducible and relatively inexpensive hepatocyte culture system could, therefore, mimic the whole animal tests in predicting hepatotoxicity of this novel group of anti-cancer drugs.

Conclusions

The need for standardisation of the preparation and culturing of hepatocytes is evident and a start has been made in this direction under the auspices of the Commission of the European Communities (CEC) through directorate-General XI and the European Center for the Validation of Alternative Methods (ECVAM). Full collaborative projects and an extensive validation of these methods will be necessary, however, before they can be deemed suitable for general acceptance.

Acknowledgements

Part of the work described in this paper has been funded by the Commission of the European Communities DG-XI (contract nos.B91/B4-3080/015951 and B92/B4-3063/11/13950) and by the Animal Procedures Committee of the Home Office.

REFERENCES

Guillouzo A., Morel F.,Fardel O., and Meunier B. (1993) Use of human hepatocyte cultures for drug metabolism studies.Toxicology 82:209-219.

Gustafsson J-A, Stenberg A (1974). Irreversible androgenic programming at birth of microsomal and soluble rat liver enzymes active on androst-4-ene-3,17-dione and 5α-androstane-3α,17β-diol.J.Biol.Chem 249:711-718.

Paine A (1990).The maintenance of cytochrome P$%) in rat hepatocyte cultures: Some applications of liver cell cultures to the study of drug metabolism, toxicity and the induction of the P450 system.Chem-biol Interact. 74:1-31.

Rogiers V.,Vercruysse A.(1993) Rat hepatocyte cultures and co-cultures in biotransformation studies of xenobiotics. Toxicology 82:193-208.

Seglen P.O. (1976) Preparation of rat liver cells.Methods Cell Biol. 13:29-83.

Skett P. (1994) Problems in using isolated and cultured hepatocytes for xenobiotic metabolism-based toxicity testing - solutions? Toxicology In Vitro (in press).

ISOLATION AND GROWTH OF HEPATOCYTES AND BILIARY EPITHELIAL CELLS FROM NORMAL AND DISEASED HUMAN LIVERS

Alastair J. Strain, Lorraine Wallace, Ruth Joplin, James Neuberger and Deirdre Kelly

Liver Units, Queen Elizabeth Hospital and Children's Hospital and School of Biochemistry, University of Birmingham, Edgbaston, Birmingham, UK

INTRODUCTION

The regulation of mammalian liver regeneration is complex. From both experimental animal studies and the use of isolated hepatocytes, numerous hormones, nutrients and growth factors have been implicated (Bucher and Strain, 1992). In keeping with all mammals, the human liver regenerates following resection or injury. More recently with the advent of liver transplantation, other situations where human liver growth occurs have also become pertinent. For example, organs from younger donors transplanted into larger, older recipients enlarge to the correct organ size for body weight (Van Thiel et al., 1987) and segmental grafts also regenerate with growth and development in pediatric recipients (Emond et al., 1989). The factors which control these responses in man are not well understood.

Understanding the regulation of human liver growth is important in several clinical contexts. The ability to enhance the capacity or the rate of the liver to regenerate would be of considerable value particularly in the treatment

NATO ASI Series, Vol. H 88
Liver Carcinogenesis
Edited by G. G. Skouteris

of acute liver disease. This would offer the possibility of an alternative to transplantation for example in patients with fulminant hepatic failure, for whom spontaneous recovery can occur if sufficient viable tissue to maintain liver function and to enable regeneration and repair of the organ remains. Thus, investigations into the control of human liver cell growth is an essential component in the consideration of future therapeutic approaches to liver diseases.

In recent years, with the development of major liver transplant programmes, the availability of normal and diseased human liver has greatly improved. As a consequence, human hepatocytes are now used in a wide variety of studies to improve our understanding of hepatic growth, physiology and pathology (Strain et al., 1991; Ismail et al., 1991; Moshage et al., 1992; Blanc et al., 1992). Sinusoidal cells from human liver have also been studied *in vitro* (Casini et al., 1993; Brouwer et al., 1988; Friedman et al., 1992) although to a lesser extent due to the greater technical difficulty in isolating purified preparations and the smaller number of cells obtained. Interest in the biology and pathology of the biliary epithelium has also received greater attention. Several established methods are now available for the isolation of rat biliary epithelial cells (BEC) from normal and cholestatic tissue (Sirica et al., 1985; Kumar and Jordan, 1986; Ishii et al., 1989). Using normal or diseased human liver tissue, intra-hepatic adult human BEC (IHBEC) can also be isolated with high purity (Joplin et al., 1989; Demetris et al., 1988). In this study we summarise the methodology available for the isolation of human hepatocytes and biliary epithelial cells and their subsequent use in growth and functional studies.

MATERIALS AND METHODS

Tissue

Normal liver was obtained from the pediatric transplant programme (excess tissue from reduced-graft segmental pediatric transplants). Organs had been harvested from brain-dead cadavers, perfused with University of Wisconsin (UW) fluid (Belzer and Southard, 1988) and stored at 4°C for 6-24h. Following surgical dissection, the surplus tissue was transferred to the laboratory in sterile UW maintained at 4°C. For diseased tissue, hepatectomies removed at the time of orthotopic liver transplantation were placed on ice immediately upon removal.

Hepatocyte Isolation

Hepatocyte were isolated from normal human liver segments (150-250 g) by enzymatic digestion as previously described (Strain et al., 1991; Ismail et al., 1991). From diseased hepatectomies, 150-200 g wedges were cut from the right lobe, ensuring the capsule surrounding this segment remained undamaged. Segments were perfused (3-4 cannulae) with Ca^{+2}/Mg^{+2}-free PBS or HBSS (+10 mM HEPES) for 15 min (60 ml/min) to warm the tissue at 37°C. This was followed by 500 ml buffer containing 0.5 mM EGTA, and finally 500 ml EGTA-free HBSS. Perfusion with enzyme solution (0.05% collagenase, 0.025% hyaluronidase, 0.05% dispase, 0.005% DNAase in HBSS + $CaCl_2$) was continued until the liver was judged to be sufficiently softened. The tissue was then minced coarsely in Dulbecco's modified Eagles medium (DMEM) containing 10% FCS, filtered sequentially through 250 μm and 60μm nylon mesh and pelleted by centrifugation at 50 g for 10 min. Cells were then washed 3x in medium containing FCS. Viability and total yield were determined by trypan blue dye exclusion and haemocytometer counting.

Hepatocyte Culture

Hepatocytes were plated at 3x 10^5 cells/ml into 35 mm tissue culture dishes (2ml) coated with rat-tail tendon collagen (McGowan et al., 1981; Strain et al., 1982). Following attachment for 2 h, cells were washed 2x PBS and were then refed with serum-free modified William's E medium (Strain et al., 1991; Ismail et al., 1991). Cells were maintained in monolayer culture for up to 7 days with 24h medium changes. DNA synthesis was determined following ^{3}H-thymidine incorporation directly in cell extracts or by autoradiography as previously described (McGowan et al., 1981; Strain et al., 1982). Thymidine incorporation was reduced by >90% in the presence of hydroxyurea indicating replicative DNA synthesis. Each experiment was carried out with a minimum of three separate cell isolates from different donors or end-stage livers. For autoradiography, a minimum of 500 cells were counted in 3 randomly chosen fields.

Biliary Epithelial Cell Isolation

IHBEC were isolated from 30-50 g segments of liver as previously described (Joplin et al., 1989, 1990). Briefly 10-30 g tissue was minced finely in 0.05% collagenase and was incubated at 37°C for 30 min. After filtration through 60μm mesh, suspensions were layered on cushions of 1.04 and 1.09 g/ml iso-osmotic Percoll and were spun at 800 g for 30 min. The layer equilibrating at the interface was removed and subjected to final immuno-magnetic purification using and antibody (HEA125) which recognises a plasma membrane glycoprotein expressed exclusively in adult human liver by biliary epithelial cells (Momberg et al., 1987). Isolates were extensively washed prior to plating in tissue culture plates and flasks.

Biliary Epithelial Cell Culture

Cultures were monitored rigorously by phase-contrast microscopy. At appropriate intervals, culture wells were washed in PBS (x2) and were fixed in ice-cold ethanol for 10 minutes. These cultures were then characterised using a variety of markers for the various liver cell types summarised as follows: Biliary Epithelial Cells HEA125 (Momberg et al., 1987), Cytokeratin 19 (Van Eyken et al., 1987), Gamma glutamyl Transpeptidase (Sirica et al., 1969); Kupffer cells Endogenous peroxidase (Smedsrod et al., 1985), EBM-11 (Kelly et al., 1988); Hepatocytes Asialoglycoprotein Receptor (Mizuno et al., 1984); Endothelial Cells Factor VIII related antigen (Scoazec et al., 1991); Fat-Storing Cells Desmin (Win et al., 1993). Immunopositivity was visualised by immuno-alkaline phosphatase (Dako) fast red (Sigma) using primary antigen and EB11 (all obtained from Dako Ltd), HEA125 (Bradsure Biologicals) and asialoglycoprotein receptor (Gift from Dr. I. McFarlane, King's College Medical School). Gamma-glutamyl transpeptidase was visualised by histochemistry (Rutenberg et al., 1969). For autoradiographic determination of the growth of cells, explant cultures were incubated with ^{3}H-thymidine for 4h, fixed and processed routinely using Ilford K5 photographic emulsion.

RESULTS

Hepatocytes and biliary epithelial cells were successfully isolated from normal human liver and from the livers of patients with end-stage Primary Biliary Cirrhosis (PBC) and primary Sclerosing Cholangitis (PSC).

TABLE 1
Viability and Yield of Hepatocytes Isolated from Normal and Diseased Human Liver

Group	Age	Sex	Tissue (g)	Viability (%)	Yield ($x10^{-6}$/g)
Normal	51	F	180	82	6.7
Normal	37	F	150	80	6.9
PBC	41	F	200	82	4.0
PBC	40	F	185	70	6.6
PSC	66	F	100	78	6.0
PSC	41	M	170	88	7.9

Despite the fact that livers were from patients with end-stage disease and macroscopically the degree of cholestasis and nodularity were at times

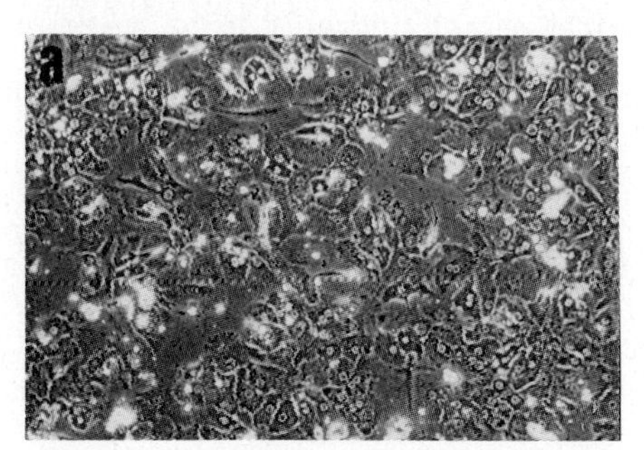

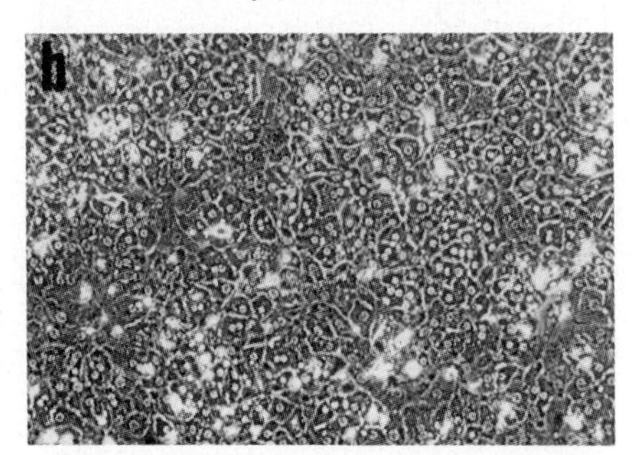

Figure 1. Phase-contrast photomicrographs of adult human hepatocyte monolayers after culture for 72h in the absence (A) or presence (B) of HGF.

extensive, the tissue digested with remarkable ease. For hepatocyte preparations, there was some variation in the time required for enzyme treatments (10-20 min) with diseased tissue. The yield and viability of hepatocytes from the two diseased groups compared very favourably with normals (Table 1) which were within the same range as previously found (Strain et al., 1991; Ismail et al., 1991). Representative hepatocyte cultures are shown in Fig. 1.

The yield of BEC from normal tissue was approximately 10^5/g tissue and from diseased tissue was often slightly higher. The morphology of both cell types from normal, PBC and PSC livers was livers was similar (not show). In early work, BEC were plated onto growth-arrested fibroblast feeder layers (Fig.2a) since they rapidly deteriorated when plated on tissue culture plastic (Fig.2b). However, in the presence of Hepatocyte Growth Factor (HGF) BEC can be cultured without feeder-layers and can be maintained for several weeks and passages (Fig. 2c and d).

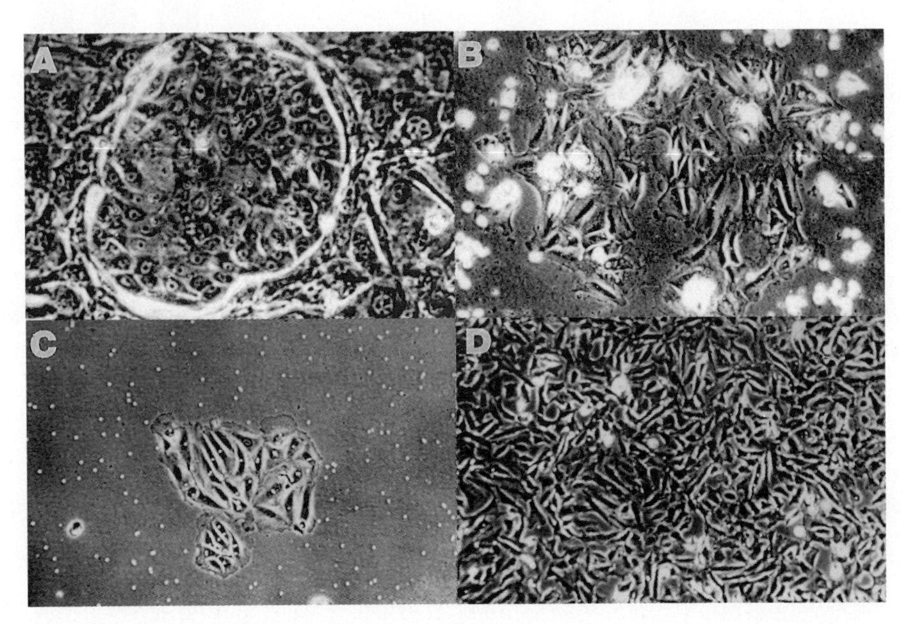

Figure 2. Phase-contrast photomicrographs of isolated human biliary epithelial cells. Cells cultured on fibroblast feeder layers (A) or on plastic (B) after 7 days without HGF. With HGF cells proliferate on tissue culture plastic from small colonies (C) to confluent monolayers (D).

We next determined the ability of cells to respond to growth factors and hormones.

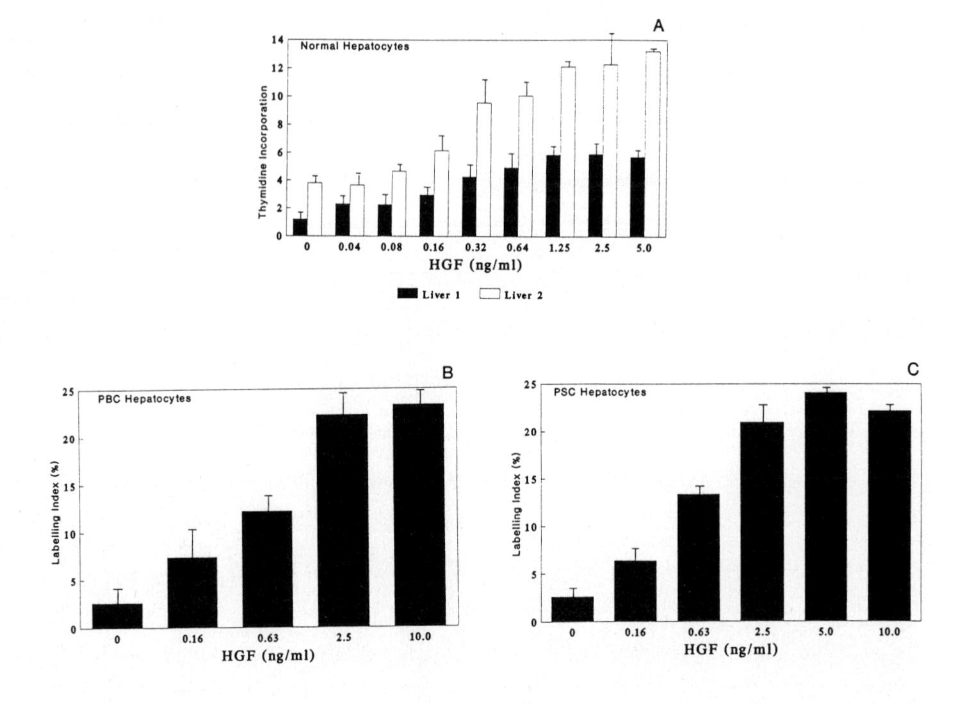

Figure 3. Stimulation of DNA synthesis by HGF in human hepatocytes isolated from normal (A), Primary Biliary Cirrhosis (B) and Primary Sclerosing Cholangitis (C) livers.

HGF has been shown to stimulate DNA synthesis in primary monolayers of adult hepatocytes with half-maximal activity of approximately 5 pM (Starin et al., 1991). Figure 3 indicates that sensitivity of cell preparations from individual normal donors to HGF does vary, since activity was evident with as little as 0.08 ng/ml (Fig. 3A). The half-maximal activity of the two cell preparations illustrated was approximately 2 and 4 pM respectively. Thymidine incorporation was also stimulated by HGF in hepatocyte cultures prepared from both PSC and PBC livers (Fig 3B and C). The dose responses, as determined both by autoradiography and scintillation counting (not shown), were in the same as found with hepatocytes from normal livers with half-maximal activity of 5 pM, TGF-β, which is known to antagonise the growth stimulatory effects of growth factors such as EGF and HGF, was found to inhibit the response of PBC and PSC derived hepatocytes (Fig. 4A and B).

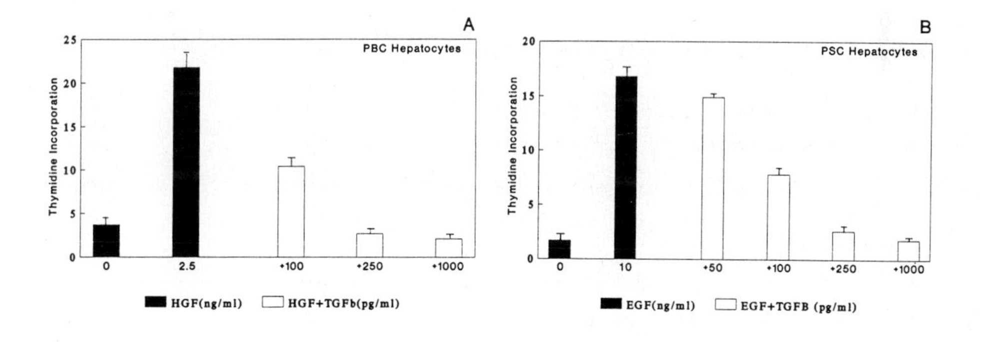

Figure 4. Inhibition of DNA synthesis in human hepatocytes from PBC (A) and PSC (B) livers by TGFβ.

We have previously shown that IHBEC proliferate in response to HGF in the presence of EGF and serum, with maintenance of markers of differentiated biliary phenotype (Joplin et al., 1992a). Using these conditions, cells from adult tissue can now be maintained in culture for up to 12-14 weeks and 7 passages. Throughout this time, they maintain expression of CK-19 and GGT (Table 2) markers of adult differentiated biliary phenotype. By contrast, they loose expression of the surface determinant HEA125 within 3-4 weeks which appears to be accompanied by a parallel expression of the intermediate filament vimentin Table 2).

TABLE 2

Expression of phenotypic markers by isolated human biliary epithelial cells

Age (wk)	Passage No. (approx)	CK19	GGT	HEA125	Vimentin
3	1	+	+	+	-
4	2	+	+	+/-	-/+
6	3	+	+	-	-/+
8	4	+	+	-	+
10	5	+	+	-	+
12	6	+	+	-	+
14	7	+	+	-	+

Several recent observations have been made regarding the biology of HGF. Firstly, it is now apparent that cells from younger donor livers failure to grow satisfactorily in the presence of HGF (Table 3). Current efforts suggest that an additional requirement for Keratinocyte Growth Factor (KGF) may be necessary. Additionally although gall bladder epithelial cells can be readily isolated using the same immuno-isolation technique, they also grow relatively poorly in HGF containing medium. Finally, we have successfully isolated IHBEC from small tissue fragments obtained from needle biopsies.

TABLE 3

Growth of BEC from range of human liver donors

AGE RANGE	0-9y (7)	10-18y (10)	19-65y (14)
Poor	71%	30%	7%
Fair	14%	20%	29%
Good	14%	50%	64%

In the presence of HGF cells slowly migrate from the explanted tissue and proliferate (Fig.5). The cell colonies which are derived in this way can be sub-cultured and expanded yielding significant cell numbers. These cells demonstrate expression of CK19 and GGT. Without HGF BEC failed to migrate and proliferate from the biopsies.

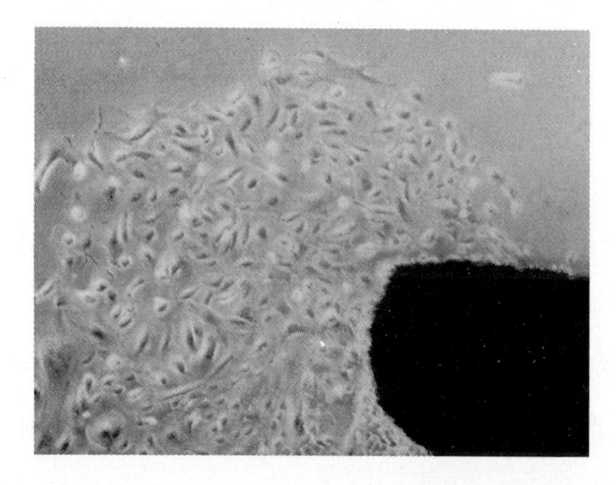

Figure 5. Phase contrast photomicrograph of human biliary epithelial cells migrating and proliferating from an explant of normal adult human liver.

DISCUSSION

The data outlined in the present study form part of an ongoing series of investigations designed to help elucidate the mechanisms which control the regeneration and repair of human liver. Largely as a result of improved access to human liver tissue through transplantation programmes, normal human liver tissue through transplantation programmes, normal human liver cell preparations of all major cell types are now routinely prepared in many laboratories and are used for a wide variety of growth and functional studies (Moshage and Yap, 1992; Strain et al., 1991; Ismail et al., 1991; Friedman et al., 1992; Sirica et al., 1985; Joplin et al., 1989).

The ability to isolate viable and functional hepatocytes from PBC and PSC end-stage livers with relative ease is perhaps surprising. Often the livers removed at time of orthotopic transplantation are highly cholestatic and fibrotic. However in PBC and PSC liver where the primary damage is targeted to the biliary epithelium, cirrhosis is often less extensive, thus enabling perfusion of the parenchymal tissue with buffers and proteolytic enzymes. These cells also demonstrated growth factor responses similar to those of hepatocytes from normal liver (Strain et al., 1991; Ismail et al., 1991; Blanc et al., 1992). This clearly indicates that these hepatocytes, under the conditions used in the present study, express functional plasma membrane receptors for HGF and TGFβ.

One implication of these studies is that the factors which have been shown to demonstrate *in vitro* bio-activity may be of relevance to human liver growth control *in vivo* both in normal and diseased conditions. Extrapolation of observations made in cell culture to the *in vivo* situation has been made in rat studies where factors such as HGF and TGFβ which regulate DNA synthesis in isolated hepatocytes (Strain et al., 1991; Zarnegar and

Michalopoulos, 1989; Strain et al., 1987; Braun et al., 1988; Francavilla et al., 1991; Ishiki et al., 1992). However, similar conclusions regarding human liver growth control can only be made with caution, since information regarding the *in vivo* biological activity of growth factors in man cannot be obtained. Another caveat is that when hepatocytes are isolated and removed from the influences imposed upon them within the diseased tissue, it is possible that they revert to a normal phenotype. Studies of cell isolates from diseased livers into other functional parameters such as phase I and phase II drug metabolising enzyme activities and plasma protein synthesis and further comparisons with hepatocytes from normal human liver would be required to resolve this question. Despite these questions, access to hepatocytes from livers of patients with diseases such as PBC and PSC may give useful insights into understanding disease processes.

Access to high quality primary BEC from human liver has opened the way for investigations of biliary pathophysiology. In our laboratory these have included studies of autoimmune antigen expression (Joplin et al., 1992b), cytotoxicity and HLA expression. The novel methodology to obtain preparations of biliary epithelial cells from human liver biopsies is particularly valuable. The method requires only very small fragments of liver and is therefore not restricted to material from transplant hepatectomies or large wedge biopsies. This development offers several advantages over the other established techniques used to isolate intra-hepatic human BEC which require >10 g of tissue (Joplin et al., 1989,1990) or perfusion of the intact biliary tree (Demetris et al., 1988). Precutaneous needle biopsies taken pre- or post-transplantation for diagnostic purposes can be utilised since only a minor fragment of the core obtained is necessary.

Thus this technique introduces the exciting possibility that using in vitro approaches, investigations of the pathophysiology of the biliary epithelium during progressive chronic liver disease can be now undertaken. Additionally, it will allow centres where there is no liver transplantation programme (and therefore where tissue availability is much more limited) to initiate studies of this nature.

The observation that BEC migrate and proliferate from adherent tissue explants was of interest since it was clearly HGF dependent. In addition to its potent mitogenicity for epithelial cells HGF, also known as "Scatter Factor" (Stoker et al., 1987; Strain 1993), induces cell motility and/or three-dimensional morphogenetic responses in a number of different cells types (Montesano et al., 1991; Rosen et al., 1991). Therefore it seems possible that the ability to generate BEC cultures from explants as described here is at least partly attributable to this property of HGF. Indeed, analogous to a more fibroblast-like morphology seen when MDCK cells "scatter" in response to HGF (Weidner et al., 1990), BEC also appear more fibroblast-like and form less compact colonies with a much less well defined border than in previous cultures (Joplin et al., 1992a), suggesting that human BEC may respond to HGF in more than one way, namely mitogenic and motogenic.

BEC from biopsies demonstrated expression of GGT and CK-19 throughout, indicating stability of these markers of adult biliary epithelial phenotype. However, there is a gradual increase in expression of the intermediated filament vimentin. We have also previously noted that in BEC cultures from collagenase tissue digests, expression of vimentin the intermediate filament originally regarded as a mesenchymal cell marker in vivo, increases with time in culture (Joplin et al.,1993) and form around 4-5 weeks (or 2-3 passages) all CK-19 and GGT positive cells co-express vimentin (manuscript in preparation). The pattern of vimentin expression by BECs in the present study is consistent with these previous observations. Co-expression of vimentin and cytokeratins by cells has been widely documented (Franke et al.,1979; Lazarides , 1982) and reflects a general response of many epithelial cells when isolated and maintained in culture. Co-expression of vimentin/cytokeratin intermediate filament also occurs in epithelia *in vivo*. For example in cells of newly formed bile ducts following bile duct ligation in the rat (Milani et al., 1989) and in damaged and regenerating renal tubular epithelium (Grone et al., 1987). Although the significance of this co-expression is not known, it has been suggested to correlate with the proliferation/differentiation status of cells. What is clear however, is that vimentin is no longer regarded as a definitive marker of mesenchymal cells, at least *in vitro*.

The growth and differentiation characteristics of human biliary epithelial cells have not been extensively characterised. Our observations suggest that during development, there is a phenotypic switch in growth factor requirements of BEC and that cells from young donors *in vitro* require additional factors. We believe that Keratinocyte Growth Factor (KGF) may be one such agent. KGF is a member of the Fibroblast Growth Factor super-family (now known to comprise 9 members). Unlike other members of the family, however, KGF is though to target its action to epithelial cells while synthesis and release appears restricted to mesenchymal cells (Rubin et al., 1989; Finch et al., 1989). This synthesis and targeting specificity is analogous to HGF. However, while HGF is mitogenic for both human hepatocytes and biliary epithelial cells (Strain et al., 1991; Joplin et al., 1992), KGF does not appear to stimulate growth of primary human hepatocytes (Strain et al.,1994). This is an important distinction and may indicate a means of directing cellular compartments proliferate with differing kinetics during tissue growth and repair. Similarly, although it is not yet clear which additional factors are required for GBEC cells to proliferate readily, the data reported here indicate further phenotypic differences in cells from different sites on the biliary tree.

In conclusion, these studies indicate that not only is it possible to isolate in relatively high yield, viability and purity cells from severely compromised livers, but that these cells continue to express functional growth factor receptors. These findings therefore help shed light on the factors which may be relevant to human liver growth in vivo. Thus, the concept of improving liver regenerative capacity through therapeutic intervention using recombinant growth factors, which are now widely available, becomes a step closer to reality. The use of cells from such tissue may also serve as a useful model for further investigations of liver pathophysiology.

Acknowledgements

We wish to thank Mr P McMaster, J Buckels and D Mayer for help in collecting biopsy samples and to Dr S Hubscher for assessment of histology.

This work was supported by grants from the Wellcome Trust and The Children's Liver Disease Foundation.

REFERENCES

Belzer FO., Southard H. (1988) Principles of solid organ preservation by cold storage. Transplantation 45:673-676.

Blanc P., Etienne H.,Daujat M., Fabre I., Zindy F., Domergue J., Astre C., Saint A., Michel H., Maurel P. (1992) Mitotic responsiveness of cultured adult human hepatocytes to epidermal growth factor, transforming growth factor a and human serum. Gastroenterol 102:1340-1350.

Braun L., Mead J., Panzica M., Mikumo R., Bell GI., Fausto N. (1988) Transforming growth factor beta mRNA increases during liver regeneration. Proc.Natl.Acad.Sci.U.S.A. 85:1539-1543.

Brouwer A., Barelds RJ., De Leeuw AM., Blauw E., Plas A., Yap SH., van den Broek AMWC (1988) Isolation and culture of Kuppfer cells from human liver: ultrastructure, enducytosis and prostaglandin synthesis. J. Hepatol. 6:36-49.

Bucher NLR, Strain AJ. (1992) Regulatory mechanisms in hepatic regenerations. In: Wright's Liver and biliary disease Millward-Sadler GH, Wright R, Arthur MJP (eds) London WB Saunders, pp 258-274.

Casini A., Pinzani K., Milani S., Grappone C., Galli C., Jezequel AM., Schuppan D., Roteaal CM., Surrenti C. (1993) Expression of extra-cellular matrix synthesis by transforming growth factor β in human fat-storing cells. Gastroenterol 105:245-253.

Demetris AJ., Markus BH., Saidman S., Fung JJ., Makowka L., Graner S., Duquesnoy R., Starzl TE. (1988) Isolation and primary cultures of human intra-hepatic biliary epithelial ductular epithelium. In Vitro Cell Dev Biol 24:464-470.

Emond JC., Whitington PF., Thistlethwaite JR., Alonso EM., Broelsch C. (1989) Reduced-size orthotopic liver transplantation: use in the management of children with chronic liver disease. Hepatol. 10:867-872.

Finch PW., Rubin JS., Miki T., Ron D., Aaronson SA. (1989) Human KGF is FGF-related with properties of a paracrine effector of epithelial cell growth. Science 245:752-755.

Francavilla A., Starzl TE., Porter K., Foglieni CS., Michalopoulos GK., Carrieri

G., Trejo J., Azzarone A., Barone M., Zeng QH. (1991) Screening for candidate hepatic growth factors by selective portal infusion after canine Eckl's fistula. Hepatol. 14:665-670.

Franke W., Schmidt E., Winter S., Osborn M., Weber K. (1979) Widespread occurrence cells from diverse vertebrates. Exp.Cell Res. 123:25-46.

Friedman SL., Rockey DC., McGuire RF., Maher JJ., Boyles JK., Yamasuki G. (1992) Isolated human lipocytes and Kupffer cells from normal human liver: morphological and functional characteristics in primary culture. Hepatol. 15:234-243.

Grone HJ., Weber E., Grone U., Osborn M. (1987) Co-expression of keratin and vimentin in damaged and regenerating tubular epithelium of the kidney. Am.J. Pathol. 129:1-8.

Ishiki Y., Ohnishi H., Muto Y., Matsumoto K., Nakamura T. (1992) Direct evidence that hepatocyte growth factor is a hepatotrophic factor for liver regeneration and has a potent anti-hepatitis affect *in vivo*. Hepatol. 16:1227-1235.

Ishii M., Vroman B., La Russo NF. (1989) Isolation and morphologic characterization of bile duct epithelial cells from normal rat liver. Gastroenterol. 97:1236-1247.

Ismail T., Howl J., Wheatley M., McMaster P., Neuberger JM., Strain AJ. (1991) Growth of normal human hepatocytes in primary culture: Effect of hormones and growth factors on DNA synthesis. Hepatol. 14:1076-1082.

Joplin R., Hart LL.,Neuberger JM. and Strain AJ. (1993) Characterization of long term biliary epithelial cell (BEC) cultures. FASEB J. 7:A496.

Joplin R., Hishida T., Tsubouchi H., Daikuhara Y., Ayeres R., Neuberger JM., Strain AJ. (1992a) Human intra-hepatic biliary cells proliferate in response to human hepatocyte growth factor. J.Clin. Invest. 90:1284-1289.

Joplin RJ., Lindsay JG., Johnson GD., Strain AJ., Neuberger J. (1992b) Membrane dihydrolipoamide acetyltransferase (E2) on human biliary epithelial cells in primary biliary cirrhosis. Lancet 339:93-94.

Joplin R., Strain AJ., Neuberger JM. (1990) Biliary epithelial cells from the livers of patients with primary biliary cirrhosis: isolation, characterisation and short-term culture. J. Pathol. 16:255-260.

Kelly PMA., Bliss E., Morton JA., Burns, McGee J. (1988) Monoclonal antibody EBM/11 : high cellular specificity for human macrophages. J.Clin Pathol. 41:510-515.

Kumar U., Jordan TW. (1986) Isolation and culture of biliary epithelial cells

from the biliary tract of normal rats. Liver 6:369-378.
Lazarides E. (1982) Intermediate filaments - a chemically heterogeneous, developmentally regulated class of proteins. Ann. Rev.Biochem. 51:219-250.
McGowan JA., Strain AJ., Bucher NLR. (1981) DNA synthesis in primary cultures of adult rat hepatocytes in a defined medium: effects of epidermal growth factor, insulin, glucagon and cyclic AMP. J.Cell Physiol. 108:353-363.
Milani S., Herbst H., Schuppan G., Niedobitek G., Kim KY., Stein H. (1989) Vimentin expression of newly formed rat bile duct epithelial cells in secondary biliary cirrhosis. Virchows Archiv A Pathol Anat 415:237-242.
Mizuno M., Kloppel TM., Nakane PK., Brown WR. (1984) Cellular distribution of the asialoglycoprotein receptor in rat liver. Gastroenterol. 86:142-149.
Momberg F., Moldenhauer G., Mammerling GH.(1987) Immunohistochemical study of a MW 34.000 human epithelium-specific glycoprotein in normal and malignant tissues. Cancer Res. 47:2883-2891.
Montesano R., Matsumoto K., Nakamura T., Orci L. (1991) Identification of a fibroblast-derived epithelial cell morphogen as hepatocyte growth factor. Cell 67:967-973.
Moshage H., Yap. SH. (1992) primary cultures of human hepatocytes : a unique system for studies in toxicology, virology, parasitology and liver pathophysiology in man. J. Hepatol. 15:404-413.
Rosen EM., Jaken S., Carley W., Luckett PM., Setter E., Bhargava M., Goldberg ID. (1991) Regulation of motility in bovine brain endothelial cells. J. Cell Physiol. 146:325-335.
Rubin JS., Osada H., Finch PW., Taylor WG., Rudikoff S., Aaronson SA. (1989) Purification and characterization of a newly identified growth factor specific for epithelial cells. Proc.Natl.Acad.Sci.U.S.A. 86:802-806.
Russell WE., Coffey RJ., Ouellette AJ., Moses HL. (1988) Type beta transforming growth factor reversibly inhibits the early proliferative response to growth factor reversibly inhibits the early proliferative response to partial hepatectomy in the rat. Proc. Natl.Acad. Sci. U.S.A. 85:5126-5130.
Rutenberg AM., Kim H., Fishbein JW., Hanker JS., Wasserkrug HL., Seligman AM. (1969) Histochemical and ultrastructural demonstration of gamma-glutamyl transpeptidase activity. J. Histochem. Cytochem. 17:517-522.
Scoazec J., Feldmann G. (1991) In situ immunophenotyping study of endothelial cells of the human sinusoid: results and functional implications. Hepatol. 14:789-797.

Sirica AE., Cihla HP. (1984) Isolation and partial characterization of oval and hyperplastic bile ductular cells enriched populations from the livers of carcinogen and non-carcinogen treated rats. Cancer Res. 44:3454-3466.

Sirica AE., Sattler CA., Cihla HP. (1985) Characterization of a primary bile ductular cell culture from the livers of rats during extra-hepatic cholestasis. Am.J. Pathol. 120:67-78.

Smedsrod B., Pertoft H., Eggertsen G., Sundstrom C. (1985) Functional and morphological characterization of cultures of Kupffer cells and liver endothelial cells prepared by means of density separation in Percoll and selective adherence. Cell Tiss Res. 241:639-649.

Stoker M., Gherardi E., Perryman M.,Gray J. (1987) Scatter factor is a fibroblast-derived modulator of epithelial cell mobility. Nature 327:239-242.

Strain AJ. (1993) Hepatocyte Growth Factor: another ubiquitous cytokine. J.Endocrinol. 137:1-5.

Strain AJ., Frazer A., Hill DJ., Milner RDG. (1987) Transforming growth factor beta inhibits DNA synthesis in hepatocytes isolated from normal and regenerating liver. Biochem. Biophys.Res.Commun. 145:436-442.

Strain AJ., Ismail T., Tsubouchi H., Arakaki N., Hishida T., Kitamura N., Daikuhara Y., McMaster P. (1991) native and recombinant human HGF are highly potent promoters of DNA synthesis in both human and rat hepatocytes. J.Clin.Invest. 87:1853-1857.

Strain AJ., McGowan JA., Bucher NLR. (1982) Stimulation of DNA synthesis in primary cultures of adult rat hepatocytes by rat platelet-associated substances. In Vitro 18:108-116.

Strain AJ., McGuiness G., Rubin JS., and Aaronson SA. (1994) Keratinocyte growth factor and fibroblast growth factor action on DNA synthesis in rat and human hepatocytes: Modulation by heparin . Exp.Cell Res. 210:253-259.

Weidner KM., Behrens J., Vanderkerkhove J., Birchmeier W. (1990) Scatter factor: molecular characteristics and effect on the invasiveness of epithelial cells. J. Cell Biol. 111:2097-2108.

Win KH., Charlotte F., Mallat A., Cherqui D., Martin N., Mavier P., Preaux A., Dhumaeux D., Rosenbaum J. (1993) Mitogenic effect of transforming growth factor beta1 on human Ito cells in culture: Evidence for the dediation by endogenous platelet-derived growth factor. Hepatol. 18:137-145.

Van Eyken PR., Sciot R., van Damme B., de Wolf-Peters VJ., Desmet V. (1987) Keratin immunohistochemistry in normal human liver. Cytokeratin pattern of hepatocytes, bile ducts and acinar gradient. Virchows Arch. 412:63-72.

Van Thiel DH., Gauder JS., Kam I. (1987) rapid growth of an intact liver transplanted into a donor larger than the recipient. Gastroenterol. 93:1414-1419.

Zarnegar R. and Michalopoulos GK. (1989) Purification and characterisation of hepatopoietin A, a polypeptide growth factor for hepatocytes. Cancer Res. 49:3314-3320.

A NOVEL STRATEGY FOR ISOLATING HEPR, A HUMAN SMALL INTESTINAL CYTIDINE DEAMINASE

Christos Hadjiagapiou, Federico Giannoni, Toru Funahashi and Nicholas Davidson.

Department of Medicine
University of Chicago
MC4076
5841 S.Maryland Ave.
Chicago, IL 60637
U.S.A.

INTRODUCTION

Apolipoprotein B (apo B)is a large hydrophobic protein which circulates in two distinct forms, each the product of a single apo B gene (Chan, 1992). The small intestine synthesizes and secretes a form of apo B referred to on a centile scale as apo B48, a protein of 264 KDa, which is associated with chylomicron remnants (Powell et al., 1987, Chen et al., 1987). Human liver produces a form of apo B referred to as apo B100. Apo B100, a 550 KDa protein, circulates in association with low density lipoproteins, the major transport vehicle for cholesterol ester in humans (Chan, 1992). This organ-specific production of apo B isophorms has as its basis a unique post-transcriptional modification to the apo B mRNA. Apo B transcription occurs in the liver and small intestine and yields a 24 Kb mRNA species which contains a CAA codon at position 2153, encoding glutamine in apo

NATO ASI Series, Vol. H 88
Liver Carcinogenesis
Edited by G. G. Skouteris

B 100 and which is specified in the genomic sequence. In the small intestine, this apo B mRNA undergoes a site-specific cytidine deamination which results in the production of a UAA codon and thereby specifies an in-frame translational stop (Powell et al.,1987, Chen et al., 1987). Work conducted over the last few years using isolated intestinal cell extracts and synthetic apo B RNA templates has enabled investigators to determine that apo B mRNA editing is a process mediated by protein(s) and with the characteristics of an enzymatic reaction (Teng and Davidson, 1992).

The recent demonstration that chicken intestinal apo B mRNA was not edited, led to the discovery that chicken enterocytes while unable to mediate apo B RNA editing alone, contain protein factor(s) which paradoxically enhance the ability of mammalian intestinal extracts to edit a synthetic apo B RNA template (Teng and Davidson, 1992). This demonstration led in turn to the recent cloning of REPR, a rat small intestinal cDNA which encodes a protein of 27 KDa and which, in association with chicken intestinal extracts, mediates *in vitro* apo B RNA editing (Teng et al.,1993). REPR contains the conserved histidine and paired cysteine residues described in other members of the cytidine deaminase gene family (Navaratnam et al., 1993a). The present report describes the use of a novel strategy to isolate the human homolog of REPR, referred to as HEPR, from adult small intestine.

MATERIALS AND METHODS

Normal adult human small intestine was obtained during the course of surgical resections at the University of Chicago Hospitals and was provided by Dr. D.Rukstalis, Department of Surgery. Total RNA was prepared using guanidinium thiocyanate purification as previously described (Teng et al., 1990) and used for reverse transcription-polymerase chain reaction (RT-PCR) amplification with two degenerate oligonucleotide primers D1 and D2 selected from a region spanning residues 86-174 of the rat intestinal apo B mRNA editing protein (Teng et al., 1993). 500 ng total RNA from normal human jejunum was annealed to 60 pmoles primer D1 and RT was carried out at 60°C for 15 minutes using rTth (Perkin-Elmer-Cetus). PCR amplification was conducted by adding chelating buffer and 60 pmoles primer D2, as described by Giannoni et al., (1994). 40 cycles of PCR were conducted with denaturation for 30 seconds at 95°C, annealing at 45C for 1 min and extension at 65°C for 1.5 min. A single PCR product of 267 bp was cloned directly into pCRII (InVitrogen, San Diego, CA) and sequenced on both strands. Using the authentic sequence, specific primers were then prepared for amplification of the 5' and 3' ends of the cDNA with overlap at a convenient restriction site (see below).

Polyadenylated human jejunal RNA was prepared and used as a source for cloning of the 5' end of the human cDNA. Reverse transcription was performed using 2 μg poly A+RNA with AMV reverse transcriptase (BRL, Gaithersburg, MD) and 10 μM gene-specific primer, GSP1. A modified 5' phosphorylated oligonucleotide anchor, as supplied by Clontech laboratories, was then ligated to the single stranded cDNA using 10 units T4 RNA ligase (Edwards et al., 1991) at 22°C for 20h as described by the supplier (Clontech). PCR amplification was conducted using an anchor primer AP, supplied by Clontech laboratories and a nested antisense (specific) primer. PCR was conducted using Hot Tub DNA polymerase (Amersham, Arlington

Heights, IL) with 35 cycles of 45 seconds denaturation at 94°C, 45 seconds annealing at 60°C and 2 minutes extension at 72°C with a final 7 minute extension after the last cycle. A single 374 bp PCR product was cloned into pCRII and subjected to DNA sequencing.

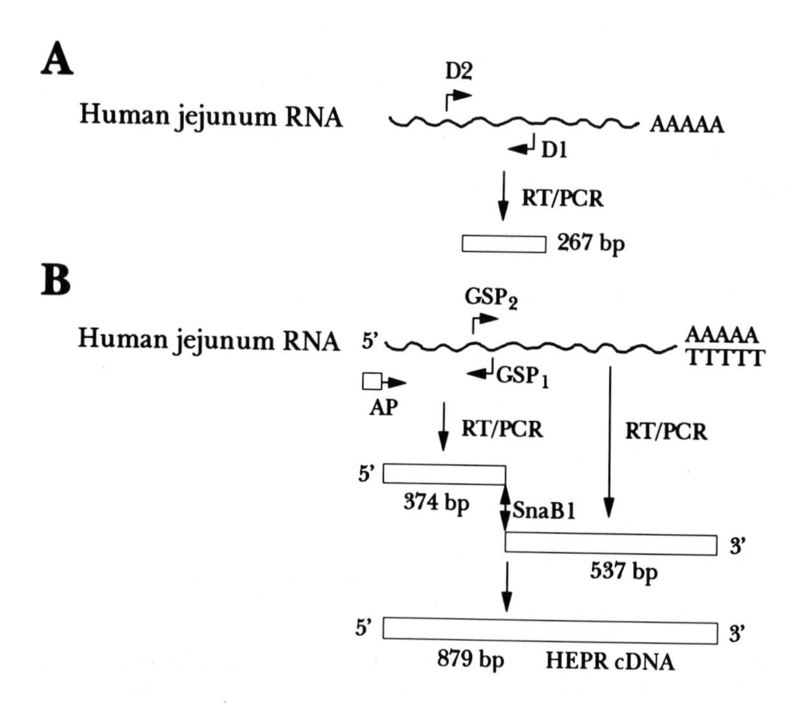

Figure 1. RT-PCR Strategy for amplification of HEPR from human small intestine. [A] Total RNA was used to amplify a conserved region of the rat apo B mRNA editing protein (REPR) using degenerate primers flanking the zinc-binding domain.[B] The 267bp amplicon was sequenced and used to design gene-specific primers GSP1 and GSP2 which were then used in separate 5 prime and 3 prime amplification reactions.These independent PCR amplification products were then ligated through a SnaB1 site present in both cDNA fragments.

The 3' end of the cDNA was independently obtained (Frohman et al.,1988) using 2 µg poly A+ RNA from human jejunum to generate first strand cDNA synthesis which was primed using an adaptor primer containing oligo dT (GIBCO-BRL, Gaithersburg, MD).

PCR amplification used an upstream gene-specific primer GSP2. PCR conditions (30 cycles) were: denaturation at 94°C for 30 seconds, annealing at 55°C for 1 min and extension at 72°C for 1.5 min with a final extension time of 15 min. A single 537 bp PCR product was cloned into pCRII and subjected to DNA sequencing. A convenient SnaB1 site present in both clones allowed the digestion and religation of a full length cDNA clone. The outline of the method is illustrated in Figure 1.

```
HEPR         60  NHVEVNFIkkftserdfhpsisctitwfls--WSPCWECSQAIREFLS  105

REPR         60  KHVEVNFIekftteryfcpntrcsitwfls--WSPCGECSRAITEFLS  105
ECCDD       101  VHAEQSAIshawlsgekalaaitvn------YTPCGHCRQFMNELNS  141
ECOCDDA     101  VHAECSAIshawlsgekalaaitvn------YTPCGHCRQFMNELNS  141
HUMDEOXDEA   83  CHAELNAImnknstdvkgcsmyva-------LFPCNECAKLIIQAGI  122
DCTD-BPT2   103  IHAELNAIlfaarngssiegatmyvt-----LSPCPDCAKAIAQSGI  143
DCTD-BPT4   103  IHAELNAIlfaarngssiegatmyvt-----LSPCPDCAKAIAQSGI  143
YSCDCD1     231  LHAEENALleagrdrvgqnatlycd------TCPCLTCSVKIVQTGI  272
BSCDDGENE    52  NCAERTALfkavsegdtefqmlavaadtpgq-VSPCGACRQVISELCT   98
S52873       63  ICAERTAIqkavsegykdfraiaiasdmqddfISPCGACRQVMREFGT  110
```

Figure 2. Alignment of HEPR with other cytidine deaminase. The deduced amino acid sequence of HEPR flanking the conserved zinc-binding region is shown in alignment with the corresponding region from REPR as recently described (Navaratnam et al., 1993a). Other members of the cytidine deaminase family are included in this alignment (see Navaratman et al., 1993a for identity of these gene products).

Results

Two degenerate primers, chosen based upon the rat apoB mRNA editing protein (REPR) cDNA sequence, were used to obtain a 267bp fragment from human jejunum by RT/PCR (see above, Figure 1). The fragment was found to have 75% nucleotide homology to REPR.

Subsequently, two overlapping specific primers GSP1 and GSP2 were designed from this fragment and used to clone the 3-prime and 5-prime ends of HEPR from human jejunum using reverse transcription and PCR. The two fragments were joined together at their unique SnaB1 site. The nucleotide sequence was 879bp encoding a 236 amino acid protein with a predicted molecular weight of 28206 daltons. The full length cDNA sequence showed 76% homology to REPR, whereas the amino acid homology was 69%. HEPR was found to have an N-glycosylation site (NTTN) at residue 57, phosphorylation sites for casein kinase 2 (STGD, SERD) at residues 8, 72, c-AMP-dependent kinase (KKFT), residue 68, and protein kinase C sites (SEK, TLR, SRK, SKG, SER, SRR) at residues 3,13,47,54,72, and 196 respectively. All residues are identified with the single letter code.

Sequence alignment revealed that HEPR contained the three previously identified conserved amino acid residues, His61,Cys93, and Cys96, shown to promote zinc binding to the active site of other known cytidine deaminases (Navaratnam et al., 1993a). This alignment is illustrated above in Figure 2.

A notable difference between the predicted amino acid sequence of HEPR and REPR is the absence of the leucine zipper from the carboxyl terminus of HEPR. Further analysis, using Northern blotting of various human tissues, showed that HEPR mRNA is expressed only in the adult small intestine with greater abundance in the jejunum (data not shown). Finally, immunocytochemical studies using a polyclonal antibody raised against a conserved region of REPR showed HEPR to be more abundant in the villi than in the crypts (data not shown).

DISCUSSION

Apo B mRNA editing is a critical factor determining the systemic catabolism of mammalian apo B (Chan, 1992). This is largely because apo B100 contains the recognition sequence for binding to the low density lipoprotein receptor, a motif which is present in the carboxyl terminus of apo B100 and is therefore missing from apo B48 (Chan, 1992). Recent evidence suggests that protein factors which have been proposed to bind to mammalian apo B RNA flanking the edited base (Harris et al., 1993, Navartnam et al., 1993b). Specifically, proteins in the range of 40-70 KDa have been reported to be associated with apo B RNA as evidenced by UV cross linking studies and these have been hypothesized to mediate binding of perhaps other factors involved with the deamination reaction (Lau et al., 1991, Harris et al., 1993, Navaratnam et al., 1993b). None of these studies, however has suggested the involvement of proteins in the size range reported for REPR or HEPR. Earlier studies by Driscoll and Casanova (1990) suggested that the native molecular size of the editing complex was approximately 125 KDa. Assuming that the REPR/HEPR cytidine deaminase is an essential component of this enzyme complex, the other components of the complex, i.e. the so-called complementation factor(s), should account for the remainder, suggesting that these additional factor(s) should have a molecular size of approximately 100 KDa. Whether this complementation factor(s) is a single or multiple gene product is currently unknown.

In conclusion, this study has demonstrated a novel RT-PCR based method to isolate a cDNA from human small intestine which is a member of the deaminase family of genes. This gene product functions in the context of the post-transcriptional editing of apo B mRNA and thereby plays an essential role in regulating the tissue-specific delivery of lipoproteins in humans.

Acknowledgements

This work was supported by National Institutes of Health grants HL-38180 and DK-42086 to NOD. The outstanding assistance of Susan Skarosi, Jennifer Ziouras and Trish Glascoff is gratefully acknowledged.

REFERENCES

Chan L. (1992) Apolipoprotein B, the major protein component of triglyceride-rich and low density lipoproteins. J.Biol.Chem. 267:25621-25624.

Chen SH., Habib G., Yang CY., Gu ZW., Lee BR., Weng S., Silberman SR., Cai SJ., Deslypere JP., Rosseneu M., Gotto AM., Li WH., and Chan L. (1987) Apolipoprotein B-48 is the product of a messenger RNA with an organ-specific in-frame stop codon. Science 238:363-366.

Driscoll DM. and Casanova E. (1990) Characterization of the apolipoprotein B mRNA editing activity in enterocyte extracts. J.Biol.Chem. 265:21401-21403.

Edwards JBDM.,Delort J. and Mallet J. (1991) Oligodeoxyribonucleotide ligation to single-stranded cDNAs: a new tool for cloning 5' ends of mRNAs and for constructing cDNA libraries by in vitro amplification

Nucleic Acids Res. 19:5227-5232.
Frohman MA.,Dush MK., and Martin GR. (1988) Rapid production of full-length cDNAs from rare transcripts: amplification using a single gene-specific oligonucleotide primer. Proc.Natl.Acad.Sci. U.S.A. 85:8998-9002.
Giannoni F.,Field FJ., and Davidson NO. (1994) An improved reverse-transcription polymerase chain reaction method to study apolipoprotein gene expression in Caco-2 cells. J. Lipid Res. 35:340-350.
Harris SG.,Sabio I.,Mayer E.,Steinberg MF., Backus JW., Sparks JD.,Sparks CE., and Smith HC. (1993) Extract-specific heterogeneity in high-order complexes containing apolipoprotein B mRNA editing activity and RNA-binding proteins. J.Biol.Chem. 268:7382-7392.
Lau PP.,Chen SH.,Wang JC., and Chan L. (1991) A 40 kilodalton rat liver nuclear protein binds specifically to apolipoprotein B mRNA around the RNA editing site. Nucleic Acids Res. 18:5817-5821.
Navaratnam N.,Morrison JR.,Bhattacharya S., Patel D.,Funahashi T.,Giannoni F.,Teng BB.,Davidson NO. and Scott J. (1993a) The p27 catalytic subunit of the apolipoprotein B mRNA editing enzyme is a cytidine deaminase. J.Biol.Chem. 268:20709-20712.
Powell LM.,Wallis SC.,Pease RJ.,Edwards YH.,Knott TJ. and Scott J. (1987) A novel form of tissue-specific RNA processing produces apolipoprotein B-48 in intestine.Cell 50:831-840.
Teng BB.,Burant CF. and Davidson NO.(1993) Molecular cloning of an apolipoprotein B mRNA editing protein. Science 260:1816-1819.
Teng BB. and Davidson NO. (1992) Evolution of intestinal apolipoprotein B mRNA editing.Chicken apolipoprotein B mRNA is not edited, but chicken enterocytes contain in vitro editing enhancement factor(s). J.Biol.Chem. 267:21265-21272.
Teng BB., Verp M.,Salomon J. and Davidson NO. (1990) Apolipoprotein B messenger RNA editing is developmentally regulated and widely expressed in human tissues. J.Biol.Chem. 265:20616-20620.

REGENERATING LIVER: ISOLATION OF UP- AND DOWN-REGULATED GENE PRODUCTS

Theodorus B.M.Hakvoort[1] and Wouter H. Lamers
Department of Anatomy and Embryology
University of Amsterdam
Meibergdreef 15
1105 AZ Amsterdam
The Netherlands

INTRODUCTION

Ageing is associated with a general loss of regenerative capacity in vertebrates. Among the few exceptions is the liver, which preserves a considerable regenerating ability. The regenerative process, i.e. the restoration of functional mass after e.g. liver injury or fulminant hepatitis, has drawn the attention of several well established laboratories in the last decade, but remains a poorly understood phenomenon, studies by Fausto et al. (1989), Michalopoulos (1990) and Mohn et al. (1991) and Haber et al. (1983). Among the identified factors are humeral factors, such as circulating hormones, growth factors and nervous input, see e.g Michalopoulos (1990) and Haber et al. (1993). However the list is still incomplete.

In order to obtain a better understanding of the reconstitution of functional liver mass during regeneration, attention is focused on the signals

[1] Part of this work was performed in the J.v.Gool Laboratory for Experimental Internal Medicine, Academic Medical Center, University of Amsterdam, Meibergdreef 9, 1105 AZ Amsterdam, The Netherlands.

NATO ASI Series, Vol. H 88
Liver Carcinogenesis
Edited by G. G. Skouteris

and mechanisms of cellular communications between the various liver cell types involved (mitogenic and morphogenic signals) and on the delicate balance of growth-stimulating and -inhibiting factors. Studies on the interactions between liver cell types appear only recently in the literature and factors that induce cell-cycling and growth, are being identified. Most of these studies have focused on the more abundant gene products. The procedures employed to isolate the regenerating gene products do not guarantee the isolation of the abundant gene products. Moreover, many of these methods require very extensive screening procedures, such as colony dot-blot hybridization, to identify the cDNAs involved in liver regeneration.

As part of our interest in hepatotrophic and hepatogenic factors we describe a method which can be used to enrich a cDNA population for up- and down-regulated gene products from a regenerating and a control rat liver. In this procedure the common liver gene products ("house-keeping" genes) have been eliminated by a subtractive hybridization method and the remaining gene products were amplified, facilitating the isolation of underrepresented gene products. The unique regenerating gene products can be isolated directly from the enriched regeneration cDNA population by means of a "display gel" and can be further characterized without having to use of a colony screening hybridization method.

MATERIALS AND METHODS

All procedures, except where indicated, were according to standard molecular biology protocols which can be found in various laboratory manuals.

RNA and cDNA from regenerating and sham-operated control rat livers. Liver tissues were obtained from 150-200 g male Wistar rats after a laparotomy or after a partial (2/3) hepatectomy according to Higgins and Anderson (1931). These preparations are denoted control and regenerating liver respectively. At 3,6,12,18,24 and 36 hours after operation the livers were taken from the animals and used to isolate mRNA. Equal portions of poy(A)+ mRNA from regenerating or control liver after 3,6 and 12 hours ("early" regenerating or control), in a separate experiment after 18,24, and 36 hours ("late" regenerating or control),were mixed to make double-stranded cDNA. First strand cDNA was obtained using oligo(dT) and the murine Superscript Reverse Transcriptase (GIBCO), second strand cDNA by RNase H (BRL) and DNA polymerase I (Boehringer). Part of the obtained cDNAs were used to construct a phage lambda cDNA library (Stratagene) for future cDNA isolation.

Enrichment of the up- and down-regulated cDNAs from a regenerating liver was performed according to the procedure described by Wang and Brown (1991) and modified in our laboratory by Hakvoort et al. (1994). Briefly, the "early" and "late" regeneration and control cDNAs population was restricted with *AluI* and the second part with *AluI* plus *RsaI*. Double-stranded, phosphorylated oligodeoxynucleotide linkers having a blunt end a 3' protruding end, were ligated onto the blunt-ended cDNA fragments. The linker used for the regeneration cDNA fragments contained the oligonucleotide sequence 5'-AATTCAGGCCCAGTCGGCCGG and the linker for the control cDNAs 5'-CTCTTGCTTGAATTCGGACTA. Using these specific sequences, the regenerating and control cDNA fragments were amplified employing Taq DNA polymerase, after removal of the excess linkers used in the ligation step with a Qiagen spin-20 column (Qiagen). The PCR amplification conditions were a one-minute denaturation step at 94°C, annealing of the primers at 58°C for 90 seconds followed by an elongation step at 72°C for two minutes. The whole procedure was repeated for 30 cycles and concluded by a 5 minutes final elongation at 72°C. The starting amplification mixture contained 200 μM dNTPs and 1.5 mM $MgCl_2$ and was kept under paraffin wax throughout the entire amplification procedure. The PCR products (referred to as tracer and driver) were purified using Qiagen spin-20 columns which were recycled by re-equilibration.

In order to achieve a more general enrichment of gene products from a regenerating liver it was decided to mix the "early" (3 through 12 hours) and the "late" (18 through 36 hours) cDNA fragments in equal ratios. The same was done for the control cDNAs. The subtractive-hybridization was started with 5μγ tracer cDNA and 100 μg photobiotinylated driver cDNA for the long hybridization step (20 hours). A 10-fold excess driver was used for the short hybridization step (20 hours). A 10-fold excess driver was used for the short hybridization step (4 hours), following the procedure exactly as described by Wang and Brown (1991). After each long and short hybridization step the biotinylated hybrids and excess driver were removed. The remaining enriched cDNA fragments, were amplified after both the long and the short hybridization steps were done, employing the above mentioned PCR conditions. Up-regulated gene products from the regenerating cDNAs (tracer) were obtained by using the control cDNA (driver) to subtract with. For the down-regulated gene products the entire procedure was reversed, that is, the control cDNA fragments were used as tracer and subtracted with biotinylated regeneration cDNA fragments (driver).

Polyacrylamide gel electrophoresis of the enriched cDNA fragments. The enriched cDNA fragments were amplified using Tap DNA polymerase in the presence of 0.5 μM [α-^{35}S]dATP (10 μCi, 1200 Ci/mmole, Amersham) and 20 μM non-radioactive dATP; all other dNTPs were kept at 200 μM in a 25μl reaction volume. The PCR product (5μl) was mixed with 3μl denaturing medium (95% formamide, 20 mM EDTA and dye solutions), heated for 5 minutes at 80°C, cooled on ice and loaded onto a standard polyacrylamide-urea DNA sequencing gel. After electrophoresis the gel was dried directly on a 3 MM Whatman paper and exposed to X-ray film with proper marking of the orientation.

Amplification and sequencing of cDNA fragments.
cDNA fragments of interest were excised from the dried DNA sequencing gel using a scalpel and guided by the developed film. After rehydration of the gel slice in distilled water (100 μl) for ten minutes, the DNA was diffused out of the gel by boiling the sample for 15 minutes. The cDNA was recovered from the extract by ethanol precipitation (70%) in the presence of 0.3 M sodium acetate, 2μg carrier tRNA, and dissolved in 10 μl water. The cDNA was reamplified using exactly the same original PCR conditions as described above (200 μM dNTPs and 4 μl of recovered cDNA per 50 μl PCR reaction). The amplified products were analyzed on an agarose gel (10μl) and directly cloned into the pCR™ cloning vector using the TA Cloning System (invitrogen). Ligation was exactly according the manufacturers instructions using one microliter from a 1:100 dilution of the non-purified PCR product.

Transformants were analyzed by partial sequence analysis using the chain-termination method with Sequenase (USB) on minipreparation plasmid DNA. Sequences were compared with other reported sequences in the GENbank data base.

RESULTS AND DISCUSSION

To isolate factors which are involved in rat liver regeneration, we prepared two cDNA populations: one from a control, sham-liver and one from a regenerating liver after partial hepatectomy. Both cDNA populations were representative for the gene products present over the first 36-hour time period after the regenerating stimulus (partial hepatectomy). mRNA was isolated from the liver samples taken from rats at regular time intervals, combined and used to make cDNA population and in addition the subtractive-hybridization scheme was simplified dramatically: instead of performing a cascade of subtractions on each of the cDNA populations isolated at given time point, the composite cDNA populations were used as starting material for the enrichment procedure. Aliquots of isolated poly(A+) mRNAs were stored at -80^{0}C in ethanol, which can be used to construct specific cDNA libraries.

Before a description is given on the isolation of up- and down-regulated gene products some more general aspects of the subtractive-hybridization procedure are given. To facilitate a more complete hybridization and amplification pattern of the double-strand cDNAs, the regeneration as well as the control cDNA fragments were each tagged with restriction enzymes. Regeneration and control cDNA fragments were each tagged with their own set of primers/adapters, thereby facilitating specific amplification of the DNA in large amounts. Mismatching of the primers, i.e. using the regenerating oligonucleotide primers together with the control cDNA fragments which contain the control primer sequence, did not result in product formation (data not shown). In order to further minimize potential artifacts by random-priming during thc apmlification cycles the annealing temperature was optimized i.e. identification of maximum annealing temperature. The average size of the (amplified) cDNA fragments was in the range of 100 to 700 nucleotides (not shown).

The main object of the subtractive-hybridization procedure is to enrich a representing cDNA population from a regenerating liver for the up- and down-regulated gene products. After three complete rounds of subtractive-hybridization a first impression on the effectiveness of the procedure was obtained by performing a screening on a Southern blot. In contrast to the display method, which will be shown later, such a hybridization screening can be laborious an often only yields a rather qualitative result. Figure 1 shows an example of such a Southern blot, using regeneration cDNA as a probe. Equal amounts of DNA (1 μg) were loaded onto an agarose gel, original "early" and "late" material and enriched material after 1,2 and 3 subtractive hybridization steps were hybridized with the original regeneration cDNA fragments and with these fragments after 3 subtractive-hybridization steps.

As expected, using the original regeneration cDNA material as a probe (panel A) an equal hybridization signal was found for the "early" and "late" regeneration and control cDNAs (lanes R-a,-b,C-c,-d). In the lanes C-1, C-2 and C-3, representing the control cDNA fragments after 1,2 and 3 rounds of

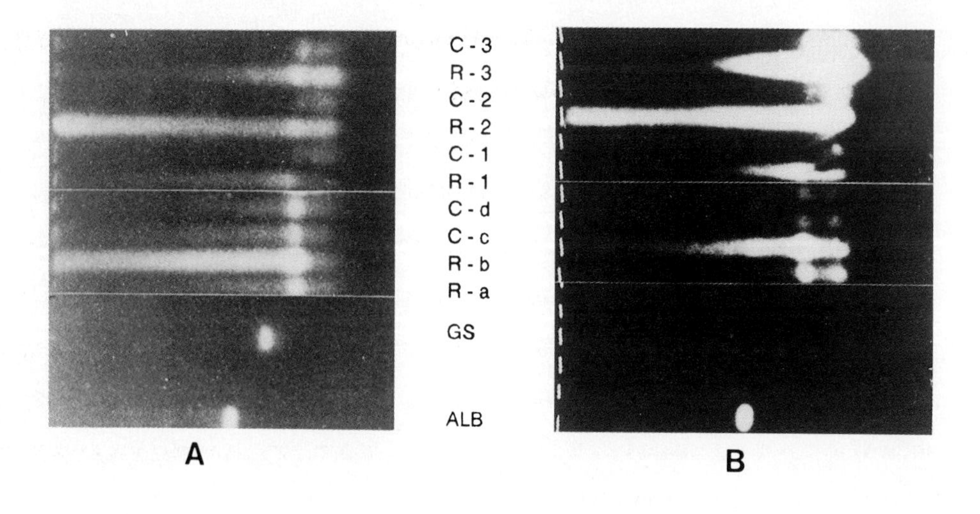

Figure 1. *Southern blot of up- and down-regulated gene products form a regenerating rat liver.* cDNA fragments from a regenerating and control liver were blotted onto Hybond and probed with random prime ^{32}P-labeled cDNA fragments from regenerating liver before (panel A) or after (panel B) enrichment by 3 rounds of subtraction. Equal amounts of cDNA (1 μg) were loaded. R-a, starting material of "early" and R-b, "late regenerating liver; C-c, starting material of "early" and C-d, "late" control liver; R-1,-2,-3 & C-1,-2,-3 subtracted one,two and three times respectively; GS:rat glutamine synthase cDNA fragment, Alb:rat albumin cDNA fragment.

subtraction with regeneration cDNA fragments, the intensity decreases, it remains constant for the regeneration cDNA lanes (lanes R-1,R-2 and R-3) whereas when the enriched regeneration DNA is used as a hybridization probe (panel B), the result is confirmed. The hybridization signal with the enriched regeneration cDNA probe in the original "early" and "late" control cDNA lanes (C c,-d) is reduced compared to the original "early" and "late" regeneration counterpart (R-a,-b). Furthermore, the hybridization intensity increases after each subtraction step for the regeneration lanes (compare lanes R-1,R-2 and R-3) whereas it is reduced in the control lanes (lanes C-1, C-2 and C-3). Also shown in Figure 1 is the elimination, by the subtraction procedure, of a known liver cDNA marker, namely glutamine synthase (GS) from the regeneration cDNA fragments. Albumin and GS were both present in the original cDNA preparation (positive hybridization signal, see panel A) whereas no hybridization signal was detected anymore in the GS-lane (see in panel B). The albumin gene product is still present in the enriched regeneration cDNA fragments, c.f. panel A & B. When the original or subtracted control cDNA material was used as a hybridization probe, complementary results were found, though less pronounced. Noteworthy, however was the elimination of both albumin and glutamine synthase cDNA from the enriched control cDNA populations (data not shown). All these results can only be taken as a qualitative indication of enrichment: a second screening procedure, performed by cloning the enriched regeneration, and separately the enriched control cDNA fragments followed by a partial sequence analysis, demonstrated that the albumin is still an abundant product in the enriched control cDNAs although less that in the regeneration counterpart (not shown).

A more elegant way to estimate the efficiency procedure was offered by the display method. With this procedure the enriched cDNA fragments are directly analyzed using the single nucleotide resolution of the polyacrylamide-urea gel. The regeneration and control cDNA fragments obtained after each round of subtractive hybridization were PCR-amplified in the presence of [α^{35}S]dATP. The experimental conditions were exactly the same as those used to generate the larger quantities of the cDNA populations needed for the subtraction procedure. To obtain a sufficiently high specific activity of the incorporated ([α^{35}S]dATP/total dATP x 100=2.5%) had no effect on the product formation as could be judged from agarose gels, in which identical amplification patterns were found after ethidium bromide staining (fragment range from 100 to 700 nt, data not shown). Decreasing the dATP concentration to 2μM ([^{35}S]dATP/total dATPx100=20%) resulted in an reduced overall yield of cDNA fragments. In figure 2 the regeneration cDNA fragments amplified in the presence of 2 (panel A) and 20 μM dATP (panel B) (the original regeneration cDNA fragments and those subtracted 2 and 4 times with control cDNA fragments) were run on a sequencing gel. The lower parts of panels A and B show that the distribution pattern of the smaller cDNA fragments was unaffected. The number of larger (>300 nt) cDNA fragments, however, was dramatically reduced by lowering the dATP concentration from 20 μM (2.5% ^{35}S-label) to 2 μM (20% ^{35}S-label) in he amplification mixture, as shown in the top part of panel A and B.

Since the 2.5% ^{35}SdATP ratio gave a reproducible distinct pattern, no further attempts were made to optimize the amplification under limiting dATP conditions (e.g. the elongation time). The original cDNA fragments from both the regeneration and control cDNA population represented all the gene products present in the liver before the enrichment procedure was started. As a result of this an almost evenly distribution of cDNA fragments over the entire separation range of the sequencing gel (see lanes denoted 0 in figure 2), was found. The resolution was improved by elimination of the common gene products by the subtractive-hybridization procedure. With an increasing number of subtractions (increasing lane number in Figure 2) the total number of gene products (cDNA fragments) remaining was reduced resulting in more distinct patterns. A known side-effect of the Taq DNA polymerase is that the enzyme has the capability to randomly add one adenosine at the 3' end of the newly synthesized DNA strand. Therefore, multiple bands, appearing as doublets or even triplets, in the display may originate from one and the same denatured cDNA fragment. The 3' terminal transferase activity of the Taq DNA polymerase can be minimized by deleting the 5 min. elongation step at the end of the amplification program, but does not eliminated it entirely. This type of multiplicity of cDNA fragments, however, did not disturb the analysis of the enrichment procedure. It should be noted that due to the restriction digestion of the original cDNAs also multiple fragments derived from one and the same gene product are present. These fragments will be distributed over the display gel according to their size. A higher resolution of the cDNA fragments in size range of 250 nt and larger can be obtained on a 5% gel as shown in Figure 2, panel C.

The enrichment for the up-regulated gene products from a regenerating liver is shown in Figure 2. Enrichment of the up-regulated gene products can be seen in those bands for which the relative amount of [^{35}S]-labeled cDNA

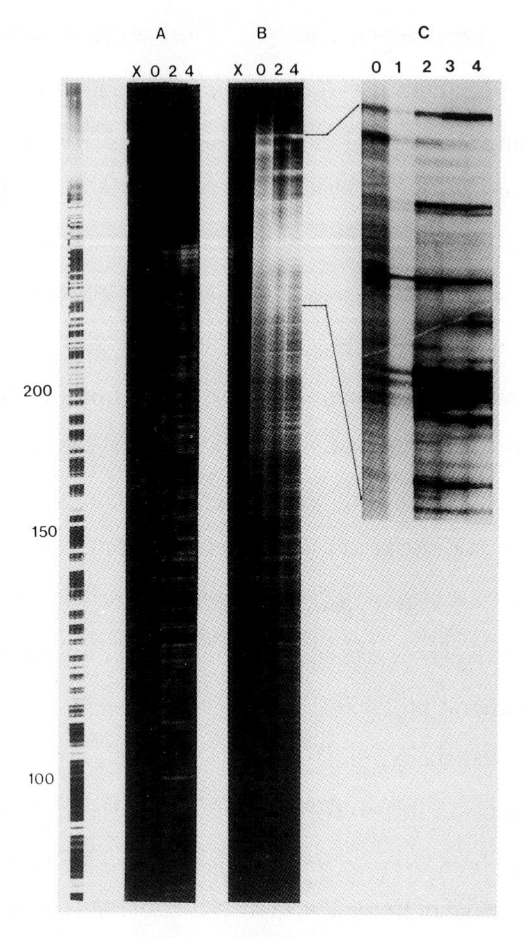

Figure 2. *Display of up-regulated gene products from a regenerating rat liver.* cDNA fragments from a regenerating liver were separated on a 6% sequencing gel. Visualized fragments were derived from regeneration cDNA fragments before and after one, two, three and four rounds of subtractive hybridization (lanes numbered 0,1,2,3 and 4, respectively). The cDNA fragments were PCR amplified in the presence of 10 μCi [α-^{35}S]-labeled dATP and autoradiography was performed by overnight exposure to an X-ray film. Panel A: cDNA fragments amplified in the presence of 2 μM and panel B: 20μM non-radioactive dATP. Panel C: cDNA fragments amplified in the presence of 20 μM non-radioactive dATP and separated on a 5% sequencing gel. Only the 250-600 nt separating range is shown. X are control lanes for the amplification reaction in which no cDNA fragments were used. As a marker for the fragment size one lane of a sequencing reaction was used.

increases with an increasing number of subtractions (compare e.g. lanes 0 through 4 in Figure 2 panel C). Also shown on the gel are [^{35}S]-cDNA bands for which no change in staining intensity was found. These cDNA fragments represent common gene products in the regeneration and control cDNA fragments and are apparently unaffected by 4 rounds of subtractive-hybridization. The down-regulated gene products can be isolated from the enriched control cDNA fragments (results not shown). Various up-regulated cDNA fragments were isolated from the gel and analyzed by partial sequencing. The results were compared with sequences in the GENbank public data domain. Among the identified cDNA fragments were α_2-macroglobulin, albumin, hemopexin, vitamin-D-binding protein, mitochondrial DNA, fibronectin, fibrinogen, α-foetoprotein, together with fragments for which no homologous counterpart could be found. These unknown cDNAs are potential new regeneration factors. The isolated up- and down-regulated gene products from a regenerating liver can be used directly e.g. in *in situ* hybridization studies on the involvement of the various liver cell-types during regeneration. An additional screening procedure is required in order to establish the i.e. function. For this, the complete cDNA of the unknown factor can be isolated from a phage lambda cDNA library using a plaque lifting procedure and the unknown cDNA fragments as a probe. As a second criterium the new regeneration factors can be screened for e.g. mitogenic activity in a proliferation assay using primary hepatocytes. These studies are currently being done in our laboratory.

ACKNOWLEDGEMENTS

The authors wish to acknowledge Dr.P.Hordijk who was involved in the initial steps of the project. Mr.A.Leegwater for technical assistance and Dr.R.Chamuleau for stimulating discussions.

REFERENCES

Fausto N., and Mead J. (1989) Regulation of liver growth: Proto-oncogenes and transforming growth factors. Lab.Invest. 60:4-13.

Haber BA.,Mohn KL.,Diamond RH. and Taub R. (1993) Induction patterns of 70 genes during nine days after hepatectomy define the temporal course of liver regeneration. J.Clin.Invest. 91: 1319-1326.

Hakvoort TBM., Leegwater ACJ., Michiels FAM.,Chamuleau RAFM. and Lamers WH. (1994) Identification of enriched sequences from a cDNA subtraction-hybridization procedure. Nucleic Acid Res. 22: In press.

Higgins GM. and Anderson RM. (1931) Experimental pathology of the liver: Restoration of the liver of white rat following surgical removal. Arch.Pathol. 12:186-202.

Michalopoulos GK. (1990) Liver regeneration: molecular mechanisms of growth control. FASEB J.4:176-187.

Mohn KL., Laz TM., Hsu J-C, Melby AE., Bravo R. and Taub R. (1991) The immediate-early growth response in regenerating liver and insulin-stimulated H-35 cells: Comparison with serum-stimulated 3T3 cells and identification of 41 novel immediate-early genes. Mol.Cell Biol. 11:381-390.

Wang Z. and Brown DD. (1991) A gene expression screen. Proc.Natl.Acad.Sci.U.S.A. 88:11505-11509.

HEPATOTROPHIC AND RENOTROPHIC ACTIVITIES OF HGF *IN VIVO*: POSSIBLE APPLICATION OF HGF FOR HEPATIC AND RENAL DISEASES

Kunio Matsumoto and Toshikazu Nakamura
Division of Biochemistry
Biomedical Research Center,
Osaka University Medical School
Suita, Osaka 565, Japan

INTRODUCTION

Hepatocyte growth factor (HGF), a ligand for *c-met* protooncogene product, was initially identified, purified and cloned as a potent mitogen for adult rat hepatocytes in primary culture (Nakamura et al.,1984; Russell et al.,1984; Nakamura et al.,1986; Nakamura et al., 1987; Nakamura et al.,1989), but it is known as a pleiotropic factor which elicit mitogenic, motogenic, and morphogenic activities (Nakamura, 1991; Matsumoto and Nakamura, 1993). Although this growth factor has been considered to be a long-sought hepatotrophic factor for liver regeneration, we recently found that HGF acts as renotrophic factor and pulmonotrophic factor for regeneration of the kidney and lung, as well as hepatotrophic factor (Kinoshita et al.,1989; Kinoshita et al.,1991; Nagaike et al.,1991; Ishiki et al.,1992; Hamanoue et al.,1992; Igawa et al., 1993; Yanagita et al.,1993; Kawaida et al.,1994). On the basis of its "organotrophic function", we expected that HGF would become useful drug for

NATO ASI Series, Vol. H 88
Liver Carcinogenesis
Edited by G. G. Skouteris

treatments of diseases and injuries of various tissues and organs. The present study was undertaken to explore the clinical usage of HGF for the treatment of hepatic or renal injuries, by administrating human recombinant HGF into experimental animals (Ishiki et al.,1992; Igawa et al.,1993; Kawaida et al.,1994). Our results indicate that HGF may well become effective drug for treatment of hepatic or renal diseases and injuries.

Enhancement of liver regeneration by administration of HGF:

First to examine whether administration of human recombinant HGF (hrHGF) into animals accelerate liver regeneration *in vivo*, the liver was injured by 30% partial hepatectomy. Fig. 1 shows distribution of labeled cells in the remnant liver of the mouse at the peak of DNA synthesis of hepatocyte after 30% partial hepatectomy. The cells undergoing DNA synthesis were detected by incorporation of 5'-bromodeoxyuridine (BrdU), and following immunohistochemical staining using monoclonal anti-BrdU antibody. Control mice injected with saline alone after 30% partial hepatectomy were killed 72 hr after surgery and mice injected with 1μg hrHGF and 5 μg hrHGF were respectively killed 60 hr and 36 hr after surgery.

In normal liver, there were few cells undergoing DNA synthesis (not shown), however, several hepatocytes underwent DNA synthesis after 30% partial hepatectomy. Intravenous injection of hrHGF remarkably increased the number of cells undergoing DNA synthesis. The number of labeled hepatocytes increased in dose-dependent fashion with exogenous HGF administration (Fig. 2). Moreover, hrHGF-injection increased the weight of remnant liver in a dose-dependent manner (not shown). Thus, exogenous HGF stimulate DNA synthesis in hepatocytes and accelerate liver regeneration after 30% partial hepatectomy.

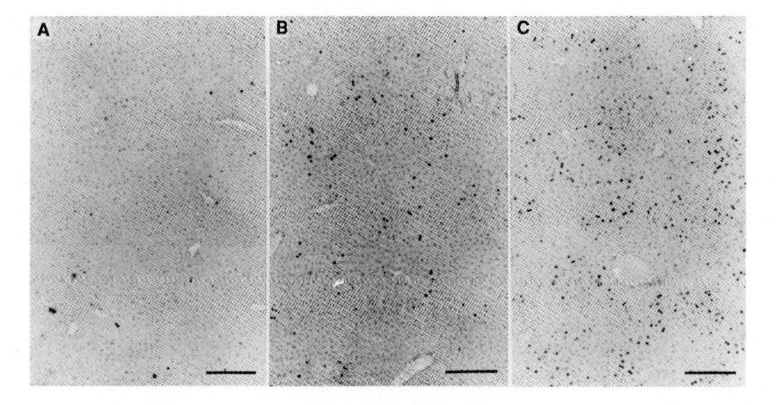

Fig. 1. Distribution of cells undergoing DNA synthesis in the remaining liver of mice with or without HGF-injection after 30% partial hepatectomy. A, control mouse; B, mouse injected with 1 μg hrHGF; C, mouse injected with 5 μg hrHGF. Bars represent 50 μm.

We next examined the effect of hrHGF on DNA synthesis of hepatocytes 36 hr after 70% partial hepatectomy. In 70% partial hepatectomy, marked DNA synthesis occurred in control mice given with saline with a labeling index of 25.9%, and exogenous HGF (1 μg/mouse) increased it to 30.8% (Fig. 2). Thus, in the case of 70% partial hepatectomy, HGF enhances DNA synthesis in the remnant liver with lessor potency than in the case of 30% partial hepatectomy. This may due to that hepatocyte DNA synthesis is well stimulated by endogenously elevated HGF in 70% partial hepatectomy (Kinoshita et al., 1991).

Carbon tetrachloride (CCl_4) is well-known hepatotoxin which induces sever experimental hepatitis. When CCl_4 was administrated into rats, HGF mRNA expression in the liver increased rapidly and markedly (Kinoshita et al., 1989; Noji et al., 1990). We analyzed the efficacy of hrHGF on DNA synthesis 48 hr after CCL_4-administration. Hepatocytes around the central veins were necrotic, and the degree of necrosis did not differ between control mice and mice given 1 µg hrHGF (not shown). However, hrHGF-injection enhanced the labeling index 2.1-fold compared with the control group (Fig.2).

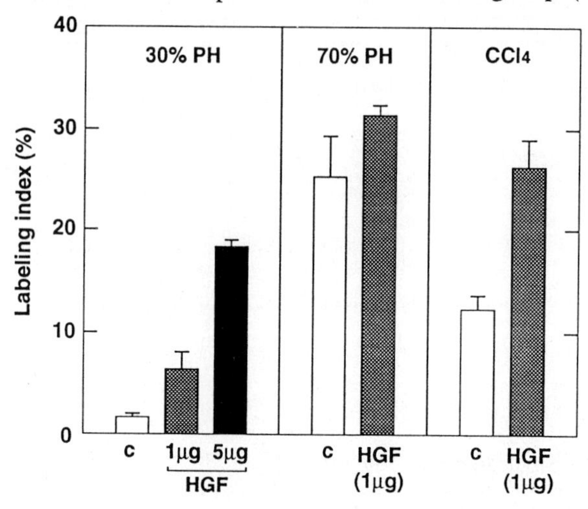

Fig. 2. Change in DNA synthesis (labeling index) of hepatocytes in mice with liver injury caused by administration of hrHGF. Livers of mice were injured with 30% partial hepatectomy (30% PH), 70% partial hepatectomy (70% PH), or CCl_4-administration (CCl_4).

HGF has anti-hepatits action *in vivo*

Administration of α-naphthylisothiocyanate (ANIT) is known to induce intrahepatic cholestasis. When ANIT was intraperitoneally administrated to mice, the epithelia of bile ducts were necrotic and hepatocytes at the portal areas were focally necrotic (Ishiki et al., 1992). We then analyzed the efficacy of hrHGF in cholestatic hepatic injury induced by ANIT. Forty-eight hours after ANIT-administration, there were few labeled cells in control mice given with saline alone after ANIT-administration, however, injectionof hrHGF markedly increased the number of cells undergoing DNA synthesis. More importantly, intravenously injected hrHGF exerted anti-hepatitis action *in vivo* (Fig. 3). The increases of cytosolic enzymes and bilirubin in the sera caused by ANIT-administration were markedly suppressed by hrHGF in a dose-dependent manner. Thus, HGF prevented the onset of severe hepatitis and cholestasis caused by ANIT-administration, which suggests anti-hepatitis action of HGF *in vivo*.

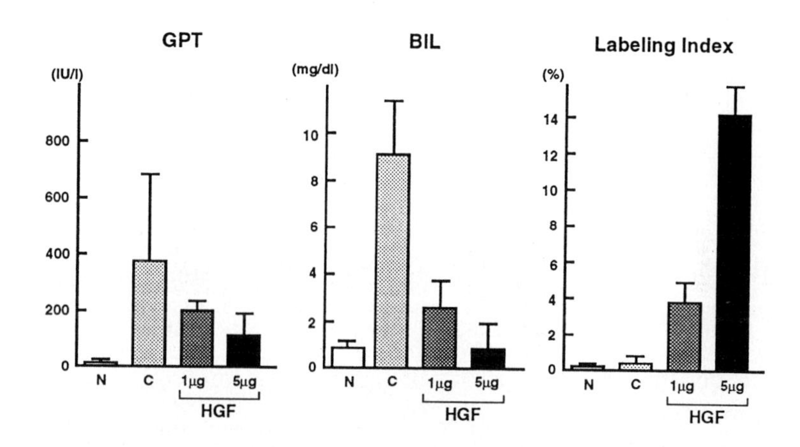

Fig. 3. Changes in GPT and bilirubin in serum and DNA synthesis (labeling index) in the liver of mice injected with hrHGF after ANIT-administration. GPT, glutamic pyruvic transaminase; BIL bilirubin; N, normal mice; C, control mice injected with saline alone; HGF, mice injected with hrHGF.

HGF suppresses hepatic fibrosis/cirrhosis:

Hepatic fibrosis/cirrhosis is a common hepatic disease characterized by hyperaccumulation of connective tissue components, and hepatic necrosis. In patients with liver cirrhosis, there is high coincidence of portal hypertension, esophageal varices, hypoalubuminemia, and hepatocellular carcinoma, though there has been no effective treatment for liver cirrhosis. Polypeptide growth factors and cytokines have a pertinent role in pathogenesis of liver fibrosis/cirrhosis. Especially, the over-expression of transforming growth factor-β1 (TGF-β1) is thought to be a predominant cause of hepatic fibrosis/cirrhosis (Czaja et al., 1989; Nakatsukasa et al., 1990). TGF-β1 is a potent growth inhibitor for hepatocytes (Nakamura et al., 1985) and also a most potent enhancer for extracellular matrix deposition (Ignotz and Massague, 1986; Weiner et al., 1990). Moreover, we recently found that TGF-β1 is a strong suppressor of gene expression of HGF (Matsumoto et al., 1992). Based on this, we assume that exogenously injected HGF acts as a potent mitogen for hepatocytes and possibly for bile duct epithelial cells in the cirrhotic liver in which the endogenous supply of HGF would be impaired. We recently obtained evidence that administration of hrHGF into rats prevents onset of hepatic fibrosis/cirrhosis. Repeated administration of low doses of dimethylnitrosamine (DMN) into rats for four weeks induced onset of hepatic fibrosis/cirrhosis. Extracellular matrix components were hyper-accumulated and the structure of the liver lobules was almost destroyed by DMN-treatment. However, when rats were given with hrHGF during DMN-treatment, extracellular matrix components were remarkably reduced and the structure of liver lobules was remained even after repeated DMN-treatment (not shown). Taken together with theses findings, possible pathogenetic mechanisms of hepatic fibrosis/cirrhosis are described in Fig. 4.

Transient hepatic injury would induce transient induction of HGF mRNA followed by TGF-β1 mRNA, which leads to normal liver regeneration. However,if hepatic injury is prolonged or repeated, as is the case in chronic hepatic diseases, TGF-β1 would suppress the induction of HGF expression, enhancing deposition of extracellular matrix and fibroblast growth, the result

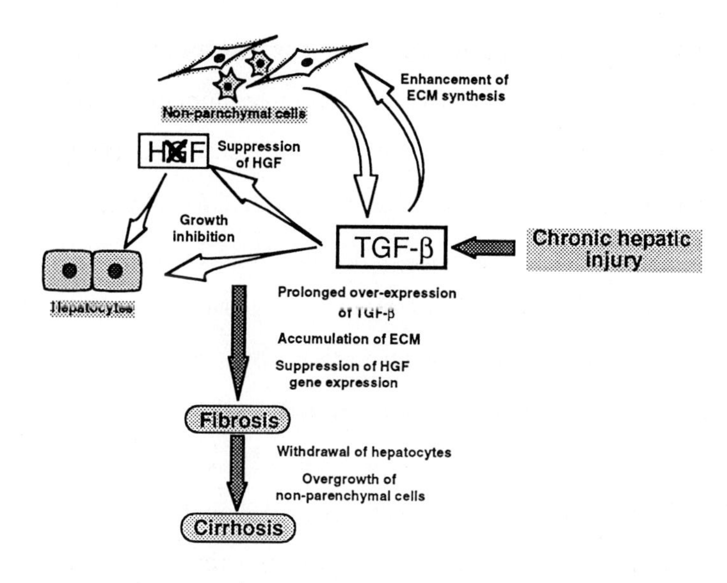

Fig. 4. Possible pathogenetic mechanisms for hepatic fibrosis/cirrhosis from the aspects of TGF-β1 and HGF.

being impairment of hepatocyte proliferation and hepatic fibrosis. Prolonged depletion of HGF as well as prolonged expression of TGF-β1 would lead to withdrawal of parenchymal hepatocytes and overgrowth of stromal fibroblasts with the accumulation of extracellular matrices, and then hepatic cirrhosis would result. Not only over-expression of TGF-β1 but also impaired HGF expression may be directly lined to hepatic fibrosis/cirrhosis, and thus exogenously administrated HGF would prevent the onset and progress of hepatic fibrosis/cirrhosis.

HGF enhances renal regeneration and prevents acute renal failure:

Onset of acute renal failure is often encountered with treatments of diseases by administration of nephrotoxic drugs, renal ischemia, or unilateral nephrectomy. Although we found that HGF may be physiologically functioning renotrophic factor for renal regeneration after acute renal injuries (Igawa et al., 1991; Nagaike et al., 1991; Igawa et al., 1993), we recently obtained evidence that administration of hrHGF accelerated renal regeneration and prevented onset of severe renal dysfunction in experimental animals (Igawa et al., 1993; Kawaida et al., 1994).

We analyzed the stimulatory effect on growth of renal cells and a possible protective action in case of acute renal dysfunction. Fig. 5A shows changes in blood urea nitrogen (BUN) and creatinine in mice with acute renal injury caused by administration of $HgCl_2$, a strong nephrotoxic compound. Both BUN and creatinine increased to 3-4 times higher level than those in normal mice at 72 hr after $HgCl_2$-administration, thereby indicating the onset of typical acute renal failure. However, intravenous injection of hrHGF at 250 μg/kg strongly suppressed the increase in BUN and creatinine (Fig. 5A). The result suggested that the intravenously injected HGF had a potent protective activity against acute renal failure induced by $HgCl_2$.

To extend the protective effect of HGF on renal function, acute renal failure was induced by administration of cisplatin, the most widely used anti-tumor drug, and efficacy of hrHGF was examined (Fig. 5B). Subcutaneous injection of cisplatin induced severe acute renal failure as shown by marked increases in BUN and creatinine. However, injection of hrHGF remarkably suppressed increases in BUN and serum creatinine levels. Thus, HGF has a strong protective effect on renal functions against cisplatin-induced acute renal dysfunction.

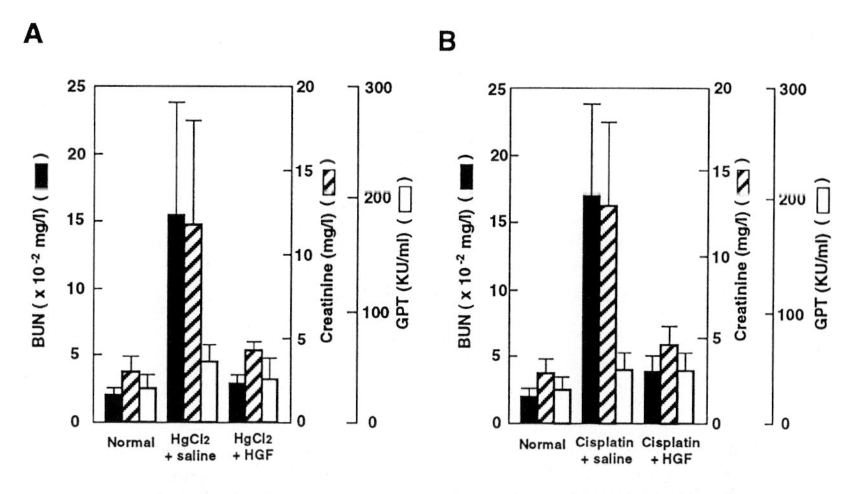

Fig. 5. Preventive effect of hrHGF on acute renal failure induced by administration of HgCl2 or cisplatin in mice. BUN, blood urea nitrogen, GPT, glutamic pyruvic transaminase.

Next, we examined effects of exogenous HGF on DNA synthesis of renal tubular cells in $HgCl_2$-treated mice (Fig. 6). In the kidney of normal mouse, there are few cells undergoing DNA synthesis (not shown), however, several cells underwent DNA synthesis at 72 hrs after $HgCl_2$-administration. Intravenously injected hrHGF stimulated DNA synthesis of renal cells (Fig. 6), and these cells were mostly renal tubular cells located in outer medulla (not shown).

Unilateral nephrectomy has served as an experimental model for compensatory renal regeneration, we examined the effect of exogenous hrHGF on DNA synthesis of renal cells after unilateral nephrectomy. Intravenous injection of hrHGF stimulated DNA synthesis of renal tubular cells *in vivo*.

Thus, HGF enhances renal regeneration after acute renal failure and unilateral nephrectomy.

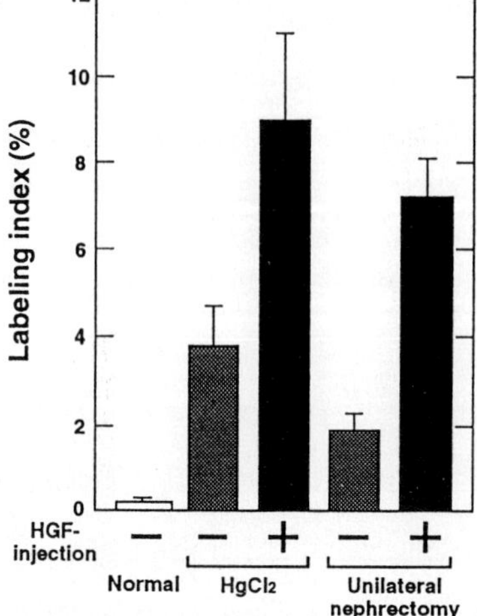

Fig. 6. Enhancement of DNA synthesis (labeling index) by hrHGF-injection in mice with acute renal failure caused by $HgCl_2$-administration or unilateral nephrectomy.

Conclusion and Perspective:

In the present paper, we showed that administration of recombinant HGF accelerated hepatic or renal regeneration, and more importantly it prevented the onset of acute hepatic or renal dysfunction. Pleiotropic functions of HGF to elicit mitogenic, motogenic, and morphogenic activities for various cells seem to be all responsible for the organotrophic roles of HGF for regeneration of organs and tissues (Nakamura, 1991; Montesano et al., 1991; Gherardi and Stoker, 1991; Johnson et al., 1993; Matsumoto and Nakamura, 1993). HGF may be effective for treatment of hepatic diseases, including ischemic injury after surgery, liver transplantation, partial hepatectomy for removal of tumor tissue, hepatitis induced by drugs or viruses, and hepatic fibrosis/cirrhosis. On the other hand, HGF may also become effective drug for treatment of renal diseases, including renal transplantation, acute renal failure induced by drugs (anti-tumor drug, immunosuppressive agent, antibiotics, etc.), and ischemic injury after surgery. Moreover, as elevated blood HGF levels were noted in patients with hepatic or lung diseases (Selden et al., 1986; Tsubouchi et al., 1991; Yanagita et al., 1993), HGF levels in blood or tissues may useful for diagnosis and prospect of hepatic or renal diseases.

Our recent study showed that HGF may function as a pulmonotrophic factor lung regeneration (Yanagita et al., 1993). Together with the findings that HGF targets a wide variety of cells, including hepatocytes, renal tubular cells (Igawa et al., 1991), alveolar and bronchial epithelial cells (Mason et al., 1994), gastric mucosal epithelial cells (Takashi et al., 1993), epidermal keratinocytes and melanocytes (Matsumoto et al., 1991a; Matsumoto et al., 1991b), corneal epithelial cells, hematopoietic stem cells (Mizuno et al., 1993), etc., HGF may function as a naturally occurring "organotrophic factor" for regeneration of various tissues and organs. We predict here that application of HGF and its gene may well become effective treatment for various diseases and injuries in various tissues and organs, as "growth factor therapy".

REFERENCES

Czaja, M.J., Weiner, F.R., Flanders, K.C., Giambrone, M., Wind, R., Biempica, L., Zern, M.A. (1989). In vitro and in vivo association of transforming growth factor-β1 with hepatic fibrosis. J. Cell. Biol. 108: 2477-2482.

Gherardi, E., Stoker, M. (1991). Hepatocyte growth factor-scatter factor: mitogen, motogen, and *met*. Cancer Cells 3: 227-232.

Hamanoue, M., Kawaida, K., Takao, S., Shimazu, H., Noji, S., Matsumoto, K., Nakamura, T. (1992). Rapid and marked induction of hepatocyte growth factor during liver regeneration after ischemic or crush injury. Hepatology 16: 1485-1492.

Igawa, T., Kanda, S., Kanetake, H., Saito, Y., Ichihara, A., Tomita, Y. Nakamura, T. (1991). Hepatocyte growth factor is a potent mitogen for cultured rabbit renal tubular epithelial cells. Biochem. Biophys. Res. Commun. 174: 831-838.

Igawa, T., Matsumoto, K., Kanda, S., Saito, Y., Nakamura, T. (1993). Hepatocyte growth factor may function as a renotropic factor for regeneration in rats with acute renal injury. Am. J. Physiol. 265: F61-69.

Ignotz, R.A., Massague, J. (1986). Transforming growth factor-b stimulates the expression of fibronectin and collagen and their incorporation into the extracellular matrix. J. Biol. Chem. 261: 4337-4345.

Ishiki, Y., Ohnishi, H., Muto, Y., Matsumoto, K., Nakamura, T. (1991). Direct evidence that hepatocyte growth factor is a hepatotrophic factor for liver regeneration and has potent antihepatitis effects *in vivo*. Hepatology 16: 1227-1235.

Johnson, M., Koukoulis, G., Matsumoto, K., Nakamura, T., Iyer, A. (1993). Hepatocyte growth factor induces proliferation and morphogenesis in nonparenchymal epithelial liver cells. Hepatology 17: 1052-1061.

Kawaida, K., Matsumoto, K., Shimazu, H., Nakamura, T. (1994). Hepatocyte growth factor prevents acute renal failure and accelerates renal regeneration. Proc. Natl. Acad. Sci. USA, in press.

Kinoshita, T., Tashiro, K., Nakamura, T. (1991). Possible endocrine control in non-parenchymal liver cells of rats treated with hepatotoxins. Biochem. Biophys. Res. Commun. 165: 1229-1234.

Kinoshita, T., Hirao, S., Matsumoto, K., Nakamura, T. (1991). Possible endocrine control by hepatocyte growth factor of liver regeneration after partial hepatectomy. Biochem. Biophys. Res. Commun. 177: 330-335.

Mason, R.J., Leslie, C.C., McCormick-Shannon, K., Deterding, R., Nakamura, T., Rubin, J.S., Aaronson, S.A., Shannon, J.M. (1994). Hepatocyte growth factor is a growth factor for rat alveolar type II cells. Am. J. Respir. Cell & Mol. Biol., in press.

Matsumoto, K., Hashimoto, K., Yoshikawa, K., Nakamura T. (1991a). Marked stimulation of growth and motility of human keratinocytes by hepatocyte growth factor. Exp. Cell Res. 196: 114-120.

Matsumoto, K., Tajima, H., Nakamura, T. (1991b). Hepatocyte growth factor is a potent stimulator of human melanocyte DNA synthesis and growth. Biochem. Biophys. Res. Commun. 176: 45-51.

Matsumoto, K., Nakamura, T. (1993). Roles of HGF as a pleiotropic factor in organ regeneration. In *"Hepatocyte Growth Factor-Scatter Factor (HGF-SF) and the C-MET Receptor"*, eds. ID Goldberg, E.M. Rosen, pp. 225-249. Birkhauser Verlag, Basel, Switzerland.

Mizuno, K., Higuchi, O., Ihle, J.N., Nakamura, T. (1993). Hepatocyte growth factor stimulates growth of hematopoietic progenitor cells. Biochem. Biophys. Res. Commun. 194: 178-186.

Montesano, R., Matsumoto, K., Nakamura, T., Orci, L. (1991b). Identification of a fibroblast-derived epithelial morphogen as hepatocyte growth factor. Cell 67: 901-908.

Nagaike, M., Hirao, S., Tajima, H., Noji, S., Taniguchi, S., Matsumoto, K., Nakamura, T. (1991). Renotropic functions of hepatocyte growth factor in renal regeneration after unilateral nephrectomy. J. Biol. Chem. 266: 22781-22784.

Nakamura, T., Nawa, K., Ichihara, A. (1984). Partial purification and characterization of hepatocyte growth factor from serum of hepatectomized rats. Biochem. Biophys. Res. Commun. 122: 1450-1459.

Nakamura, T., Tomita, Y., Hirai, R., Yamaoka, K., Kaji, K., Ichihara, A. (1985). Inhibitory effect of transforming growth factor-β on DNA synthesis of adult rat hepatocytes in primary culture. Biochem. Biophys. Res. Commun. 133: 1042-1050.

Nakamura, T., Teramoto, Y., Hirai, R., Yamaoka, K., Kaji, K., Ichihara, A. (1985). Purification and characterization of hepatocyte growth factor from rat platelets for mature parenchymal hepatocytes in primary cultures. Proc. Natl. Acad. Sci. USA 83: 6489-6493.

Nakamura, T., Nawa, K., Ichihara, A., Kaise, N., Nishino, T. (1987). Purification and subunit structure of hepatocyte growth factor from rat platelets. FEBS Lett 224: 311-318.

Nakamura, T., Nishizawa, T., Hagiya, M., Seki, T., Shimonishi, M., Sugimura, A., Tashiro, K., Shimizu, S. (1989). Molecular cloning and expression of human hepatocyte growth factor. Nature 342: 440-443.

Nakamura, T. (1991). Structure and function of hepatocyte growth factor. Prog. Growth Factor Res. 3: 67-86.

Nakatsukasa, H., Nagy, P., Everts, R.P., Hsia, C-C., Marsden, E., Thorgeirsson, S.S. (1990) Cellular distribution of transforming growth factor-β1 and procollagen types I, III, and transcripts in carbon tetrachloride-induced rat liver fibrosis. J. Clin. Invest. 85: 1833-1843.

Noji, S., Tashiro, K., Koyama, E., Nohno, T., Ohyama, K., Taniguchi, S., Nakamura, T. (1990). Expression of hepatocyte growth factor gene in endothelial and Kupffer cells of damaged rat liver, as revealed by in situ hybridization. Biochem. Biophys. Res. Commun. 173: 42-47.

Selden, C., Johnstone, R., Darby, H., Gupta, S., Hodgson, H.J.F. (1986). Human serum does contain a high molecular weight hepatocyte growth factor: studies pre- and post-hepatic failure. Biochem. Biophys. Res. Commun. 139: 261-366.

Takahasi, M., Ota, S., Terano, A., Yoshiura, K., Matsumura, M., Niwa, Y., Kawabe, T., Nakamura, T., Omata, M. (1993). Hepatocyte growth factor induces mitogenic reaction to the rabbit gastric epithelial cells in primary culture. Biochem. Biophys. Res. Commun. 191: 528-534.

Tsubouchi, H., Niitani, Y., Hirono, S., Nakayama, H., Gohda, E., Arakaki, N., Sakiyama, O., Takahashi, K., Kimoto, M., Kawakami, S., Setoguchi, M., Shin, S., Arima, T., Daikuhara, Y. (1991). Levels of the human hepatocyte growth factor in serum of patients with various liver diseases determined by enzyme-linked immunosorbent assay. Hepatology 13: 1-5.

Yanagita, K., Matsumoto, K., Sekiguchi, K., Ishibashi, H., Niho, Y., Hakamura, T. (1993). Hepatocyte growth factor may act as a pulmotrophic factor on lung regeneration after acute lung injury. J. Biol. Chem. 268: 21212-21217.

Weiner, F.R., Giambrone, M-A., Czaja, M.J., Shah, A., Annoni, G., Takahashi, S., Eghbali, M., Zern, M.A. (1990). Hepatology 11: 111-117.

TARGETED GENE DELIVERY AND EXPRESSION

George Y. Wu
Department of Medicine
Division of Gastroenterology-Hepatology
University of Connecticut School of Medicine
263 Farmington Avenue, Room AM-044
Farmington, CT, USA.

INTRODUCTION

A unique characteristic of parenchymal liver cells is the presence of large numbers of high affinity cell-surface receptors that can recognize galactose-terminal (asialo-)glycoproteins. The physical characteristics and properties of the asialoglycoprotein receptor have recently been reviewed (Ashwell *et al.*, 1974, Wall *et al.*, 1980, and Stockert *et al.*, 1991). Binding of an asialoglycoprotein by the receptor leads to invagination of the plasma membrane and internalization of the ligand-receptor complex in membrane-bound vesicles, endosomes. Subsequently, acidification of the endosomes results in segregation of the ligand from the receptor. The receptor is cycled back to the plasma membrane, while the glycoprotein is usually transported to lysosomes where degradation occurs (Schwartz, 1991).

Asialoglycoproteins have been used previously to deliver a variety of biologically active compounds to liver cells (Wu *et al.*, 1985, Wu *et al.*, 1983,

NATO ASI Series, Vol. H 88
Liver Carcinogenesis
Edited by G. G. Skouteris

and Wu and Wu, 1988a) and others. In these studies, materials were coupled to asialoglycoproteins by covalent links (Jung *et al.*, 1981). We were interested in the possibility of targeted delivery of DNA to liver cells via asialoglycoprotein receptors. To avoid modification of DNA that might result from covalent binding to the asialoglycoprotein, a targetable conjugate was developed consisting of two components: 1) a cell targeting component consisting of an asialoglycoprotein ligand covalently bound to 2) a DNA binding component consisting of a polymer containing multiple positive charges (polycation).

Methods and Results:

To test this carrier system *in vitro*, a plasmid carrying the chloramphenicol acetyltransferase gene (CAT) under the control of an SV-40 viral promoter was added to the conjugate to form a soluble complex (Wu and Wu, 1988b). The complex was transfected into Hep G2, an asialoglycoprotein (AsG) receptor (+) cell line, and SK-Hep 1, an AsG receptor (-) cell line. After incubation, no CAT activity could be demonstrated in AsG receptor (-) cells. Similarly, CAT activity could not be observed in AsG receptor (+) cells treated with separate components of the complex. However, CAT activity was detected in complex-treated AsG receptor (+) cells (Wu and Wu, 1988b). CAT gene expression was completely inhibited by the addition of an excess of asialoglycoprotein to compete with the complex for receptor uptake (Wu and

Wu, 1988b). This indicated that recognition of the complex by Hep G2 cells was directed by the asialoglycoprotein component.

Targeted gene delivery in hepatocytes was examined *in vivo* (Wu and Wu, 1988a). ^{32}P-labeled DNA alone or as a complex with AsOR-PL conjugate was injected intravenously into groups of rats. After 10 min, in rats treated with DNA alone, most of the radioactivity, 55%, was found still circulating in the blood. Seventeen percent of the injected counts were detected in the liver. In contrast, in rats injected with labeled DNA in the form of a complex, 85% of the counts were detected in the liver while only 5 % of the radioactivity remained in the blood. This organ distribution of the complex was similar to that of ^{125}I-asialoorosomucoid suggesting that the complex retained its ability to be recognized by AsG receptors *in vivo* (Wu and Wu, 1988a).

To determine whether any of this targeted DNA remained hybridizable after uptake by the liver in vivo, complexed DNA was injected intravenously, and after 24 hours, liver DNA was extracted and subjected to dot-blot analysis using a ^{32}P-labeled CAT cDNA probe. CAT DNA sequences were detected in transfected liver, but not in control liver DNA (Wu and Wu, 1988a). To determine whether the targeted gene was functional, livers were assayed for CAT gene activity. CAT gene expression was detected only in rats treated with complexed DNA (Wu and Wu, 1988a). No CAT activity was detected in kidney, spleen, or lungs of those animals (Wu and Wu, 1988a).

Using the Watanabe rabbit model for familial hypercholesterolemia, the feasibility of targeting the gene for the low density lipoprotein (LDL) receptor was determined *in vivo* (Wilson *et al.*, 1991). A plasmid carrying a full-length cDNA for the human LDL receptor was constructed. Transcriptional regulatory elements from the mouse albumin gene were inserted to drive the LDL receptor gene and provide liver-specific expression. The plasmid was complexed with the AsG-PL conjugate and injected intravenously into rabbits. Cellular DNA extracted from livers removed 10 min after injection, demonstrated approximately 1000 copies of plasmid per cell by Southern blot (Wilson *et al.,* 1992). These levels declined progressively with time, and by 48 hours plasmid DNA was less than 0.1 copies/cell (Wilson *et al.,* 1992). Total RNA was extracted and analyzed by RNase protection assays to quantitate levels of human LDL receptor transcripts. Exogenous LDL receptor mRNA was detected at 4 hours, reached a peak at 24 hours and decreased to undetectable levels by 72 hours after transfection (Wilson *et al.,* 1992). Maximal levels of recombinant LDL receptor mRNA achieved at 24 hours were estimated to be 2-4% of normal endogenous levels. In order to determine the metabolic effects of hepatocyte-directed gene transfer *in vivo*, WHHL rabbits injected with complexed LDL receptor gene or CAT gene were analyzed for changes in total serum cholesterol. After administration of the LDL receptor gene complex, there was a rapid, but definite decline in serum cholesterol which lasted 6 days (Wilson *et al.,* 1992). The drop in cholesterol levels was maximal at 2 days post-injection and was 25% - 30% of pretreatment values (Wilson *et al.,* 1992). Treatment with complexed CAT gene had no effect on serum cholesterol under identical conditions.

To prolong the observed transient expression, efforts were focused on stimulation of hepatocyte replication to enhance the probability of integration of the foreign gene into the genome of foreign cells. In the normal adult liver, few hepatocytes are dividing at any given time. However, hepatocytes can replicate in response to injury resulting in regeneration, e.g., partial hepatectomy (Fabrikant 1968). To determine if gene expression could be made to persist in hepatocytes stimulated to replicate by partial hepatectomy, rats were injected with complexed DNA. Thirty min later, 66% hepatectomies were performed. In these rats, CAT activity was detected for at least 11 weeks post-transfection (Wu *et al.,* 1989). Analysis of liver DNA 11 weeks post-surgery demonstrated the presence of CAT sequences (Wu *et al.,* 1989).

The possibility of using the DNA delivery system combined with hepatocyte replication to provide long term correction of a metabolic dysfunction in genetically defective animals was studied using the Nagase analbuminemic rat as a model (Wu *et al.,* 1991). A plasmid containing the structural gene for the human serum albumin driven by mouse albumin enhancer-rat albumin promoter sequences was constructed and complexed with the targetable conjugate. The resulting soluble complex was injected intravenously into Nagase rats and 66% partial hepatectomies were performed. Two weeks after injection, densitometric quantitation revealed an average copy number of 1000 copies of the plasmid/diploid genome (Wu *et al.,* 1991). Most of the foreign DNA was in plasmid form. Evidence of transcription of the targeted DNA using RNase protection assay estimated the level of human albumin mRNA to be between 0.01% and 0.1% of the normal level of rat albumin in normal rats (Wu *et al.,* 1991). The presence of circulating human albumin was determined by Western blot analysis which showed the appearance of human albumin in the serum within 48 hours. Human serum albumin was first detectable at a level of 0.05 mg/ml, but increased in concentration to a peak level of 34 mg/ml by 2 weeks post-injection. This level remained stable through 4 weeks. No antibodies directed against the human albumin were detected during the 4 weeks of the study.

We conclude that DNA in the form of genes can be complexed in a non-covalent fashion to an asialoglycoprotein based DNA carrier system for targeted delivery and expression specifically in hepatocytes.

Acknowledgements:

We thank Rosemary Pavlick and Martha Schwartz for secretarial assistance in the preparation of the manuscript.

This work was supported in part by the United States Public Health Service, NIDDK Grant DK-42182, Grant 91-0786 from the March of Dimes Birth Defects Foundation, and a grant from TargeTech, Inc./Immune Response Corp. Dr. Wu holds equity in the Immune Response Corp.

REFERENCES

Ashwell G, Morell A (1974) Rule of surface carbohydrates in the hepatic recognition and transport of circulating glycoproteins. Adv Enzymol 41: 99-128.

Fabrikant JI (1968) The kinetics of cellular proliferation in regenerating liver. J Cell Biol 36: 551-565.

Jung G, Kohnlein W, Luders G (1981) Biological activity of the anti tumor protein neocarzinostatin coupled to a monoclonal antibody by N-succinimydyl 3-(2-pyridylthio)-propionate. Biochem Biophy Res Comm 101: 599-606.

Schwartz A (1991) Trafficking of asialoglycoproteins and the asialoglycoprotein receptor. *In*: Wu G, Wu C, ed. Liver Diseases. Targeted Diagnosis and Therapy Using Specific Receptors and Ligands. New York: Marcel Dekker, 3-40.

Stockert RJ, Morell AG, Ashwell G (1991) *In*: Wu G, Wu C, ed. Liver Diseases. Targeted Diagnosis and Therapy Using Specific Receptors and Ligands. New York: Marcel Dekker, 41-64.

Wall D, Wilson G, Hubbard A (1980) The galactose specific recognition system of mammalian liver: The route of ligand internalization in rat hepatocytes. Cell 21: 79-83.

Wilson JM, Grossman M, Wu CH, Chowdhury NR, Wu GY, Chowdhury JR (1992) Hepatocyte directed gene transfer *in vivo* leads to transient improvement of hypercholesterolemia in LDL receptor-deficient rabbits. J Bio Chem 267: 963-967.

Wu CH, Wilson JM, Wu GY (1989) Targeting genes: delivery and persistent expression of a foreign gene driven by mammalian regulatory elements *in vivo*. J Biol Chem 264: 16985-16987.

Wu G, Keegan-Rogers V, Franklin S, Midford S, Wu CH (1988a) Targeted antagonism of galactosamine toxicity in normal rat hepatocytes *in vitro*. J Biol Chem 263: 4719-4723.

Wu G, Wu C, Rubin M (1985) Acetaminophen hepatotoxicity and targeted rescue: A model for specific chemotherapy of hepatocellular carcinoma. Hepatology 5:709-5713.

Wu G, Wu C, Stockert R (1983) Model for specific rescue of normal hepatocytes during methotrexate treatment of hepatic malignancy. Prod Natl Acad Sci USA 80: 3078-3080.

Wu G, Wu C (1988b) Evidence for targeted gene delivery to HepG2 hepatoma cells *in vitro*. Biochemistry 27:887-892.

Wu GY, Wilson JM, Shalaby F, Grossman M, Shafritz DA, Wu CH (1991) Receptor-mediated gene delivery *in vivo*: Partial correction of genetic

analbuminemia an Nagase rats. J Biol Chem 266: 14338-14342.
Wu GY, Wu CH (1988a) Receptor-mediated gene delivery and expression *in vivo*. J Biol Chem 263: 14621-14624.

THE GENERATION OF TUMOR VACCINES BY ADENOVIRUS-ENHANCED TRANSFERRINFECTION OF CYTOKINE GENES INTO TUMOR CELLS

G. Maass, K. Zatloukal, W. Schmidt, M. Berger, M. Cotten, M. Buschle, E. Wagner and M.L. Birnstiel
Research Institute of Molecular Pathology
Dr. Bohrgasse 7, 1030-Vienna, Austria

INTRODUCTION

The major obstacle to immunotherapy of cancer is the poor immunogenicity of tumors, especially in man. This is somewhat surprising since the large number of genetic alternations found in advanced cancers should give rise to peptide neo-epitopes capable of being recognized in the context of MHC-I molecules by cytotoxic lymphocytes (Lurquin et al., 1989). Indeed, tumor-associated (Groen et al., 1987) and tumor-specific antigens (van der Bruggen and van den Eynde, 1992) which should be targets for immunological attack have recently been identified. This leads to the suspicion that foreign antigens are indeed present on most, or at least many, tumor cells and that tumor cells are not rejected as foreign because the response of the immune system to the presented foreign antigen is inadequate (Fearon et al., 1990).

If tumor-specific antigens can be identified on all or most kinds of tumors, it is conceivable that someday antibodies against the tumor antigens may be used to combat cancer of that recombinant antigens can be used to elicit

NATO ASI Series, Vol. H 88
Liver Carcinogenesis
Edited by G. G. Skouteris

predominant form of this DNA was found to be episomal, based on restriction analyses. That this episomal DNA in liver was intact plasmid was confirmed by rescue from total hepatocyte DNA by transformation in *E. coli*. The DNA recovered from livers of transfected rats also retained a bacterial pattern of methylation suggesting that the plasmid had not replicated *in vivo*. These results are consistent with the conclusion that the majority of transgene sequences are retained as stabilized plasmids. The specific form of DNA which is transcriptionally active was not identified in these studies. There was no evidence of integration of the foreign DNA into the host genome. This represents a new mechanism for retaining foreign DNA in eukaryotic cells *in vivo* and has implications both for the development of somatic gene therapies and the pathogenesis of viral diseases (Wilson *et al.*, 1992).

Methods and Results:

If the transgene had not integrated, why did the targeted gene expression persist? To determine the mechanism by which partial hepatectomy resulted in persistence of transgene expression, the intracellular fate of the DNA was assessed as a function of time. A plasmid containing the gene for bacterial chloramphenicol acetyltransferase (CAT) was complexed with an asialoorosomucoid-polylysine (AsOR-PL) conjugate and injected intravenously into groups of rats. At various time points, the internalized DNA was quantitated by Southern blot analysis. To confirm that the DNA had targeted to hepatocytes, parenchymal and non-parenchymal cells were isolated by collagenase perfusion (Seglen, 1976) followed by Percoll gradient centrifugation (Page *et al.*, 1979). Twenty minutes after administration, 80-85% of the plasmid appeared in the liver, 80% of which was within hepatocytes, representing 12000 to 18000 copies per hepatocyte. In sham-operated rats, the transgene concentration decreased to 8-12% and 2-4% of the initial levels in 4 and 24 hr, respectively, and became undetectable at 7 days. In rats subjected to 66% hepatectomy 20 min after DNA administration, approximately 20%, 9% and 7% of the plasmid in the residual liver persisted at 4 hr, 24 hr and 7 days, respectively. Liver homogenates were centrifuged at 750 X g for 20 min. The pellet was subfractionated into nuclear and membranous fractions (Lipman *et al.*, 1990); of these, the internalized plasmid was present mainly in the membranous fraction. Plasmid in the 750 X g membrane fraction was DNase sensitive and became undetectable in 24 hr. The plasmid in isolated nuclei was undetectable by Southern blot, but was demonstrated by polymerase chain reaction (PCR). The 750 X g supernatant was subfractionated on a Percoll gradient. Twenty minutes after administration, the internalized plasmid was distributed bimodally in a plasma membrane/endosome-enriched fraction and a lysosome-enriched fraction. In sham-operated controls, the plasmid in all fractions was degraded by 24 hr. In the 66% hepatectomy group, plasmid in the lysosomal fraction became undetectable in 24 hr. In contrast, the plasmid in the plasma membrane/endosome fraction was DNase resistant and persisted undegraded, in a transfection-competent from throughout the experiment (7 days), indicating that cytoplasmic vesicles are the main site of persistence of the endocytosed DNA (Roy Chowdhury *et al.*, 1993). Previous studies have been directed toward to objective of increased duration of targeted gene expression. In an effort to increase the levels of transgene expression, adenovirus type 5 was modified by coupling an asialoorosomucoid-polylysine conjugate (AsOR-PL). The strategy was based on several facts: 1) from the above analysis of the fate of targeted DNA, the data indicate that the delivery system is efficient. However, following internalization within ensdosomes, the normal degradative pathway results in destruction of the vast majority of the DNA 2) adenovirus like many non-enveloped viruses have evolved a mechanism by which it escapes from the endosome prior to fusion with lysosomes (Svensson *et al.*, 1984 and Pastan *et al.*, 1987) 3) the specificity of infection of adenovirus for host tissue is dictated by recognition of sites on the fiber components of the virus and receptors on the surface of target cells 4) only the virus fibers contain carbohydrate. We hypothesized that if reactions that activate carbohydrate residues were used in the linkage procedures, the recognition sites of the virus might be obscured.

Wild-type virus and replication defective dl312 adenovirus were exposed separately to $NaIO_4$ to activate the carbohydrate, then treated with $NaAsO_2$ to destroy the oxidant, and finally coupled to an asialoorosomucoid-polylysine conjugate. To assess possible changes in infection specificity, modified viruses were exposed to various cells in culture and the viability determined by microscopically counting cells capable of trypan blue exclusion as a function of time, Table 1.

Table 1
Effect of Modification of Adenovirus on Infection Specificity[+]

Addition	HeLa S3*	% Control	Huh 7*	% Control	SK Hep1*	% Control
None	1.1 ± .10	100	1.4 ± .13	100	1.4 ± .10	100
Wild-type	.11 ± .01	10	.43 ± .15	31	.20 ± .02	14
Modified wild-type	1.0 ± .05	91	.26 ± .02	19	1.2 ± .30	86
Modified wild-type + AsOR	1.0 ± .05	91	1.3 ± .02	93	1.4 ± .30	100
Wild-type + AsOR	.10 ± .05	9	.37 ± .03	27	.23 ± .20	16
dl312	1.2 ± .20	109	.98 ± .15	70	1.2 ± .11	86
Modified dl312	1.0 ± .11	91	1.3 ± .10	93	1.2 ± .11	86
Modified dl312 + AsOR	1.0 ± .10	91	1.4 ± .12	100	1.3 ± .12	93

*. number of viable cells (x 10^{-6}) determined by trypan blue exclusion.
[+] reprinted with permission from J. Biol. Chem. 269:1-5 (1994)

Table 1 shows that modified wild-type virus had greatly *decreased* infectivity toward normally susceptible HeLa S_3 [asialoglycoprotein receptor (-)] and SK Hep1 [asialoglycoprotein receptor (-)] cells leaving 91% and 86% viable, respectively. However, with Huh 7 [asialoglycoprotein receptor (+)] cells, **modified** virus **retained** its infectivity leaving only 19% of cells viable under identical conditions (Wu, *et al.* 1994).

To assess foreign gene expression, modified virus was complexed to a plasmid, pSVHBV, containing the gene for hepatitis B virus surface antigen, as a marker. Huh 7, receptor (+), cells treated with modified wild-type, and modified replication-defective dl312 virus complexed to DNA raised antigen levels by approximately 13- and 30-fold, respectively, compared to asialoglycoprotein-polylysine DNA complex alone, Table 2.

Table 2
Effect of Modified Virus on Targeted Gene Expression[+]

Addition	HBV surface antigen production (pg/10^6 cells/24 hrs)
Untreated control cells	0
AsOR-PL-DNA complex alone	5.4 $\pm$ 1.1
Complex + wild-type virus	14 $\pm$ 2.5
Complex + dl312 virus	15 $\pm$ 5.5
Modified wild-type virus-DNA complex	70 $\pm$ 28
Modified dl312 virus-DNA complex	160 $\pm$ 15
Modified dl312 virus-DNA complex + 1000-fold AsOR	4.6 $\pm$ 1.0
Modified dl312 virus-DNA complex + dl312 virus (1:100)	150 $\pm$ 25

[+] reprinted with permission from J. Biol. Chem. 269:1-5 (1994)

Competition with a large excess of an asialoglycoprotein blocked the enhancement by more than 95%. Finally to determine the number of cells that can be transfected (transfection efficiency), modified replication-defective dl312 virus was used to avoid possible cytotoxic effects of wild-type virus during the incubations. Modified dl312 virus was complexed with DNA containing the gene for a *nuclear localizing* b-galactosidase. By microscopically counting the number of cells with blue nuclei, cells transfected by modified virus was found to be 200-fold higher than the complex without virus. Yet, specificity was retained exclusively for asialoglycoprotein receptor-bearing cells (Wu *et al.,* 1994).

We conclude that natural agents that have evolved mechanisms for entering the cytoplasm of cells without degradation of their nucleic acids, can be incorporated into an asialoglycoprotein-based DNA carrier system that retains the specificity for asialoglycoprotein receptors on hepatocytes.

Other investigators have used the adenovirus to enhance the introduction of agents into the internal compartments of cells. For example, FitzGerald *et al.* have shown that co-administration of adenovirus and polypeptides to cells in culture increased the delivery of those peptides into the cytosol of exposed cells (FitzGerald *et al.,* 1983). More closely related experiments (Curiel *et al.,* 1992 and Cristiano *et al.,* 1993) have shown that co-addition of adenovirus and protein-polylysine conjugates resulted in more than a 1,000-fold increase in the efficiency foreign gene expression when plasmid DNA was delivered to cells. Co-localization by direct chemical linkage of the complex to the virus has been used to advantage (Wagner *et al.,* 1992) to increase the likelihood of adenoviral-mediated escape of transgene. They chemically coupled adenovirus to polylysine and formed transferrin-based ternary complexes with foreign DNA. This resulted in further enhancement of targeted gene expression (Wagner *et al.,* 1992). In these experiments, polylysine was coupled to adenovirus by random attachment of polycation to exposed glutamyl or lysyl residues of the virus. However, Curiel demonstrated that in this system, **both** transferrin *and* adenoviral receptors were involved in the gene delivery (Curiel *et al.,* 1991). In the currently described experiments, we deliberately sought to eliminate the contribution of the adenovirus receptors in the cell entry process in order to produce a transfection exclusively mediated by the asialoglycoprotein ligand. The lower enhancement of expression is an expected result of the increased specificity. The infectivity and gene expression data support the conclusion that coupling of conjugates to adenovirus via a procedure that promotes carbohydrate-mediated linkages does result in alteration of viral infectivity, targeted gene expression exclusively directed by the attached ligand, and substantially increased foreign gene expression. Furthermore, because two different markers genes, b-galactosidase, an intracellular enzyme, and hepatitis B virus surface antigen, a secreted protein was used and found in appropriate locations after synthesis by target cells, the data suggest that the use of modified virus as a delivery system may be generally applicable for delivery and expression of a variety of foreign genes and regulatory elements.

Although we have focused on the use of a virus as a means for *enhancement* of receptor-mediated gene delivery, the data indicate that similar methods could be used to modify and target viruses *themselves* as vectors for gene delivery to specific cell types.

Acknowledgement:

We thank Rosemary Pavlick and Martha Schwartz for secretarial assistance in the preparation of the manuscript, Dr. Michiaki Hirayama for technical help, Dr. Claire Bonnerot for the pTZ bact nls lacZ A1 plasmid, and Dr. Nicolas Ferry for kind help and advice.

This work was supported in part by the United States Public Health Service, NIDDK Grant DK-42182, Grant 91-0786 from the March of Dimes Birth Defects Foundation, and a grant from TargeTech, Inc./Immune Response Corp. Dr. Wu holds equity in the Immune Response Corp.

REFERENCES

Cristiano RJ, Smith LC and Woo SLC (1993) Hepatic gene therapy: Adenovirus enhancement of receptor-mediated gene delivery and expression in primary hepatocytes. Proc Natl Acad Sci USA 90:2122-2126.

Curiel DT, Agarwal S, Wagner E, Cotten M (1991) Adenovirus enhancement of transferrin-polylysine-mediated gene delivery. Proc Natl Acad Sci USA 88:8850-8854.

Curiel DT, Agarwal S, Romer MU, Wagner E, Cotten M, Birnstiel ML and Boucher RC (1992) Gene transfer to respiratory epithelial cells via the receptor-mediated endocytosis pathway. Am J Respir Cell Mol Biol 6:247-252.

FitzGerald D, Padmanaban JPR, Pastan I and Willingham MC (1993) Adenovirus-induced release of epidermal growth factor and pseudomonas toxin into the cytosol of KB cells during receptor-mediated endocytosis. Cell 32:607-617.

Lipman BJ, Silverstein SC and Steinberg TH (1990) Organic anion transport in macrophage membrane vesicles. J Biol Chem 265:2142-2147.

Pastan I and Seth P (1987). Concepts In Viral Pathogenesis 141:46.

Page DT and Garvey JS (1979) Isolation and characterization of hepatocytes and kupffer cells. J Immunol Methods 27:159-173.

Roy Chowdhury N, Wu CH, Wu GY, Yerneni PC, Bommineni, VR and Roy Chowdhury JR (1993) Fate of DNA targeted to the liver by asialoglycoprotein receptor-mediated endocytosis *in vivo.* J Biol Chem 268:11265-11271.

Seglen PO (1976) Preparation of isolated rat liver cells. Methods Cell Biol 13:29-83.

Svensson U. and Persson R (1984) Entry of adenovirus 2 into Hela cells. J Virol 51:687-694.

Wagner E, Zatloukal K, Cotten M, Kirlappos H, Mechtler K, Curiel DT and Birnstiel ML (1992) Coupling of adenovirus to transferrin-polylysine/DNA complexes greatly enhances receptor-mediated gene delivery and expression of transfected genes. Proc Natl Acad Sci USA 89:6099-6103.

Wilson JM, Grossman M, Cabrera JA, Wu CH and Wu GY (1992) A novel mechanism for achieving transgene persistence in vivo following somatic gene transfer into hepatocytes. J Biol Chem 267:11483-11489.

Wu GY, Zhan P, Sze LL, Rosenberg AR and Wu CH (1994) Incorporation of adenovirus into a ligand-based DNA carrier system results in retention of original receptor specificity and enhances targeted gene expression J. Biol Chem 269:1-5.

STUDIES ON PERSISTENCE AND ENHANCEMENT OF TARGETED GENE EXPRESSION

George Y. Wu
Department of Medicine
Division of Gastroenterology-Hepatology
University of Connecticut School of Medicine
263 Farmington Avenue, Room AM-044
Farmington, CT, USA.

INTRODUCTION

Intravenous injection of DNA-protein complexes into rats has been shown by us previously to result in transient recombinant gene expression in liver, lasting 4-5 days. The eventual deterioration of gene expression is due in part to instability of the targeted DNA. However, we noted retention of transgene sequences in liver and persistent recombinant gene expression, lasting 2-4 months, when the animals were subjected to partial hepatectomy immediately following *in vivo* gene transfer. Therefore, in an attempt to determine the mechanism(s) responsible for persistent gene expression following partial hepatectomy, we characterized the molecular state of the retained, liver-associated transgenes.

Southern blot analysis of DNA from liver tissues harvested various times after *in vivo* gene transfer and partial hepatectomy (10 min to 11 weeks) demonstrated high levels of transgene DNA (100-10,000 copies/cell). The

NATO ASI Series, Vol. H 88
Liver Carcinogenesis
Edited by G. G. Skouteris

cellular immunity against tumors. Since to date the distribution and nature of tumor antigens is mainly unknown, attempts are being made to use the entire tumor cells, which were transfected with cytokine-expressing plasmids as a source of antigen and to rely on the immune system to seek out and to mount a response against such foreign antigens.

It was found empirically (Fearon et al., 1990) that immunological tolerance of tumors can be broken by transfecting tumor cells with IL-2-gene expression vectors. When such cells are transplanted into syngeneic mice, a powerful systemic response based on T-lymphocytes is mounted by the organism leading to the destruction of both the IL-2 expressing tumor cells as well as parental tumor cells (not expressing cytokines) injected at distant sites. The systemic response does not derive from an increased systemic level of IL-2 as a consequence of implantation of cytokine producing tumor cells, rather it is the high level of cytokines arising locally, which is thought to have a dramatic effect on reprogramming the immune system.

This initial observation has led to an avalanche of rather unsystematic studies for many mouse tumor models using different cytokine genes (reviewed by Zatloukal et al., 1993) in which rejection of parental tumor cells was reproduced using the above mentioned strategy. It is hoped that such findings will ultimately lead to clinical protocols in which the primary tumor will be removed from patients, the tumor cells set to culture, transfected with cytokine expression vectors, inactivated with X-rays and implanted as tumor vaccines back into patient. If a strong immune response results, one can be hopeful that this procedure will lead to an eradication of distant micrometastases over a long time which have arisen previously from disseminated tumor cells.

Transferrinfection of Tumor Cells for the Generation of Tumor Vaccines

Since the tumor tissue may be heterogeneous due to tumor progression and selection by the host immune system, it may be desirable not to use transfected or transduced cell clones, as has been done in most experiments reported up to now (Zatloukal et al., 1993). This procedure, besides being time consuming and delaying unnecessarily the application of the "tumor vaccine" may inadvertently lead to selection of an unrepresentative cell clone. Furthermore, extensive culturing and expansion of cells may lead to a loss of the tumor antigen. A procedure by which the bulk of the tumor cells can be transfected soon after removal and culture would seem desirable.

We believe that our recently developed adenovirus-augmented, receptor-mediated DNA transfer technique "Adenovirus-Enhanced Transferrinfection" (AVET) (Cotten et al., 1992, Curiel et al., 1992, Wagner et al., 1992, Zatloukal et al., 1992) has many advantages over retroviral transduction or DNA transfection followed by clonal expansion of the genetically modified cells. AVE is a new transfection protocol in which the plasmid DNA to be transported into the tumor cells is reacted with transferrin-polylysine to form highly condensed round particles with a diameter of approx. 100 nm, referred to as "donuts" (Zatloukal et al., 1992). These "donuts" are linked to adenovirus Ad5d11014 which, owing to its endosomolytic property, greatly enhances the receptor-mediated transfer of genes into cells and ultimately into the cell nucleus (Cotten et al., 1992, Curiel et al., 1992, Wagner et all, 1992, Zatloukal et al., 1992).

One of the outstanding features of AVET is that with this technique a multiplicity of DNA plasmids (per cell) can be introduced into a large fraction of cells both from cell lines and primary cell culture, including freshly prepared murine and human melanoma cells. Followed transferrinfection, extraordinarily high levels of cytokine or reporter gene expression can be achieved routinely in freshly isolated tumor cells. Cytokine-expressing cells can then be easily mixed with non-transfected irradiated cells to obtain any desired cytokine-expressing level. This seems to be a good starting point for the generation of a tumor vaccine by the procedure described above.

To test the concept of a tumor vaccine, we chose a murine melanoma skin cancer model. For several reasons, melanoma cancer seems to be a particularly interesting target: There is ample evidence that melanomas are subject to immunological control in humans. For instance, there is a (transient) regression of skin cancers in about 25% of the patients and complete remission in about 0.5%, and tumor-specific antigens (MAGE 1, MAGE 3) have been identified (van der Bruggen and van den Eynde, 1992; Gaugler et al., 1994). In addition, cytotoxic T-lymphocytes directed against the tumor can be obtained from patients (Herin et al., 1987; Topallan et al., 1989).

Immunization with IL-2 Transfected Melanoma Cells Protects Mice from Tumor Development

The murine melanoma cell line Cloudman S91 (clone M3) was obtained from ATCC (No. CCL53.1). M3 cells which were established from a spontaneously developed melanoma in DBA/2 mice (Cloudman, 1941) express low level of MHC-I antigens (unpublished observation) and are only moderately immunogenic.

Groups of 6 mice were immunized with IL-2 transfected (24,000 Units/10^6 cells / 24 h), or non transfected tumor cells. Prior to immunization, cells were irradiated (2,000 rad) to avoid further cell proliferation. Mice were immunized twice subcutaneously with 1 x 10^5 cells into the left flank in a weekly interval. After an additional week, animals were challenged with 1 x 10^5 viable M3 melanoma (30-fold tumorigenic dose) cells into the right flank and tumor growth was scored weekly.

Control animals which only received a challenge dose developed tumors within 2 weeks, whereas M3/IL-2 vaccinated mice were completely protected (fig. 1). 5 out of 6 animals which were immunized with non transfected melanoma cells developed tumors. Our results show that irradiation of tumor cells led to a somewhat slower tumor growth, but was not sufficient for the generation of an adequate immune response in our model.

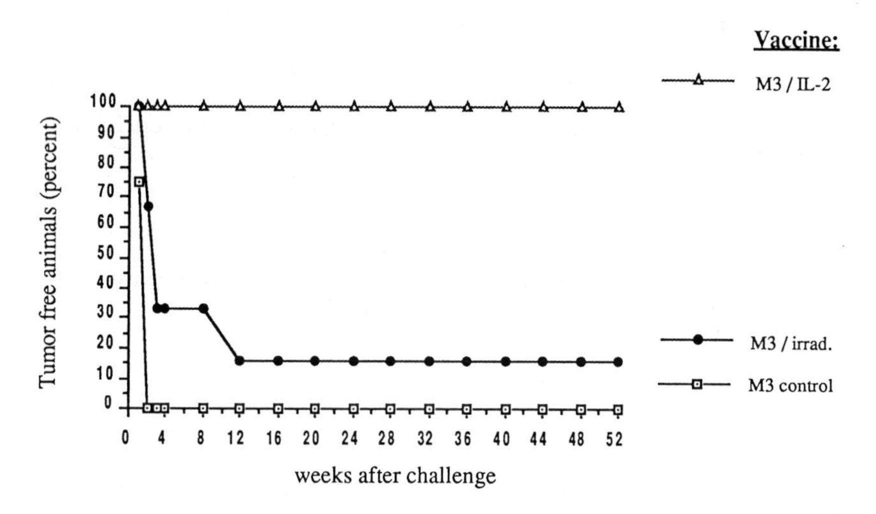

Fig. 1. Groups of DBA/2 mice (each 6 animals) were vaccinated twice in a one week interval with 1 x 10^5 IL-2 transfected, irradiated M3 cells and challenged with 1 x 10^5 viable M3 cells one week later.

AVET leads to very high cytokine expression levels with IL-2 Production rates in M3 cells in vitro of more than 60.000 Units/10^6 cells / 24 h. Since other groups used much lower cytokine levels for vaccination of animals, we were interested whether the high IL-2 production provides any advantage for vaccine efficiency. Vaccines which produced different amounts of IL-2 were generated by using transfection complexes in which the IL-2 expression vector (pWS2m) was mixed with plasmid missing the IL-2 coding region (pSP) in different ratios. This procedure allowed to adjust the IL-2 expression to 400 Units/ 10^6 cells/ 24 h without changing the complex formation.

Table 1. IL-2-dose dependent protection of immunized mice

Challenge/ IL-2-production	1 w	2 w	3 w	4 w	5 w	6 w	8 w
1 x 10^5 M3 IL-2 high	0/6	0/6	0/6	0/6	0/6	0/6	0/6
1 x 10^5 M IL-2 low	0/5	0/5	1/5	1/5	1/5	1/5	1/5
3 x 10^5 M3 IL-2 high	0/8	0/8	1/8	1/8	1/8	1/8	1/8
3 x 10^5 M3 IL-2 low	0/8	1/8	5/8	5/8	5/8	5/8	5/8

Table 1: Mice were immunized twice with 1 x 10^5 IL-2 transfected, irradiated (2000 rad) M3 cells producing either 20.000-30.000 Units IL-2/ 10^6 cells / 24 h (IL-2 high) or 400-500 Units IL-2/ 10^6 cells / 24 h (IL-2 low). One week after the second immunization, the mice were challenged with 1 x 10^5 or 3 x 10^5 parental (tumorigenic) M3 cells. Numbers represent tumor bearing animals/total number of animals at the indicated periods after challenge.

Using this approach, mice were immunized twice with high and low level IL-2 producing vaccines as described above. One week after the second immunization animals were challenged with either 1 x 10^5 or 3 x 10^5 wild-type M3 cells.

We find that there is a dose-dependent relationship between IL-2 expression in vitro and vaccine efficiency in vivo. The low level IL-2 vaccine is less effective than the high level IL-2 vaccine: a lower cancer cell numbers are rejected, tumor development occurs earlier, and fewer animals show protection (Table 1).

Fates of Cytokine-Transfected Tumor Cells In Vivo

The Polymerase Chain Reaction (PCR) amplification technique with an appropriate internal standard for quantitative evaluation was adopted to determine the survival time of IL-2 plasmid and adenovirus DNA (another component of the transfection complex) of subcutaneously injected, transferrinfected and irradiated M3 cells. The persistence of the DNAs was taken as a measure for the survival of the cells at the vaccination site. Three groups of DBA/2 mice (two animals each) were injected with 3 x 10^5 IL-2 transfected, irradiated M3 cells. PCR analysis of the samples taken at day one, two and five after immunization shows a fast destruction of the IL-2 plasmid after subcutaneous injection into the back of the mouse. While still detectable after 24 h and 48 h, the IL-2 plasmid could only be amplified from one DNA sample prepared from skin specimens than had been injected 5 days previously, the second DNA sample from day 5 was negative for IL-2. Adenovirus DNA could be successfully amplified from day 1 and day 2. However, both DNA samples from day 5 were completely negative (Figure 2).

The possible transfer of recombinant or viral DNAs to nearby lymph nodes, to different organs of the animals as well as monocytes and macrophages from the peripheral blood was investigated at high sensitivity. 12 male and 12 female DBA/2 mice were injected with 1 x 10^6 IL-2 transfected M3 cells. 24 h and 48 h after injection, DNA from six animals of each group was prepared from peripheral blood cells and several organs. PCR analysis fails to disclose amplifiable IL-2 adenovirus DNA in all DNA samples from the various mouse organs including draining lymph nodes, spleen, kidney, lung, liver, colon, testis, ovaries, and the peripheral blood mononuclear cells.

In order to determine systemic IL-2 levels following immunization, groups of 3 mice were injected with 1 x 10^6, 3 x10^5 and 1 x 10^5 IL-2 transfected M3 cells which produced 46.000 Units IL-2/10^6 cells/24h in vitro. After 24 and 48 h blood of animals was collected and IL-2 levels in sera were determined by ELISA (Genzyme, Cambridge, MA). As a result, after 24 h systemic levels became detectable in all three mice receiving 1 x 10^6 cells, and in 1 out of 3 mice receiving 3 x 10^5 cells. Animals treated with 1 x 10^5 IL-2 transfected cells, the dose which generates systemic protection, showed no systemic IL-2 levels in the sera. At 48 h after cell implantation, systemic IL-2 was not detectable in any animal. These results are in line with the PCR analysis above, suggesting that M3 cells are rapidly eliminated after injection.

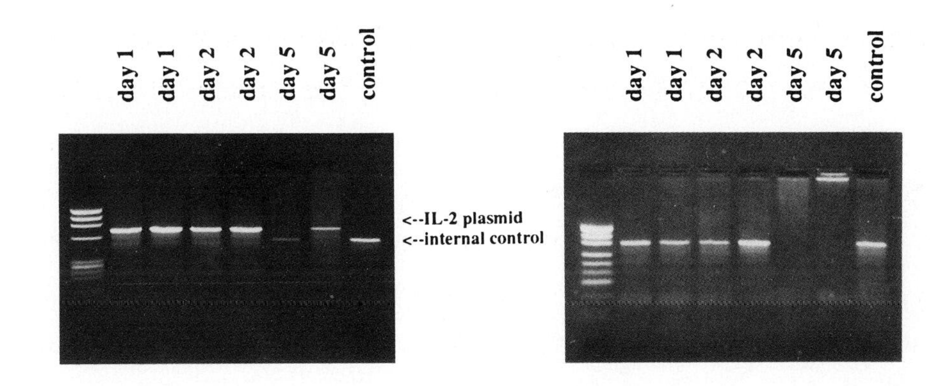

Fig. 2. PCR amplification of IL-2 plasmid from injection site: 3 x 10^5 IL-2 transfected cells were injected into DBA/2 mice. One, two and five days later, mice were sacrificed and immunization sites were excised. Tissues were incubated overnight in proteinase K buffer (50 mM Tris-HCl, 100 mM EDTA, 100 mM NaCl, 1% SDS, 0.5 mg/ml proteinase K) at 55° C. DNA was extracted twice with phenol /$CHCl_3$ followed by isopropanol precipitation. The PCR cocktail contained 1 µg DNA, 1 x PCR buffer (Boehringer Mannheim), 3 Units Taq-Polymerase (Boehringer Mannheim), 1 mM of each dNTP, and 25 pmol of specific primers. The conditions of standard PCR reaction were: 5 min denaturation at 95° C, followed by 40 cycles each 30 sec at 94°, 30 sec at 60° and 1 min at 72° C.

(Fig. 2a): As a control of IL-2 plasmid amplification (450 bp fragment), 1000 copies of a deleted version of the IL-2 plasmid were added to each PCR reaction, resulting in lower bands on the gel (280 bp fragment). Primers were 5' -GTCAACAGCGCAC-CCACTTCAAGC -3';5'-GCTTGTTGAGATGATGCTTTGACA -3'.

(Fig. 2b): Adenovirus amplification was proved by adding 1000 copies of adenovirus genome to the control reaction. Primers were 5' - GGTCCTGTG-TCTGAACCTGAG-3'; 5' -TTATGGCCTGGGGCGTTTACA-3' (317 bp fragment).

DISCUSSION

Our results clearly demonstrate that application of modified tumor cells expressing IL-2 at high levels leads to systemic and long-lasting protection against a challenge with highly tumorigenic cells. For modifying cancer cells with IL-2 genes, we have used a new gene delivery technique i.e. adenovirus-enhanced transferrinfection (AVET), which combines the receptor-mediated endocytosis uptake mechanism with the endosome disruption activity of adenovirus. This technique allows to test the efficiency of cytokine-production by the modified tumor cell over a wide range of expression of cytokines as demonstrated in our murine model. Most studies with cytokine-expressing tumor cells so far were performed with either stably transfected cell line or retrovirally transduced cells producing significantly lower levels (100-3.500 Units IL-2 /10^6 cells/24 h) than we routinely obtain. By using transferrinfection, transfected melanoma cells secrete up to 60.000 Units IL-2/10^6 cells/24 hrs. Our results show, that low levels of IL-2 (around 400 Units/ 10^6 cells/ 24 h) secreted from irradiated melanoma cells failed to protect completely. In contrast, the high production by the vaccine allowed to reject 1×10^5 parental melanoma cells in 100% of the immunized animals (see Table 1).

In contrast to stable integration of genes which is obtained by retrovirus-mediated transduction, AVET leads to long-term but transient expression of the delivered gene from episomally located DNA (Wagner et al., 1992). Besides persistence of gene expression, the survival of the cells after vaccination might be a limiting factor for application in humans since lethally irradiated cells are applied. We therefore determined how long IL-2 has to be secreted and for how long the cells have to survive in vivo to induce an adequate immune response. Using PCR we detected amplifiable IL-2 or adenovirus DNA for 2 to animal 5 days after injection at the immunization site. These data indicate that neither transient cytokine expression nor irradiation is limiting for the induction of the immune system. Another important finding is that no IL-2 DNA nor adenovirus DNA is detectable in draining lymph nodes or other tissues. These data also show that modified tumor cells do not migrate and exclude the possibility of direct gene transfer into the germ line by the applied transfection and immunization procedure.

AVET for the generation of "tumor vaccines" is not only restricted to melanoma cancer or to IL-2. Other specific targets like modified colon, liver, or renal tumor cells should also lead to enhanced immunogenicity against developing metastases when applied under the appropriate conditions. The general concept for the stimulation of the host immune system through elevated levels of cytokines produced at the immunization site seems to be more dependent on the ctyokine level than on the time period of exposure. Further experiments need to be done to investigate the host immune response to the induced cytokine expression and the resulting characterization of cells invading the tumor and the resulting change in the specific cytokine response at the immunization site will lead to further understanding the role of the induced ctyokine and the importance of high expression levels.

REFERENCES

Cloudman, A.M.: The effect of an extra-chromosomal influence upon transplanted spontaneous tumors in mice (1941). Science 93: 380-381.

Cotten, M., Wagner, E., Zatloukal, K., Phillips, S., Curiel, D.T., and Birnstiel, M.L. (1992). High efficiency receptor-mediated delivery of small and large (48 kb) gene constructs using the endosome disruption activity of defective or chemically-inactivated adenovirus particles. Proc. Natl. Acad. Sci. USA 89:6094-6098.

Curiel, D.T., Wagner, E., Cotten, M., Birnstiel, M.L., Agarwal, S., Li, C.- M., Loechel, S., and Hu, P.C. (1992). High-efficiency gene transfer mediated by adenovirus coupled to DNA-polylysine complexes. Hum. Gene. Ther. 3: 147- 154.

Fearon, E.R., Pardoll, D.M., Itaya, T., Golumbek, P., Levitsky, H.I., Simons, J.W., Karasuyama, H., Vogelstein, B., and Frost, P. (1990). Interleukin-2 production by tumor cells bypasses T helper function in the generation of an antitumor response. Cell 60: 397-403.

Gaugler, B., Van den Eynde, B., van der Bruggen, P., Romero, P., Gaforio, J.J., De Plaen, E., Lethe, B., Brasseur, F., and Boon, T. (1994). Human Gene MAGE-3 Codes for an Antigen Recognized on a Melanoma by Autologous Cytolytic T Lymphocytes. J. Exp. Med. 179: 921-930.

Groen, T.P. (1987): Tumor-associated antigens (TAA). In: den Otter, W., and Ruitenberg, E.J. (Eds.), Tumor Immunology. Elsevier, Amsterdam, 1987, pp. 13-28.

Herin, M., Lemoine, C., Weynants, P., Vessiere, A., Van Pel, A., Knuth, A., Devos, R., and Boon, T., (1987). Production of stable cytolytic T-cell clones directed against autologous human melanoma. Int. J. Cancer. 39: 390.

Lurquin, C., Van Pel, A., Mariame, B., De Plaen, E.D., Szikora, J.-P., Janssens C., Reddehase, M.J., Lejeune, J., and Boon, T. (1987). Structure of the gene of Tum-transplantation antigen P91A: the mutated exon encodes a peptide recognition with L^d by cytolytic T cells. Cell 58: 293-303.

Topalian, S.L., Solomon, D., and Rosenberg, S.A. (1989). Tumor-specific cytolysis by lymphocytes infiltrating human melanomas. J. Immunol. 142: 3714- 3725.

Wagner, E., Zatloukal, K., Cotten, M., Kirlappos, H., Mechtler, K., Curiel, D.T., and Birnstiel, M.L. (1992). Coupling of adenovirus to polylysine-DNA complexes greatly enhances receptor-mediated gene delivery and expression of transfected genes. Proc. Natl. Acad. Sci. USA 89: 6099-6103.

Van der Bruggen, P., and Van den Eynde, B. (1992). Molecular definition of tumor antigens recognized by T lymphocytes. Curr. Opin. Immuno. 4: 608-612.

Zatloukal, K., Wagner, E., Cotten, M., Phillips, S., Plank, C., Steinlein, P., Curiel, D.T., and Birnstiel, M.L. (1992). Transferrinfection: a highly efficient way to express gene constructs in eukaryotic cells. Ann. N. Y. Acad. Sci. 660: 136-153.

Zatloukal, K., Schmidt, W., Cotten, M., Wagner, E., Stingl, G., and Birnstiel, M.L. (1993). Somatic gene therapy for cancer: the utility of transferrinfection in generating "tumor vaccines". Gene 135: 199-207.

SUBJECT INDEX

Adenovirus 460,463-464,467,469,474,476-477
Aflatoxin B1 359,366-367
Amines
  aromatic 231,234-235
Apolipoprotein B 411-412,416-418
Apoptosis 182-190,199,200,210,211
Bromodeoxyuridine 198,201-203
C-fos 347,352
C-myc 2-5,109-110,113,115,119,124,306
  antibody, 115,117,120-121,123
Calcium
  channel 251
  channel blocker, 251,255
cAMP 71,74-76,78-79,81,85,342-346
Carcinogenicity 216-217
cDNA 422-427,429-433
  -downregulated 422
  -upregulated 422-427
Cell
  biliary epithelial 390,392-393,395,401-403
  interactions 287,297
  proliferation 197,199,203,208-209,211
  oval 2-3,129,131-135,149,151-152, 157-158,163,168,170,172-175,177
  stem 128-130, 132-135
  transformation 308
Cirrhosis 275-277,283,439-441
Conjugates 463
Cyclosporine A 250-253,255-256
Cytokeratins 135,170,173,175-176
Cytotoxicity 303,305,307-308
Dimethylnitrosamine 276,283
DNA synthesis 113,118,124,392,396,400
Electrophoresis 145
  two dimensional 147,150-151

Epigenetic 215, 224, 226
Food restriction 189
G Proteins 70-71,74,76,84-85
Gene 450-452
expression 232,452-454,457,462-464
Glucocorticoids 382,386
Glutamyl transpeptidase,
gamma 249,253,255,256
Growth Factor
Epidermal, 3-5,14,18,22,71-73,85,88,115-118,122-123,288,295-297
Hepatocyte, 1-6,14-24,32-46,54-60,62-66,164-165,172,176
Hepatocyte,antibody 57,62,64
Transforming-α, 1-4,14, 23-24,
Transforming-β1, 44-46,184-186,440-441
Growth Factor, receptor
Hepatocyte 33,38-39,41-42,44-45
Hepatectomy
partial, 201
Hepatitis
B, virus 301,303-308,311-313,315,318,320,322-326,329-331,341-343,349,351-353,359-360, 365-366
Duck 363-364, 366-367
enhancer 312-315, 318, 322-326, 330-331
Hepatocarcinogenesis 261
Hepatocellular carcinoma 353,359-367
Hepatocyte 1-6,70-85,287-289,291-297,373-377,380-381, 386,451-453
culture medium 376
human 301-302,305,308,390,404
microinjection 112, 118, 119, 121-123
hsp 27 261-262, 264-265, 267-271
Initiation 187
tumor 232
Infection 301, 303-306, 308
Injurin 2,21,38,40-44, 46
Intestine small 411-412, 415-418
Liver 218-220,224,251-252,254-256,275-278, 283
biopsies 399,401

growth factor 275-278, 280, 282-284
injury 283
regeneration 12-13,15-21,280
Matrix degradation 24
Metabolism 218
Methapyrilene 216-220, 224-225
Microscopy
Electron 200,202
Mitochondrial proliferation 219
Necrosis 199-200, 205, 207
Non-genotoxicity 215, 217
Norepinephrine 72,78,82,84
p53 360,362-367
PCR
amplification 413, 415
Phenotype
resistant 271
Phospholipase C 4-5,71,80-81,83,85
Prostaglandins 3-4,71-73,80-85
Protein Tyrosine Phosphatase 93, 103
Protein
Sequencing 425
Tyrosine Phosphatase 93,103
Receptor
adrenergic 71,76-80
α1 76-79
β2 76-78
Targeting 450-451
Therapeutics 282, 283
Transcription factors 342-344,352
Transferrinfection 469-470,476
Translation, In Vitro 234,236,242
Tumor
initiation 232
regression 182
Vaccines 469-470,477
Vimentin 176

NATO ASI Series H

Vol. 1: Biology and Molecular Biology of Plant-Pathogen Interactions. Edited by J.A. Bailey. 415 pages. 1986.

Vol. 2: Glial-Neuronal Communication in Development and Regeneration. Edited by H.H. Althaus and W. Seifert. 865 pages. 1987.

Vol. 3: Nicotinic Acetylcholine Receptor: Structure and Function. Edited by A. Maelicke. 489 pages. 1986.

Vol. 4: Recognition in Microbe-Plant Symbiotic and Pathogenic Interactions. Edited by B. Lugtenberg. 449 pages. 1986.

Vol. 5: Mesenchymal-Epithelial Interactions in Neural Development. Edited by J. R. Wolff, J. Sievers, and M. Berry. 428 pages. 1987.

Vol. 6: Molecular Mechanisms of Desensitization to Signal Molecules. Edited by T M. Konijn, P J. M. Van Haastert, H. Van der Starre, H. Van der Wel, and M.D. Houslay. 336 pages. 1987.

Vol. 7: Gangliosides and Modulation of Neuronal Functions. Edited by H. Rahmann. 647 pages. 1987.

Vol. 8: Molecular and Cellular Aspects of Erythropoietin and Erythropoiesis. Edited by I.N. Rich. 460 pages. 1987.

Vol. 9: Modification of Cell to Cell Signals During Normal and Pathological Aging. Edited by S. Govoni and F. Battaini. 297 pages. 1987.

Vol. 10: Plant Hormone Receptors. Edited by D. Klämbt. 319 pages. 1987.

Vol. 11: Host-Parasite Cellular and Molecular Interactions in Protozoal Infections. Edited by K.-P. Chang and D. Snary. 425 pages. 1987.

Vol. 12: The Cell Surface in Signal Transduction. Edited by E. Wagner, H. Greppin, and B. Millet. 243 pages. 1987.

Vol. 13: Toxicology of Pesticides: Experimental, Clinical and Regulatory Perspectives. Edited by L.G. Costa, C.L. Galli, and S.D. Murphy. 320 pages. 1987.

Vol. 14: Genetics of Translation. New Approaches. Edited by M.F. Tuite, M. Picard, and M. Bolotin-Fukuhara. 524 pages. 1988.

Vol. 15: Photosensitisation. Molecular, Cellular and Medical Aspects. Edited by G. Moreno, R. H. Pottier, and T. G. Truscott. 521 pages. 1988.

Vol. 16: Membrane Biogenesis. Edited by J.A.F Op den Kamp. 477 pages. 1988.

Vol. 17: Cell to Cell Signals in Plant, Animal and Microbial Symbiosis. Edited by S. Scannerini, D. Smith, P. Bonfante-Fasolo, and V. Gianinazzi-Pearson. 414 pages. 1988.

Vol. 18: Plant Cell Biotechnology. Edited by M.S.S. Pais, F. Mavituna, and J. M. Novais. 500 pages. 1988.

Vol. 19: Modulation of Synaptic Transmission and Plasticity in Nervous Systems. Edited by G. Hertting and H.-C. Spatz. 457 pages. 1988.

Vol. 20: Amino Acid Availability and Brain Function in Health and Disease. Edited by G. Huether. 487 pages. 1988.

NATO ASI Series H

Vol. 21: Cellular and Molecular Basis of Synaptic Transmission.
Edited by H. Zimmermann. 547 pages. 1988.

Vol. 22: Neural Development and Regeneration. Cellular and Molecular Aspects.
Edited by A. Gorio, J. R. Perez-Polo, J. de Vellis, and B. Haber. 711 pages. 1988.

Vol. 23: The Semiotics of Cellular Communication in the Immune System.
Edited by E.E. Sercarz, F. Celada, N.A. Mitchison, and T. Tada. 326 pages. 1988.

Vol. 24: Bacteria, Complement and the Phagocytic Cell.
Edited by F. C. Cabello und C. Pruzzo. 372 pages. 1988.

Vol. 25: Nicotinic Acetylcholine Receptors in the Nervous System.
Edited by F. Clementi, C. Gotti, and E. Sher. 424 pages. 1988.

Vol. 26: Cell to Cell Signals in Mammalian Development.
Edited by S.W. de Laat, J.G. Bluemink, and C.L. Mummery. 322 pages. 1989.

Vol. 27: Phytotoxins and Plant Pathogenesis.
Edited by A. Graniti, R. D. Durbin, and A. Ballio. 508 pages. 1989.

Vol. 28: Vascular Wilt Diseases of Plants. Basic Studies and Control.
Edited by E. C. Tjamos and C. H. Beckman. 590 pages. 1989.

Vol. 29: Receptors, Membrane Transport and Signal Transduction.
Edited by A. E. Evangelopoulos, J. P. Changeux, L. Packer, T. G. Sotiroudis, and K.W.A. Wirtz. 387 pages. 1989.

Vol. 30: Effects of Mineral Dusts on Cells.
Edited by B.T. Mossman and R.O. Begin. 470 pages. 1989.

Vol. 31: Neurobiology of the Inner Retina.
Edited by R. Weiler and N.N. Osborne. 529 pages. 1989.

Vol. 32: Molecular Biology of Neuroreceptors and Ion Channels.
Edited by A. Maelicke. 675 pages. 1989.

Vol. 33: Regulatory Mechanisms of Neuron to Vessel Communication in Brain.
Edited by F. Battaini, S. Govoni, M.S. Magnoni, and M. Trabucchi. 416 pages. 1989.

Vol. 34: Vectors asTools for the Study of Normal and Abnormal Growth and Differentiation.
Edited by H. Lother, R. Dernick, and W. Ostertag. 477 pages. 1989.

Vol. 35: Cell Separation in Plants: Physiology, Biochemistry and Molecular Biology. Edited by D. J. Osborne and M. B. Jackson. 449 pages. 1989.

Vol. 36: Signal Molecules in Plants and Plant-Microbe Interactions.
Edited by B.J.J. Lugtenberg. 425 pages. 1989.

Vol. 37: Tin-Based Antitumour Drugs. Edited by M. Gielen. 226 pages. 1990.

Vol. 38: The Molecular Biology of Autoimmune Disease.
Edited by A.G. Demaine, J-P. Banga, and A.M. McGregor. 404 pages. 1990.

NATO ASI Series H

Vol. 39: Chemosensory Information Processing.
Edited by D. Schild. 403 pages. 1990.

Vol. 40: Dynamics and Biogenesis of Membranes.
Edited by J. A. F. Op den Kamp. 367 pages. 1990.

Vol. 41: Recognition and Response in Plant-Virus Interactions.
Edited by R. S. S. Fraser. 467 pages. 1990.

Vol. 42: Biomechanics of Active Movement and Deformation of Cells.
Edited by N. Akkas. 524 pages. 1990.

Vol. 43: Cellular and Molecular Biology of Myelination.
Edited by G. Jeserich, H. H. Althaus, and T. V. Waehneldt. 565 pages. 1990.

Vol. 44: Activation and Desensitization of Transducing Pathways.
Edited by T. M. Konijn, M. D. Houslay, and P. J. M. Van Haastert. 336 pages. 1990.

Vol. 45: Mechanism of Fertilization: Plants to Humans.
Edited by B. Dale. 710 pages. 1990.

Vol .46: Parallels in Cell to Cell Junctions in Plants and Animals.
Edited by A. W Robards, W. J . Lucas, J . D. Pitts, H . J . Jongsma, and D. C. Spray. 296 pages. 1990.

Vol. 47: Signal Perception and Transduction in Higher Plants.
Edited by R. Ranjeva and A. M. Boudet. 357 pages. 1990.

Vol. 48: Calcium Transport and Intracellular Calcium Homeostasis.
Edited by D. Pansu and F. Bronner. 456 pages. 1990.

Vol. 49: Post-Transcriptional Control of Gene Expression.
Edited by J. E. G. McCarthy and M. F. Tuite. 671 pages. 1990.

Vol. 50: Phytochrome Properties and Biological Action.
Edited by B. Thomas and C. B. Johnson. 337 pages. 1991.

Vol. 51: Cell to Cell Signals in Plants and Animals.
Edited by V. Neuhoff and J. Friend. 404 pages. 1991.

Vol. 52: Biological Signal Transduction.
Edited by E. M . Ross and K . W. A. Wirtz. 560 pages. 1991.

Vol. 53: Fungal Cell Wall and Immune Response.
Edited by J. P. Latge and D. Boucias. 472 pages. 1991.

Vol. 54: The Early Effects of Radiation on DNA.
Edited by E. M. Fielden and P. O'Neill. 448 pages. 1991.

Vol. 55: The Translational Apparatus of Photosynthetic Organelles.
Edited by R. Mache, E. Stutz, and A. R. Subramanian. 260 pages. 1991.

Vol. 56: Cellular Regulation by Protein Phosphorylation.
Edited by L. M. G. Heilmeyer, Jr. 520 pages. 1991.

NATO ASI Series H

Vol. 57: Molecular Techniques in Taxonomy.
Edited by G . M . Hewitt, A. W. B. Johnston, and J. P. W. Young .
420 pages. 1991.

Vol. 58: Neurocytochemical Methods.
Edited by A. Calas and D. Eugene. 352 pages. 1991.

Vol. 59: Molecular Evolution of the Major Histocompatibility Complex.
Edited by J. Klein and D. Klein. 522 pages. 1991.

Vol. 60: Intracellular Regulation of Ion Channels.
Edited by M. Morad and Z. Agus. 261 pages. 1992.

Vol. 61: Prader-Willi Syndrome and Other Chromosome 15q Deletion Disorders.
Edited by S. B. Cassidy. 277 pages. 1992.

Vol. 62: Endocytosis. From Cell Biology to Health, Disease and Therapie.
Edited by P. J. Courtoy. 547 pages. 1992.

Vol. 63: Dynamics of Membrane Assembly.
Edited by J. A. F. Op den Kamp. 402 pages. 1992.

Vol. 64: Mechanics of Swelling. From Clays to Living Cells and Tissues.
Edited by T. K. Karalis. 802 pages. 1992.

Vol. 65: Bacteriocins, Microcins and Lantibiotics.
Edited by R. James, C. Lazdunski, and F. Pattus. 530 pages. 1992.

Vol. 66: Theoretical and Experimental Insights into Immunology.
Edited by A. S. Perelson and G. Weisbuch. 497 pages. 1992.

Vol. 67: Flow Cytometry. New Developments.
Edited by A. Jacquemin-Sablon. 1993.

Vol. 68: Biomarkers. Research and Application in the Assessment of Environmental Health. Edited by D. B. Peakall and L. R. Shugart.
138 pages. 1993.

Vol. 69: Molecular Biology and its Application to Medical Mycology.
Edited by B. Maresca, G. S. Kobayashi, and H. Yamaguchi. 271 pages.
1993.

Vol. 70: Phospholipids and Signal Transmission.
Edited by R. Massarelli, L. A. Horrocks, J. N. Kanfer, and K. Löffelholz.
448 pages. 1993.

Vol. 71: Protein Synthesis and Targeting in Yeast.
Edited by A. J. P. Brown, M. F. Tuite, and J. E. G. McCarthy. 425 pages.
1993.

Vol. 72: Chromosome Segregation and Aneuploidy.
Edited by B. K. Vig. 425 pages. 1993.

Vol. 73: Human Apolipoprotein Mutants III. In Diagnosis and Treatment.
Edited by C. R. Sirtori, G. Franceschini, B. H. Brewer Jr. 302 pages. 1993.

NATO ASI Series H

Vol. 74: **Molecular Mechanisms of Membrane Traffic.** Edited by D. J. Morré, K. E. Howell, and J. J. M. Bergeron. 429 pages. 1993.

Vol. 75: **Cancer Therapy. Differentiation, Immunomodulation and Angiogenesis.** Edited by N. D'Alessandro, E. Mihich, L. Rausa, H. Tapiero, and T. R.Tritton. 299 pages. 1993.

Vol. 76: **Tyrosine Phosphorylation/Dephosphorylation and Downstream Signalling.** Edited by L. M. G. Heilmeyer Jr. 388 pages. 1993.

Vol. 77: **Ataxia-Telangiectasia.** Edited by R. A. Gatti, R. B. Painter. 306 pages. 1993.

Vol. 78: **Toxoplasmosis.** Edited by J. E. Smith. 272 pages. 1993.

Vol. 79: **Cellular Mechanisms of Sensory Processing. The Somatosensory System.** Edited by L. Urban. 514 pages. 1994.

Vol. 80: **Autoimmunity: Experimental Aspects.** Edited by M. Zouali. 318 pages. 1994.

Vol. 81: **Plant Molecular Biology. Molecular Genetic Analysis of Plant Development and Metabolism.** Edited by G. Coruzzi, P. Puigdomènech. 579 pages. 1994.

Vol. 82: **Biological Membranes: Structure, Biogenesis and Dynamics.** Edited by Jos A. F. Op den Kamp. 367 pages. 1994.

Vol. 83: **Molecular Biology of Mitochondrial Transport Systems.** Edited by M. Forte, M. Colombini. 420 pages. 1994.

Vol. 84: **Biomechanics of Active Movement and Division of Cells.** Edited by N. Akkaş. 587 pages. 1994.

Vol. 85: **Cellular and Molecular Effects of Mineral and Synthetic Dusts and Fibres.** Edited by J. M. G. Davis, M.-C. Jaurand. 448 pages. 1994.

Vol. 86: **Biochemical and Cellular Mechanisms of Stress Tolerance in Plants.** Edited by J. H. Cherry. 616 pages. 1994.

Vol. 87: **NMR of Biological Macromolecules** Edited by C. I. Stassinopoulou. 616 pages. 1994.

Vol. 88: **Liver Carcinogenesis. The Molecular Pathways** Edited by G. G. Skouteris. 502 pages. 1994.